Adaptations for Foraging in Nonhuman Primates

Adaptations for Foraging in Nonhuman Primates

Contributions to an Organismal Biology of Prosimians, Monkeys, and Apes

PETER S. RODMAN
and
JOHN G. H. CANT,
editors

New York Columbia University Press *1984*

Library of Congress Cataloging in Publication Data
Main entry under title:

Adaptations for foraging in nonhuman primates.

Includes bibliographies and index.
1. Primates—Food. 2. Primates—Behavior.
3. Adaptation (Biology) I. Rodman, Peter S. II. Cant, John G. H.
QL737.P9A35 1984 599.8'0453 83-18954
ISBN 0-231-05226-X
ISBN 0-231-05227-8 (pbk.)

Columbia University Press
New York Guildford, Surrey

Printed in the United States of America

Clothbound editions of Columbia University Press Books are Smyth-sewn and printed on permanent and durable acid-free paper

CONTENTS

Preface *vii*

Introduction: From Comparative Morphology and Socioecology to an Organismal Biology of Primates. PETER S. RODMAN AND JOHN G. H. CANT *1*

1. On the Use of Anatomical Features To Infer Foraging Behavior in Extinct Primates. RICHARD F. KAY *21*

2. Motion Economy Within the Canopy: Four Strategies for Mobility. THEODORE I. GRAND *54*

3. Foraging, Habitat Structure, and Locomotion in Two Species of *Galago*. R. H. CROMPTON *73*

4. Use of Habitat and Positional Behavior in a Neotropical Primate, *Saguinus oedipus*. PAUL A. GARBER *112*

5. Foraging and Social Systems of Orangutans and Chimpanzees. PETER S. RODMAN *134*

6. Feeding Ecology and Sociality of Chimpanzees in Kibale Forest, Uganda. MICHAEL P. GHIGLIERI *161*

7. Ecological Differences and Behavioral Contrasts Between Two Mangabey Species. PETER M. WASER *195*

8. Body Size and Foraging in Primates. L. ALIS TEMERIN, BRUCE P. WHEATLEY, AND PETER S. RODMAN *217*

9. The Role of Food-Processing Factors in Primate Food Choice. KATHARINE MILTON *249*

10. Is Optimization the Optimal Approach to Primate Foraging? DAVID G. POST *280*

11. A Conceptual Approach to Foraging Adaptations in Primates. JOHN G. H. CANT AND L. ALIS TEMERIN *304*

Index *343*

PREFACE

This book grew out of discussions the two of us have had over several years, beginning when J. G. H. C. was a graduate student at the University of California, Davis, working under P. S. R. We found that consideration of a published paper often caused us to wonder why the author had not considered some other kind of adaptation to explain more fully the topic of initial focus. We naturally addressed the same questions to our own work, as well as that of others. Such questioning can be fruitful only if one moves ahead to propose and attempt to adopt solutions. It seems logical to us to propose that to understand animal adaptations, one must look at whole organisms whenever possible because it is these that must deal with obstacles to reproductive success. Ultimately, though as a matter of convenience it may be necessary to consider traits of a whole organism independently, the same traits should make best sense when considered as part of the whole organism in which they must be integrated. Thus our emphasis, explained more fully in chapters 1 and 12, is on adaptation of whole organisms.

We organized a symposium entitled "Foraging in Nonhuman Primates" for the 1980 annual meetings of the American Association of Physical Anthropologists, and most of the contents of this volume developed from contributions to that symposium. Chapters by Kay, Milton, and Rodman are on related but somewhat different topics from their papers at the symposium. Ted Grand, David Post, and Michael Ghiglieri were not members of the symposium but luckily had manuscripts in hand that dealt with appropriate topics in foraging and adaptations of whole animals.

We have found working with the contributors to be educational and are grateful to all of them for the privilege of editing their papers and thereby gaining early access to the insights that other readers will now find therein. L. Alis Temerin has debated questions about adaptations with each of us for about a decade, and her influence on our views and motivation in arranging this volume cannot be overestimated.

We thank Vicki Raeburn, formerly of Columbia University Press, for initial encouragement and advice on development of this project, Susan Koscielniak, the Associate Executive Editor of the Press, for her continuing interest, and Leslie Bialler, manuscript editor, for his conscientious efforts in the copy editing of the volume.

J. G. H. C. has been supported by NSF Grant BNS 8111631 during part of the preparation of this volume, and P. S. R. has received partial support from PHS Grant RR00169 throughout the project.

John G. H. Cant
Peter S. Rodman

Adaptations for Foraging in Nonhuman Primates

Introduction: From Comparative Morphology and Socioecology to an Organismal Biology of Primates

PETER S. RODMAN AND
JOHN G. H. CANT

> What I am prepared to defend is a broader thesis: there are many special disciplines and several different approaches in biological research; all are important and their integration is most important; to be effective and to be in fact biological all approaches must take into account the *organization* of *organisms,* and so must both depart from and lead to the level of whole organisms.
>
> G. G. Simpson (1964)
> *This View of Life*

THE RATIONALE for this book lies in three words of its title: *adaptations, foraging,* and *primates.* When we cast scientific eyes on an animal or plant, we tend to do so in either of two ways. First, we may ask how the organism works, leading to questions about proximate causes or mechanisms that maintain the organism. Or we may ask why the plant or animal is as it is—in other words, why it has the traits it shows. This sort of inquiry seeks to define ultimate evolutionary causes in an effort to understand the *adaptations* of organisms. In compiling this volume we have chosen to concentrate on adaptations, though we by no means intend to imply that study of proximate causes is any less valuable an investment of research effort.

We may group adaptations of animals into three more or less comprehensive classes: those concerned with avoiding predation; those that help an animal to acquire food; and those associated with reproduction. While all animals must avoid predators (and some are more constrained by this need than others), and while an individual who does not reproduce is very rarely of evolutionary importance, it is the "hunt" and capture of food that seems to select most consistently for organismal design in primates. *Foraging* to us names this class of adaptations. It is necessarily the task of the whole organism, and it occupies most of the waking time of a primate. Foraging requires the integrated use of nerves, muscles, and bone, coordinated by the central nervous system and guided by all the senses. Physiological potentials determine the suitability of any food to a specific forager, and physical structure of the food determines the potential of the forager to reduce the food to useful form. There is essentially no aspect of the organism that does not impinge on its ability to forage, and all characteristics of the organism must interact to result in the forager's behavior. This pervasive importance of foraging may be difficult for us human primates to grasp, accustomed as we are to obtaining more food than we need in too few minutes each day with too little effort, and it is important to develop an understanding of the fact that organisms other than ourselves (and even some living humans) face periodic stresses in foraging that require foraging to work well.

Given its pervasive nature, foraging draws us to the center of *organismal biology,* for to understand foraging, and particularly to understand variations among animals with respect to foraging, we must begin with the whole organism rather than its parts. Furthermore, foraging adaptations can really be understood only in natural settings, and we are thus compelled to examine our subjects *naturalistically.*

Finally, why focus on adaptations of *primates?* When humans look at other animals, comparisons with ourselves are virtually inevitable. Indeed, comparisons are seldom purposeless (as in "Have you noticed that humans have two legs, butterflies six? Another glass of sherry?"). Instead comparisons normally aim for better understanding of humans. In many cases, a comparison benefits from similarity of the organisms, and consequently, as we discuss in detail below, historically primates have received attention disproportionate to their representation in the animal kingdom. Humans are, after all, primates, and studies aiming to understand humans better naturally compare humans with the primates. A result of this is that now primates are the best known mammalian order with respect to foraging adaptations, the diversity of species studied, the diversity of adaptations studied, and the depth of the research. Thus even without current continuing interest in primates because of their relevance for understanding humans, the state of existing knowledge about them would make them an appropriate choice for a volume that aims to integrate knowledge and promote research on a highly heterogeneous topic like *foraging adaptations*.

In the rest of this introduction, we aim to introduce the work that follows by reviewing what we see as the intellectual roots of the work as a whole. The papers sort roughly into two groups with two related but somewhat independent histories: those derived from a tradition of study of comparative morphology and those derived from a newer tradition of study of comparative socioecology, which has recently developed into a theoretically sophisticated area best named "comparative behavioral ecology and sociobiology."

Comparative Morphology

It is amusing to think of the world in which our primary intellectual traditions developed as one in which nonhuman primates did not exist. Brief reflection on the fact that philosophers and theologians of Europe lived in a world populated by wet-nosed livestock and a rather impoverished natural fauna, no members of which bore any resemblance to humans, is enough to suggest why it was possible to view man as created in a mold different from any other living things, with a tremendous gap between man and the rest of animal life. The nonhuman primates, our links with the other animals, lived happily out of view to the south and east, and just what we were related to was left to the imagination. If a representative array of mammals had surrounded our gentle theologians and philosophers, any of their children could have sorted humans into a morphological group with most of the other primates. No doubt they would have had some difficulty with *Daubentonia* but little difficulty with *Pan* or even with *Cacajao*.

Of course, specialists such as Linnaeus (1758) made a point of being acquainted with all known living forms, and monkeys and apes were known to the European world. These were treated as curious caricatures of humans (Morris and Morris 1966). Perhaps it was embarrassing to early systematists to recognize the compelling morphological similarities of humans to apes and other primates, but the similarity was so clear that in their classifications, the primates emerged as an order including *Homo*. Despite the recognition of resemblance, there was as yet no need to recognize *kinship* with "lower" primates.

An evolutionary view of the living world immediately suggested that humans were not only similar to the nonhuman primates but were also related to them by common descent. The Darwinian paradigm thus generated much detailed comparison of human with nonhuman, exemplified, by Huxley's (1863) careful morphological demonstration of the extreme similarity of humans to the African apes. Anatomical description, a well-developed procedure because of its importance to medical education, became a major tool of the post-Darwinian analysis of phylogeny. Philosophically the demonstration of man's close resemblance to apes, and by evolutionary deduction his kinship with them, was an important step in intellectual history, but for the anatomist this was only a beginning. A concept of phylogeny transformed the whole array of primates (and all living things) into a genealogical puzzle to be solved by what are now the traditional methods of systematics and taxonomy. Comparative anatomists examined the primates in encyclopedic detail, as shown in the works of, for example, Elliot (1912), and this process continues to the present (e.g., Hershkovitz 1977; Hill 1953–1974).

Predictably, many works on the comparative morphology of the primates attempted to prove various hypotheses about the origin of humans in relation to the general descent of the primates (e.g., Elliot-Smith 1930; Gregory 1922; Keith 1931; Jones 1929). A genealogical concept of relations among animals suggested the possibility that fossils might be found to be "missing links" between living phyla, particularly between humans and the apes. The latter search produced not only important fossil primates, which offered more puzzles to fit together with the living primates, but also one important fraud, Piltdown Man, cleverly constructed to fill the gap between ape and human. Given the persistence of this ersatz fossil through four decades, so strong that it overshadowed the importance of real fossil hominids from South Africa as they gradually came to light, it would be interesting to analyze its effect on the entire enterprise of the study of human origins, but such analysis is not within our present scope. The interpretation of fossil primates as they appeared fell within the scope of comparative anatomists and gradually to a few primate paleontologists, such as Le Gros Clark (1959).

All of the early work—say, the first century of the Darwinian paradigm—dealt with morphology isolated from natural settings. Structural affinities were examined much as a child might sort colored glass from a sea beach, and such affinities were, as they perhaps still are, sufficient to elucidate phylogenetic relations. For fossils, only morphological comparisons were possible. But the description of primate morphology inevitably produced a large number of questions about *why* different primates were different, and a comparative morphological approach could not answer the questions alone.

Why two related species differ in structure is an evolutionary question. Darwin's initial formulation of natural selection as a process explained how the differences arose by pointing out that various organisms interact differently with their environments and that the differential success of individuals in that interaction explained how species characters change and thus how species differentiate. But the implication of this formulation for research in biology was not realized until long after Darwin explained the process: that to understand "adaptation" or function, the biologist must examine variation in organisms' interactions with the environments in which they have evolved.

Preoccupation with human descent necessarily precluded naturalistic investigation of the functional basis of critical morphological differences between humans and apes; cultural evolution seems to have removed all living humans from the environmental influences that initially shaped us into upright, flat-faced, large-brained apes. Explanations of the basic human adaptations have consequently relied on "common sense" and speculation, even to the present (e.g., Lovejoy 1981) with some enlightened exceptions (e.g., McHenry 1982).

Relevant naturalistic observation of nonhumans was possible, however, though it gained momentum slowly, beginning for our purposes with the study of gorillas by Bingham (1932) and of chimpanzees by Nissen (1931). These studies of the African apes were clearly motivated by recognition of the phylogenetic affinity of humans and the apes. Though heroic, neither study was particularly successful at describing normal behavior of apes.

C. R. Carpenter accomplished truly pioneering work not only in naturalistic observations of primates, but also in comprehensive studies of any animals in natural habitats. Each of his early studies—howler monkeys (1934); spider monkeys (1935); gibbons (1940)—was an attempt to give as nearly complete a description of the natural lives of the animals as possible under the constraints of time and observational conditions. In this work Carpenter addressed questions primarily about behavior and social relations rather than questions about function and adaptation in morphology, and we describe his role relative to the tradition of comparative socioecology below.

Carpenter's third field study of primates was part of the Asiatic Primate Expedition on which he was accompanied by Adolf Schultz, a comparative anatomist, and Sherwood Washburn, then a graduate student in anthropology at Harvard with broad interests in primate biology and evolution. Washburn was thus exposed to the approach of an extremely knowledgeable morphologist and to the methods and rationale (see below) of perhaps the most experienced field worker in the world. From his work on the expedition, Washburn produced his doctoral dissertation on comparative skeletal anatomy of langurs and macaques. This research was an important precursor to experimental and observational studies of locomotion and anatomy by Washburn's students beginning some two decades later (Avis 1962; Grand 1967; Ripley 1967; Tuttle 1967; Morbeck 1975,

1977a, b) and perhaps an even more important impetus for Washburn's own work relating naturalistic studies to human social behavior.

Numbers of field studies of primates increased in the 1950s and 1960s, but the developments in comparative morphology were oddly independent of the naturalistic observations until quite recently. Erikson (1963), a highly accomplished anatomist, dissected spider monkeys and other primates and invented a term, *semi-brachiation,* to describe the shoulder of the spider monkey. This name implied something about the locomotor behavior of the spider monkey but was based on morphological observations alone. Though common sense suggests that the morphological differences would be reflected in true behavioral differences, the behavioral measures were not taken in the laboratory or in nature. A similar approach to classification of locomotion characterized the work of Napier (e.g., 1963). The approach generated useful comparative description but inferred, by "common sense," function from structure or relied on minimal knowledge of true behavior, rather than attempting to observe both structure and behavior to learn the relationship.

Despite an exponential increase in field studies of primates through the 1960s, naturalistic studies aimed at explaining morphological patterns in nonhuman primates were rare, with the exception of work by Ripley (1967) and Mendel (1976), until the work of Fleagle (1977a, b; 1978). Instead functional analyses depended on increasingly sophisticated anatomical technique (Grand 1967, 1968a, b) and on technical development in the use of electromyography (Tuttle 1967, 1969; Tuttle and Basmajian 1974a, b; 1977) and in morphometric methods (e.g., Oxnard 1973; McHenry 1972, 1975; McHenry and Corruccini 1975, 1976, 1978).

Fleagle worked at Harvard under the influence of I. DeVore (a student of Washburn who has generated numerous naturalistic studies of nonhuman primates); C. R. Taylor (who studies the energetics of locomotion); F. A. Jenkins (whose research concerns details of the locomotion of vertebrates); and W. W. Howells, who provided education in the patterns of primate and human evolution. Fleagle carried his background to the field in Malaysia, where he took observations that showed the association of morphological differences between sympatric leaf monkeys with differences in gross aspects of their locomotor and feeding behavior in natural habitat.

This review brings us to the present in studies of comparative morphology of primates, which has proceeded from phylogenetic puzzle solving to naturalistic observations aimed at understanding the relations between form and function. Articles in this volume by Kay, Grand, Crompton, and Garber descend from this tradition. Kay's is the most nearly comprehensive attempt we know of to bring together neontological observations to interpret the relationship between structure and environment. Grand's paper offers an example of his uniquely interesting view of the manner in which a whole organism interacts with its immediate surrounding during locomotion and how differences in structure and behavior effect varying solutions to the same problems. Garber and Crompton both provide detailed descriptions of the specific interactions of organisms with their microhabitats. Each paper shows clearly the utility or necessity of examining organismal function in natural settings to approach understanding of adaptations.

Comparative Socioecology

Shortly after completing his initial field studies, which he carried out and reported ". . . with as much objectivity and as little theorizing as possible," Carpenter (1934) revealed his rationale for such work in two general papers (1942a, 1942b) by defining his assumptions as follows:

> (1) That human behavioral characteristics have anlagen, analogies and homologies in non-human invertebrate and vertebrate forms. That consequently, continuities, homologies, convergencies, variations and specializations of behavior and social relations can be discovered and systematically compared among species of the phylogenetic series. (2) That nonhuman types which have the greatest degree of morphological similarity to man, or which show anthropoid specializations or convergencies, when studied from the viewpoint of behavior or social relations, are likely to yield data most useful in understanding and interpret-

> ing problems of human behavior. (3) That for the valid investigation of some problems in comparative behavior it is obligatory not only to study "animals as wholes" but to observe whole animals in natural, organized, undisturbed groups living in that environment which operated selectively on the species and to which the species is fittingly adapted. This long recognized, but inadequately considered assumption, does not conflict with the fact that the results of field investigations and those of the laboratory should be systematically *co*-related (Carpenter 1942a).

These assumptions introduce the origins of study of comparative socioecology of primates, which has in turn generated papers such as those in the second group of this volume.

Several currents in the 1930s prepared the way for increased field research after World War II. First was the appreciation of the Darwinian paradigm that suggested the way to understand form and adaptation was to observe animals in naturalistic settings. This view is clearly enunciated among Carpenter's assumptions. Second was the growing field of ethology in Europe and England, which developed on the assumption that behavior, like morphological structures, has evolved. Carpenter's assumptions clearly state this view of behavior and, additionally, clearly state the view that this method of comparative study can usefully be applied to the study of *human* behavior. With Carpenter's early field studies and rationale in place, and with ethology developing, the war intervened.

During these days, E. A. Hooton wrote prolifically about human evolution. In his widely read books (e.g., 1946), which were essentially the lectures he gave introducing physical anthropology to students at Harvard, Hooton asserted that the proper way to understand the origins of human social behavior was by comparative study of societies of monkeys and apes. Among those exposed to Hooton's view was Sherwood Washburn, and in retrospect, the lesson seems not to have been lost on Washburn.

During the 1950s several investigators were introduced to field research on Barro Colorado (Altmann 1959, Collias and Southwick 1952). At the same time the Piltdown fossil was finally exposed as a fraud (Weiner 1955), and interest increased in understanding the remaining known protohominids, the australopithecines of South Africa. Quite naturally under the influence of the ethological point of view, it seemed logical to search for the roots of human behavior in the behavior of the protohominids. How to do this without fossil behaviors was the trick, and comparative study of the living primates seemed to be the answer. Schaller (1963) initiated a field study of gorillas in 1958, in part with the rationale that these were among man's closest relatives; others followed a similar rationale in studying chimpanzees (Reynolds 1965; Goodall 1965). Washburn observed baboons in East Africa and encouraged a social anthropologist, Irven DeVore, to study social behavior of baboons explicitly as a model for the behavior of protohominids. The rationale for the study of baboons was simple: early hominids were primates who were thought to have lived in savannas much like those occupied by baboons in East Africa today; consequently their behavior should be analogous to that of early man. This pattern of interpretation of the relationship between environment and social behavior was to persist through to the 1970s, as is described below. Washburn and DeVore published two papers on the relationship of social behavior of baboons to the social life of early man (1961a, b), and considerable interest in the possibility of reconstructing early human social behavior developed among anthropologists generally.

The important point to be drawn here relative to papers included in this volume some twenty years later is that the rationale and objective for research on nonhuman primates were both simple: social patterns must depend on descent (thus gorillas and chimpanzees) and environment (thus baboons). The subjects of the research were *social* patterns rather than individual behavior of individuals, probably because the focus of anthropological research on humans had been social structures rather than individual behaviors that made up the social patterns. Existing social theory about animals (e.g., Allee 1951; Wynne-Edwards 1962; Etkin 1964) viewed cooperativeness as the motivating force in animal social systems, the harmony of the society dominating the self-interest of individuals. Though theory contradicting this view developed shortly (Hamilton 1963, 1964), the importance of selection operating on individuals as a structuring force in the evolution of social behavior was not truly appreciated until the early 1970s.

At the transition from the 1950s to the 1960s, human evolution, and particularly the evolution of human behavior, gained wide popular and scientific interest, sparked in large part by the Leakeys' spectacular discoveries in East Africa. Robert Ardrey turned attention of the public to comparative animal behavior as the tool for understanding the origins of human behavior (1963), and several primate studies reached initial publication. Southwick (1963) published a small edited volume of papers on primate social behavior, including chapters by Japanese authors on previously little-known research on social behavior of Japanese macaques, and shortly afterward, DeVore (1965) published a larger volume with a wide range of recent results of naturalistic primate research.

As observations accumulated, there developed a need for some framework that might organize the diverse information on social behavior of the primates. In an early attempt to collate and correlate information on phylogeny and ecology of the primates, DeVore showed the relationships among a variety of variables of behavior and ecology in a short paper in 1963. In England J. H. Crook (1964, 1965) had developed a framework for comparison of avian societies, and in 1966, Crook and Gartlan published a short paper organizing known information about habitats, diets, activity patterns, group sizes, migration, sexual dimorphism (in body size and behavior), and pattern of dispersion (whether territorial or not) in the primates. Their review classified primate species into a series of grades that were explicitly argued to be evolutionary grades related to successive radiations of ancestral primates into different habitats. The paper presented an explicit model of how habitat factors maintain a social system. The model fitted with existing social theory by showing direct routes of causation from environment to social structure without considering contributions of individuals to the social structures or the influence of environment on individual behavior. This framework developed into comparative socioecology of primates in Crook's later work (1970). Similar approaches to the relation of social system to environment appeared in the works of Struhsaker (1969), Denham (1971) and Kummer (1971), and the socioecological approach has been carried to its extreme by Clutton-Brock in a series of increasingly sophisticated cross-species multivariate analyses (e.g., Clutton-Brock and Harvey 1977). Eisenberg et al. (1972) contributed to this stage in theoretical thought in their paper proposing a scheme of evolution of social structures in the primates. Once again, the analysis dealt with societies at the level of social structures themselves, rather than examining the evolution of individual behavior that produced the social patterns.

A benefit of the socioecological approach was the realization that explanations of variation in social systems would be found in the ecology of animals and that study of social behavior and social systems could not provide explanations of the systems themselves. Consequently students of primates began gathering increasingly detailed descriptions of the ecology of their subjects. Fruits of this process of the early 1970s appear most obviously in Clutton-Brock's (1977) volume on primate ecology.

Comparative socioecology ultimately has failed to provide satisfactory explanations of social systems, providing at best a more or less successful series of correlations between environmental variables and the social variables identified by Crook and Gartlan (1966). Two new currents of theory and thought have developed recently, with origins in the 1960s. Although it is difficult to specify what led to a shift of focus in primate studies to individuals rather than societies (to the extent there has been a shift), an emerging body of ideas and interpretations under the headings *behavioral ecology* and *sociobiology* have been influential. In addition, methodological improvements and a rising level of intellectual comfort in studies of behavior have caused observers to focus on individuals as the units of study, regardless of whether they ultimately aimed to examine the role of individuals in social systems or the social systems themselves.

A key paper in the emerging theoretical area of behavioral ecology was Schoener's (1971) "Theory of feeding strategies," which introduced a mode of economic analysis of foraging behavior that persists in one way or another to the present. Schoener identified a series of variables that were to be measured on individuals, and at the same time discussed, among other things, the relationship of *group living* to those variables. This sort of analysis implied a relationship between environment, individual, and *group* with the focus on the effects of various environmental variables on the individual's feeding efficiency. Stuart Altmann (1974) offered a valuable example to those interested in social behavior of primates, in which he applied Schoener's sort of analysis to understanding the nature of foraging and social behavior of savanna-living baboons.

A second important influence leading to focus on the individual social patterns was the emergence of the elements of what was to become the "new" sociobiology. R. L. Trivers, who studied under the influence of William Drury and Ian Nisbet of the Massachusetts Audubon Society in the late 1960s, was introduced to David Lack's view of life. Lack (1968) clearly rejected Wynne-Edwards' theory of group selection purporting to explain numerous social phenomenona, referring to Fisher's (1958) point that processes of group selection must necessarily proceed more slowly than processes of individual selection. Trivers' two early papers (1971, 1972) applied economic analyses to consequences of individual actions for evolution of social behaviors. At the same time, attention was drawn to W. D. Hamilton's important earlier formulation (1963, 1964) of an evolutionary theory of social bchavior integrated with the dominant mode of analysis of genetic evolution: the spread of specific alleles in populations.

A third influence on studies of primate social behavior was the methodological paper of Jeanne Altmann (1974), in which she spelled out clearly the advantages of focal animal sampling in the description of behavior. As a consequence, numerous subsequent observers have used the technique, which necessarily produces observations of individuals rather than of social phenomena directly. With this lesson in hand, the intellectual standards of research necessarily rose, and quantification rather than subjective description became the dominant mode of analysis. Laboratory techniques appropriate to ethological analyses certainly contributed to, and preceded, this change in naturalistic studies, but it seems fair to say that Altmann's paper was a prime mover in transformation of many of our techniques. As a consequence of the observational technique, it is necessary to construct descriptions of social systems out of data gathered on a set of individuals who constitute the society. This focus on individual behavior is, whether intentionally or not, consonant with the Darwinian paradigm and has helped bring the study of primate social behavior more into the realm of biology than previously. The study of social systems has been drawn from the level of the systems themselves to the more appropriate level of the individual organisms that constitute them. At the level of the organism, all influences that produce the social phenomena—proximate or ultimate, internal or external—converge and interact.

Chapters of this volume by Rodman, Ghiglieri, Waser, Temerin et al., Milton, and Post may be seen as more or less directly derived from the history discussed under comparative socioecology. Waser is a biologist who takes this opportunity to discuss the relative roles of phylogeny and environment in the behavior of two congeneric species of mangebeys. Rodman, in another comparative analysis, offers a heavily sociobiological interpretation of the similarities and differences between chimpanzees and orangutans, beginning with analysis of differences in ecological variables. Ghiglieri discusses the social behavior of a population of chimpanzees within the framework of behavioral ecology. Temerin et al. examine the effects of body size on foraging patterns of orangutans and crab-eating macaques; Temerin's work lies between the traditions of comparative morphology and comparative socioecology, dealing as it does with morphological variables and their influence on behavioral ecology of the two species. Post's contribution reconsiders a topic, optimal foraging, that is central to ideas about behavioral ecology. Milton's work seems at first to be a bit more remote from the lines of thought discussed so far, but the connection is quite clear. As concern about ecological influences on individual animals has developed, it has become increasingly necessary to consider specific relations between organism and environment rather than general characteristics of relations between a class of animals and their environments. Milton shows how variable digestive morphology affects the appropriate food choices for different primate species and thus how the same basic set of foods available to a group of species may in fact be very different for each as a function of physiological/morphological constraints. The concluding chapter by Cant and Temerin offers a general framework for designing research on animal foraging by focusing on traits of individuals and individual solutions to the multiple problems of foraging.

In conclusion, we add that the historical view presented is not necessarily that of the other contributors. Each person has a singular history and numerous sources of practical and intellectual motivation that may have been missed in our review of the roots of the studies. At least the history presented shows how we have arrived at the interests that motivated assembling this set of papers. Finally, we are aware that there may be other important contributors to the emergence of ideas represented in the volume whose contributions have been omitted. We trust these individuals will accept that constraints of space prevent a comprehensive acknowledgment of all the work that has preceded and informed us. We are grateful for their forbearance and for the benefit of all their contributions to our education.

References

Allee, W. C. 1951. *Cooperation Among Animals.* New York: Henry Schuman.

Altmann, J. 1974. Observational study of behaviour: Sampling methods. *Behaviour* 49:227–67.

Altmann, S. A. 1959. Field observations on a howling monkey society. *J. Mammal.* 40:317–30.

—— 1974. Baboons, space, time, and energy. *Amer. Zool.* 14:221–28.

Ardrey, R. 1963. *African Genesis.* New York: Delta.

Avis. V. 1962. Brachiation: the crucial issue for man's ancestry. *Southwestern J. Anthrop.* 18:118–48.

Bingham, H. C. 1932. Gorillas in a native habitat. *Carnegie Inst. Wash. Publ.* 426:1–66.

Carpenter, C. R. 1934. A field study of the behavior and social relations of howling monkeys. *Comp. Psychol. Monogr.* 10 (2):1–68.

—— 1935. Behavior of red spider monkeys in Panama. *J. Mammal.* 16:171–80.

—— 1940. A field study in Siam of the behavior and social relations of the gibbon. *Comp. Psychol. Monogr.* 16(5):1–212.

—— 1942a. Societies of monkeys and apes. *Biological Symposia* 8:177–204.

—— 1942b. Characteristics of social behavior in nonhuman primates. *Trans. N.Y. Acad. Sci.* 4:248–58.

Clark, W. E. Le Gros. 1959. *The Antecedents of Man.* Edinburgh: Edinburgh University Press.

Clutton-Brock, T., ed. 1977. *Primate Ecology: Studies of Feeding and Ranging Behaviour in Lemurs, Monkeys and Apes.* London: Academic Press.

Clutton-Brock, T., and P. H. Harvey. 1977. Species differences in feeding and ranging behaviour in primates. In Clutton-Brock, ed. 1977:557–84.

Collias, N., and C. Southwick. 1952. A field study of population density and social organization in howling monkeys. *Proc. Amer. Phil. Soc.* 96:143–56.

Crook, J. H. 1964. The evolution of social organisation and visual communication in the weaver birds (Ploceinae). *Behaviour,* Suppl. No. 10.

—— 1965. The adaptive significance of avian social organisations. *Symp. Zool. Soc. Lond.* 14:181–218.

—— 1970. The socio-ecology of primates. In Crook, ed. *Social Behaviour in Birds and Mammals: Essays on the Social Ethology of Animals and Man,* pp. 103–66. New York and London: Academic Press.

Crook, J. H. and J. S. Gartlan. 1966. Evolution of primate societies. *Nature* 210:1200–3.

Denham, W. W. 1971. Energy relations and some basic properties of primate social organization. *Amer. Anthrop.* 73:77–95.

DeVore, I., 1963. Comparative ecology and behavior of monkeys and apes. In S. L. Washburn, ed. *Classification and Human Evolution,* pp. 301–19. Viking Fund Publications in Anthropology no. 37. New York: Wenner-Gren Foundation.

DeVore, I., ed. 1965. *Primate Behavior: Field Studies of Monkeys and Apes.* New York: Holt Rinehart Winston.

Eisenberg, J. F., N. A. Muckenhirn, and R. Rudran. 1972. The relation between ecology and social structure in primates. *Science* 176:863–74.

Elliot, D. G. 1912. A review of the primates. Monograph Series, Amer. Mus. Nat. Hist., New York. Vols. 1–3.

Elliot-Smith, G. 1930. *Human History.* London: Cape.

Erikson, G. E. 1963. Brachiation in the new world monkeys. *Symp. Zool. Soc. Lond.* 10:135–64.

Etkin, W. 1964. Cooperation and competition in social behavior. In W. Etkin, ed. *Social Behavior and Organization Among Vertebrates,* pp. 1–34. Chicago: University of Chicago Press.

Fisher, R. A. 1958. *The Genetical Theory of Natural Selection.* 2d rev. New York: Dover.

Fleagle, J. G. 1977a. Locomotor behavior and muscular anatomy of sympatric Malaysian leaf-monkeys (*Presbytis obscura* and *Presbytis melalophos*). *Amer. J. Phys. Anthrop.* 46:297–307.

—— 1977b. Locomotor behavior and skeletal anatomy of sympatric Malaysian leaf-monkeys (*Presbytis obscura* and *Presbytis melalophos*). *Yearbook of Phys. Anthrop.* 1976. *Yearbook Series* 20:440–53.

—— 1978. Locomotion, posture and habitat utilization in two sympatric Malaysian leaf-monkeys (*Presbytis obscura* and *Presbytis melal-*

ophos). In G. G. Montgomery, ed. *Ecology of Arboreal Folivores,* pp. 243–51. Washington, D.C.: Smithsonian Press.

Goodall, J. 1965. Chimpanzees of the Gombe Stream Reserve. In DeVore, ed. 1965:425–73.

Grand, T. I. 1967. The functional anatomy of the ankle and foot of the slow loris (*Nycticebus coucang*). *Amer. J. Phys. Anthrop.* 26:207–18.

—— 1968a. The functional anatomy of the lower limb of the howler monkey (*Aloutta caraya*). *Amer. J. Phys. Anthrop.* 28:163–82.

—— 1968b. Functional anatomy of the upper limb. In M. R. Malinow, ed. *Biology of the Howler Monkey* (*Alouatta caraya*). *Bibl. Primat.* 8:104–25. Basel and New York: S. Karger.

Gregory, W. K. 1922. *The Origin and Evolution of Human Dentition.* Baltimore: Williams and Wilkins.

Hamilton, W. D. 1963. The evolution of altruistic behavior. *Amer. Natur.* 97:354–56.

—— 1964. The genetical evolution of social behaviour I, II. *J. Theoret. Biol.* 7:1–52.

Hershkovitz, P. 1977. *Living New World Monkeys* (*Platyrrhini*) *With an Introduction to Primates.* Chicago and London: University of Chicago Press.

Hill, W. C. Osman. 1953–1974. *Primates: Comparative Anatomy and Taxonomy.* Vols. 1–8. Edinburgh: Edinburgh University Press.

Hooton, E. A. 1946. *Up from the Ape.* Rev. ed. New York: Macmillan.

Huxley, T. H. 1863. *Man's Place in Nature.* London: Macmillan. Reprinted 1959. Ann Arbor: University of Michigan Press.

Jones, F. Wood. 1929. *Man's Place Among the Mammals.* London: Arnold.

Keith, A. 1931. *New Discoveries Relating to the Antiquity of Man.* London: Williams and Norgate.

Kummer, H. 1971. *Primate Societies: Group Techniques of Ecological Adaptation.* Chicago: Aldine.

Lack, D. 1968. *Ecological Adaptations for Breeding in Birds.* London: Methuen.

Le Gros Clark, W. E. 1959. *The Antecedents of Man.* Edinburgh: Edinburgh University Press.

Linnaeus, C. 1758. *Systema Naturae. Regnum Animale.* 10th ed. tomus I. L. Salvii, Holminae.

Lovejoy, C. O. 1981. The origin of man. *Science* 211:341–50.

McHenry, H. M. 1972. The postcranial anatomy of early Pleistocene hominids. Doctoral dissertation. Cambridge, Mass.: Harvard University.

—— 1975. Biomechanical interpretation of the early hominid hip. *J. Hum. Evol.* 4:343–56.

—— 1982. The pattern of human evolution: Studies on bipedalism, mastication and encephalization. *Ann. Rev. Anthropol.* 1982. 11: 151–73.

McHenry, H. M., and R. S. Corruccini. 1975. Multivariate analysis of early hominid pelvic bones. *Am. J. Phys. Anthrop.* 43:263–70.

—— 1976. Fossil hominid femora and the evolution of walking. *Nature* 259:657–58.

—— 1978. Analysis of the hominoid os coxae by Cartesian coordinates. *Amer. J. Phys. Anthrop.* 48:215–26.

Mendel, F. 1976. Postural and locomotor behavior of *Alouatta palliata* on various substrates. *Folia primatol.* 26:36–53.

Morbeck, M. E. 1975. Positional behavior of *Colobus guereza:* A preliminary quantitative analysis. In S. Kondo, M. Kawai, A. Ehara, and S. Kawamura, eds. *Symposia of the 5th Congress of the International Primatological Society,* pp. 331–43. Tokyo: Japan Science Press.

—— 1977a. Positional behavior, selective use of habitat substrate and associated nonpositional behavior in free-ranging *Colobus guereza* (Ruppel 1835). *Primates* 18:35–58.

—— 1977b. Leaping, bounding and bipedalism in *Colobus guereza:* A spectrum of positional behavior. *Yearbook of Phys. Anthrop.* 1976. *Yearbook Series* 20:408–20.

Morris, R. and D. Morris. 1966. *Men and Apes.* New York: McGraw-Hill.

Napier, J. R. 1963. Brachiation and brachiators. *Symp. Zool. Soc. Lond.* 10:183–95.

Nissen, H. W. 1931. A field study of the chimpanzee. *Comp. Psychol. Monogr.* 8:1–122.

Oxnard, C. E. 1973. Form and pattern in human evolution: Some mathemetical, physical, and engineering approaches. London and Chicago: University of Chicago Press.

Reynolds, V., and F. Reynolds. 1965. Chimpanzees of the Budongo Forest. In DeVore, ed. 1965:368–424.

Ripley, S. 1967. The leaping of langurs: A problem in the study of locomotor adaptation. *Amer. J. Phys. Anthrop.* 26:149–70.

Schaller, G. 1963. *The Mountain Gorilla.* Chicago: University of Chicago Press.

Schoener, T. W. 1971. Theory of feeding strategies. *Ann. Rev. Ecol. Syst.* 2:369–404.

Southwick, C. H. 1963. *Primate Social Behavior.* Princeton, N.J.: Van Nostrand.

Struhsaker, T. T. 1969. Correlates of ecology and social organization Among African cercopithecines. *Folia primatol.* 11:80–118.

Trivers, R. L. 1971. The evolution of reciprocal altruism. *Quart. Rev. Biol.* 46:35–57.

—— 1972. Parental investment and sexual selection. In B. Campbell, ed. *Sexual Selection and the Descent of Man, 1871–1971,* pp. 136–79. Chicago: Aldine.

Tuttle, R. H. 1967. Knuckle-walking and the evolution of hominoid hands. *Amer. J. Phys. Anthrop.* 26:171–206.

—— 1969. Quantitative and functional studies on the hands of the Anthropoidea. I. The hominoidea. *J. Morphol.* 128:309–64.

Tuttle, R. H., and J. V. Basmajian. 1974a. Electromyography of brachial muscles in *Pan gorilla* and hominoid evolution. *Amer. J. Phys. Anthrop.* 41:71–90.

—— 1974b. Electromyography of forearm musculature in gorilla and problems related to knuckle-walking. In F. A. Jenkins, ed. *Primate Locomotion,* pp. 293–347. New York: Academic Press.

—— 1977. Electromyography of hominoid shoulder muscles and hominoid evolution I. Retractors of the humerus and rotators of the scapula. *Yearbook of Phys. Anthrop.* 1976. *Yearbook Series* 20: 491–97.

Washburn, S. L., and I. DeVore. 1961a. Social behavior of baboons and early man. In S. L. Washburn, ed. *Social Life of Early Man,* pp. 91–105. Viking Fund publications in Anthropology No. 31. New York: Wenner-Gren Foundation.

Weiner, J. S. 1955. *The Piltdown Forgery.* London and New York: Oxford University Press.

Wynne-Edwards, V. C. 1962. *Animal Dispersion in Relation to Social Behaviour.* London: Oliver and Boyd.

1 On the Use of Anatomical Features To Infer Foraging Behavior in Extinct Primates

RICHARD F. KAY

THE BEHAVIORIST studying adaptations for foraging in a nonhuman primate has available a living, breathing subject. The species on which study focuses is readily recognized from other sympatric species, and with luck and perseverance a wealth of observational data can be obtained concerning its behavior, physiology, and ecology. From this information may be derived important conclusions about the animal's adaptations. Such a situation is in marked contrast to that facing the paleontologist who wishes to reconstruct the adaptations of an extinct species. Most avenues of information available to the behaviorist are blocked. Fossil assemblages are made of a mixture of animals that may or may not have lived together in a single community. The remains of fossil animals are very fragmentary, and plant remains are even more so. Only rarely can we iden-

RICHARD KAY studies primate and other dentitions and their functional differences from comparative and experimental perspectives. His work provides an empirically sound basis for interpreting function and inferring dietary patterns in fossil dentitions. The following paper summarizes his own work thus far and reviews other relevant research to present a concise statement of the current state of understanding of form, function, and diet through observation of body parts normally preserved in fossils.

tify what an animal ate, where it ate, and with what other species it may have been in competition. Indeed, recognition of the distinctiveness of closely related sympatric species may be very difficult. Virtually the only sources of information on which a species' adaptations may be estimated are bones and teeth.

Information about an animal's adaptations based on anatomy are of two sorts, being derived either from the structure of the animal's bones and teeth or from direct information about the uses to which these structures were put in life. For example, from the dentition we may derive information about an animal's feeding ecology based on either the unworn morphology of its teeth or the wear to which these teeth were subjected during life. Either source of information can be interpreted only on the basis of analogy with living species.

The interpretation of probable adaptations of an extinct species from morphology is possible only by means of analogy because a direct correlation between structure and function can be made only for living organisms. Kay and Cartmill (1977) proposed four criteria that need to be satisfied before a morphological trait of an extinct animal may be assumed to have particular adaptive meaning:

1. There must be some living species that has the morphological trait. If the morphological trait of the extinct species is unique, there is no analogy possible.
2. In all extant species that possess the morphological trait, the trait has the same adaptive role.
3. There must be no evidence that the trait evolved in the lineage before it came to have its present adaptive role.
4. The morphological trait in question must have some functional relationship to an adaptive role.

Although these four rules make common sense, it is remarkable how often those who reconstruct the adaptations of extinct species fail to adhere to them, with disturbing consequences. Rule one is least violated, but numerous examples may be cited when rules two through four have been overlooked. Three examples will suffice.

1. Living Old World colobine monkeys have two well-developed cross-lophs on their teeth (the bilophodont condition).

These and other crests on their teeth aid in cutting up fiber in their leaf diet. The bilophodont dentition, however, is seen also among fruit-eating species (e.g., species with low-fiber diets). Therefore an inference about dietary adaptation is not warranted simply because an animal has bilophodont molars per se. For whatever reason bilophodonty evolved, it has served as a basis for many sorts of dietary variations. The equation bilophodonty = folivory violates rule two above.

2. Living humans, and their Plio-Pleistocene forerunners, the australopithecines, have cheek teeth covered with an extremely thick layer of enamel, much thicker than typically found among living apes or Old World monkeys. Furthermore, *Homo* and *Australopithecus* are fully erect bipeds and therefore terrestrial. It has been suggested frequently that thick enamel evolved in hominids as an adaptation for feeding on various hard foods obtained by foraging terrestrially in semiarid, open woodlands. Thick enamel was taken as a marker indicating that the ramapithecines (possible ancestors of australopithecines), for which adequate postcranial evidence was lacking, were also terrestrial foragers if not actually bipedal. Thickened enamel, however, has evolved several times in parallel among arboreal primates that eat hard food, and to claim that thick enamel implies terrestriality violates rule two, which demands that all species with thick enamel also be terrestrial foragers. (Some of the environments where thick-enameled ramapithecines lived may have been more forested than has been thought previously. For example, tree shrews and lorises have been recovered from ramapithecine-bearing sediments in Asia [Jacobs 1981].) Therefore, although thick enamel may have been adaptively important for Plio-Pleistocene hominids foraging for hard food items in more open country, this morphological trait may have evolved in the hominid lineage *before* it came to have such an adaptive association. Inferring terrestriality from thick enamel is a violation of rule three.

3. Multivariate statistical analyses that are used to assess phenetic similarities among species abound with instances where no effort was made to establish *functional* links between morphological traits and adaptive roles. R. Smith's (1980) study of craniofacial morphology and its association with dietary adaptations in Miocene hominoids is an example of this. Smith takes a series

of cranial measurements on living species; assigns his species to groups by dietary, phylogenetic, and body-size criteria; and develops a mathematical model to predict diet, weighting those measurements most heavily that are best correlated with dietary pattern. Other major deficiencies in his study being ignored (e.g., use of genera rather than species as unit taxa, use of one female specimen only for each taxonomic unit, etc.), Smith makes no effort to establish plausible functional links between his measurements and the diets with which they apparently correlate, a violation of rule number four. One cannot tell whether some or all of the features are spuriously highly correlated with the dietary pattern. Smith's model is equivalent logically to the following argument: all primates have a middle-ear cavity roofed over by the petrosal bone. All primates are arboreal. Therefore, the presence in a fossil species of a middle-ear cavity covered by the petrosal bone indicates that species must have been arboreal.

Anatomical Models Relating to Foraging

Several aspects of foraging strategy are associated with anatomical peculiarities that might be preserved in the fossil record. In this section I consider anatomical parameters that may be indicative of what an animal eats, how and where its foods are obtained, and at what time of day it forages.

There has been little agreement among behaviorists or anatomists about the best way to characterize the kinds of foods available to primate consumers. Part of the problem is that an adequate characterization of food types depends, to a large degree, on what aspect of an animal's adaptations is being dealt with. If we are concerned with the masticatory system, we might be most interested in the physical and nutritive properties of the ingested food. On the other hand, if we are studying locomotor adaptations, we might wish to know from what part of the environment the food derives (e.g., high canopy, shrub layer, etc.) and how widely it is distributed in the environment. A still different sort of information would be useful to those who wish to look at adaptations of the visual system, namely, the part of the animals' daily cycle during which they forage. Inevitably, not all the sorts of information useful for understanding adaptive patterns are available. But some information is obtainable from the behavioral literature to analyze several aspects of foraging adaptations of primates based on cranial and dental structure, including the kinds of food eaten, the vertical distribution in the forest, the daily activity cycle, the means of locating prey items, and once located, how the foods are "subdued" and how a piece is separated before it is actually chewed and swallowed.

Primate Diets

A division into three gross dietetic types—insectivores, folivores, and frugivores—is commonly employed by behaviorists (see Clutton-Brock and Harvey 1977), and this system has also been employed by anatomists (e.g., Kay 1975; Kay and Hylander 1978). It is generally recognized that there is wide variation in diet within each group. Thus, the frugivore category incorporates species that may feed extensively on invertebrates and so on. These labels should be recognized for what they are—extreme simplifications of a very diverse spectrum of food types. Here and throughout, the term *frugivore* signifies an animal that eats a high percentage of fruit, nuts, or other plant foods that tend to have rather low proportions by dry weight of structural carbohydrate and low proportions of protein (except for nuts). Similarly, a folivore eats a lot of leaves, shoots, stems, or buds that, grossly speaking, tend to have much more fiber and protein than fruit does. Insectivores eat primarily insects, but this category may also signify a diet containing a large amount of noninsect animal food. Insect foods contain large amounts of readily accessible protein but are often invested with tough exoskeletons made of chitin, which, in chemical and physical structure, resembles plant structural carbohydrate.

Certain types of dietary specialization fit less readily into one of these categories. Thus, plant exudates (gums) are similar to extremely soft fruit in physical consistency but present a unique set of difficulties with respect to accessibility in the arboreal environment and digestibility once swallowed (Bearder and Martin 1980; Charles-

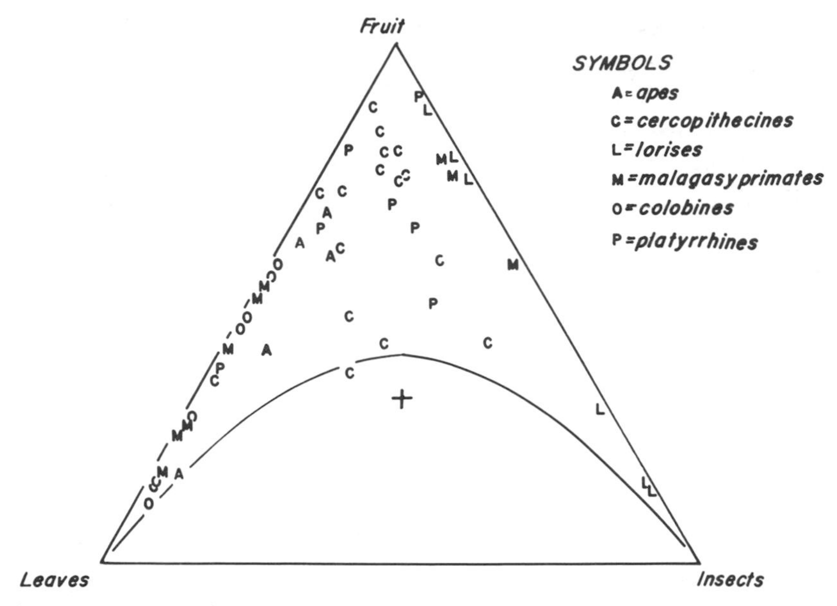

Figure 1.1. Ternary plot of the diets of 50 primates.

Dominique 1977). Similarly, grass resembles tree leaves in containing large amounts of structural carbohydrates but contains silica, which abrades the teeth of the species that feed on it to a degree not seen in species that eat tree leaves (Walker et al. 1978).

Figure 1.1 summarizes the dietary patterns of the fifty best known nonhuman primates. Foods are assigned to one of three categories—leaves, fruit, or insects, as discussed above. The diet of each species, represented by a symbol, is plotted on the figure according to the percentage of food types it eats; 100% of any of the three types would be plotted in one of the three corners of the diagram. The cross in the center is the point where a species would fall if it ate 33% of each food type.

Two generalizations emerge from this schematic representation. First, very few of the well-known species are generalists—70% of them eat 60% or more of a single food type, and in 96% of the cases examined, the first two food types (fruit plus leaves or fruit plus insects) add up to at least 80% of the diet. The rarest kind of primate in terms of its dietary pattern is the generalist; it is more common for a species to be quite highly specialized in only a single dietary category. Second, one never encounters a species that complements a primarily folivorous diet with insects as the second preferred food type and vice versa. Are there behavioral or physiological reasons for why there are no leaf-insect omnivores? The answer lies in an assessment of the implications of body size for an insectivorous or folivorous foraging strategy and in the nature of plant and animal foods available to arboreal consumers.

Diet and Body Size

Insect-eaters must be small because insects are small and hard to catch. Consider the following model, which predicts the number of insects an insect-forager must obtain per hour as a function of body weight (figure 1.2). Allow for each species a standard metabolic rate,

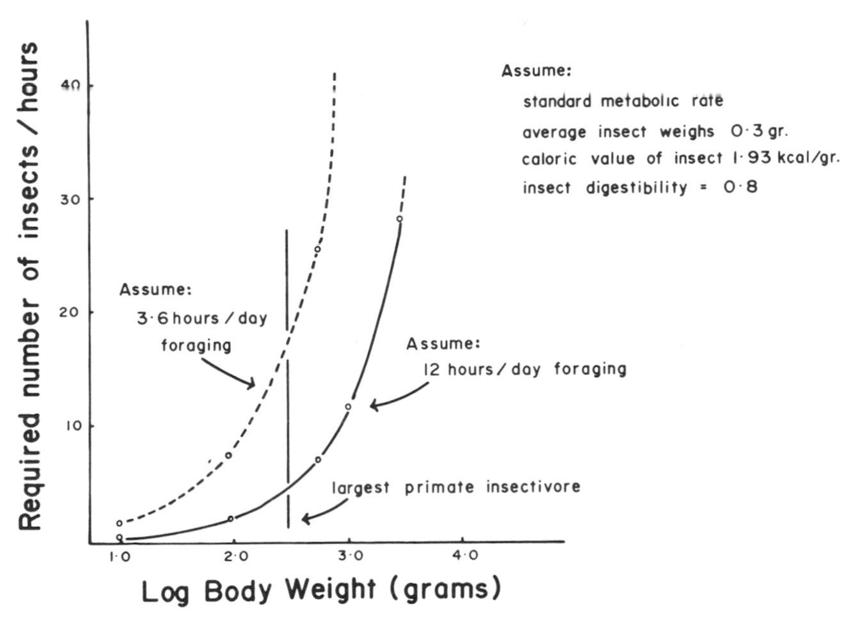

Figure 1.2. Model predicting the number of insect captures per hour as a function of body weight under two conditions: 12 hours/day foraging (solid line) and 3.6 hours/day foraging (dashed line).

such that an animal's total daily energy requirements (M) may be calculated from its body weight (W) by the equation (Kleiber 1961):

$$\text{Log } M = 1.83 + 0.756 \text{ Log } W \quad (1)$$

Assume that (1) a primate forager has 12h per day to forage; (2) on average, insects are cricket-sized; (3) the caloric value of an insect is about the same as for human flesh (1.93 Kcal/gr [Kleiber 1961]); and (4) 80% of the energy locked up in an insect can be added to the energy budget of the animal that consumes it. Given a typical primate foraging pattern where each insect is observed, captured, and eaten separately, this model predicts that a 1-kilogram primate insectivore would have to make about 150 separate cricket captures per day to survive. Actually, this number is probably on the conservative side, for the digestibility of a cricket may be somewhat less, and I have not figured in the energy cost of the foraging behavior (the metabolic values are calculated for animals at rest). Furthermore, insectivorous species more likely forage about 3.6 hours per day, as does *Galago senegalensis* (Bearder and Martin 1980). Clearly, above about 250–300 grams, a primate eating only insects finds itself running steeply uphill energetically. (Mammalian species that have beaten the energy race have found ways of locating huge numbers of insects in a limited amount of time. Termite-eaters and anteaters can do this. Once they breach the outer walls of the homes of their prey, vast quantities of insects are available.)

Leaf-eating presents a different set of problems for small mammals. Leaves have a comparatively high percentage of their calories locked up in structural carbohydrates, mainly cellulose and hemicelluloses. Aided by microorganisms in their digestive tracts, folivores can break down a small percentage of this material into assimilable form. This process is slow, and to be effective, requires considerable time; *in vitro* digestibility trials with sheep rumen fluid suggest that after 12 h, 35% to 70% of the digestible parts of grasses have been digested (McLeod and Minson 1969). Digestibility declines precipitously with less than 12 h allowed for digestion. As animals get smaller, their relative or weight-specific metabolic rates increase (see above), and they must process proportionately more food per gram body weight to fulfill these needs. But the faster they process structural carbohydrates, the lower the digestibility and the smaller the energy gain. Empirical observation suggests that the point below which folivores cannot process enough food fast enough to keep up with energy needs is about 700 g, the approximate body size of *Lepilemur mustelinus,* a folivorous malagasy prosimian, or similar-sized *Pseudocheirus peregrinus,* a phalangeroid marsupial.

All of this has an important application to the study of extinct primates. Animals that achieved an adult body weight of greater than about 350 g could not have been *primarily* insectivorous, meaning that no more than 30% to 40% of their energy needs could be filled by insect-eating. By the time 1 kg is reached, this percentage would be much less. At small size, an animal supplementing its insectivorous diet with leaves would compound its difficulties; below about 700 g body weight, folivory is difficult to sustain energetically. We should expect to find extinct folivores only at or above 700 g body weight. All of this is illustrated in figure 1.3, in histograms of $\log_{10}$ body weight for extant primate species. Insectivorous species are smaller than folivorous ones, meaning that it is possible to distinguish all extant primate insectivores from extant folivores *by body size alone.*

Three peaks are noted in the body size distribution of fruit-eating arboreal primates (fig. 1.3). This distribution may be explained by the limitation of fruit as a protein source. According to Hladik et al. (1971) and Hladik (1977), the fruits eaten by some groups of primates contain large amounts of readily available carbohydrates but small amounts of protein. Primate fruit-eaters must eat other kinds of foods to get their protein. Two such protein sources are leaves and insects, each of which contains more than 20% protein by dry weight (Hladik et al. 1971; Hladik 1977; Boyd and Goodyear 1971). The choice of leaves or insects as a source for protein in the diet of frugivorous primates is an important element in selection acting on body size. Frugivores that concentrate on insects as a source for their protein tend to be relatively small, whereas frugivores that obtain their protein from leaves tend to be relatively large. The intermediate peak in the histogram for frugivores in figure 1.3 is made up of fruit-eaters with a secondary specialization in insects and fruit-eaters with a secondary specialization in leaves.

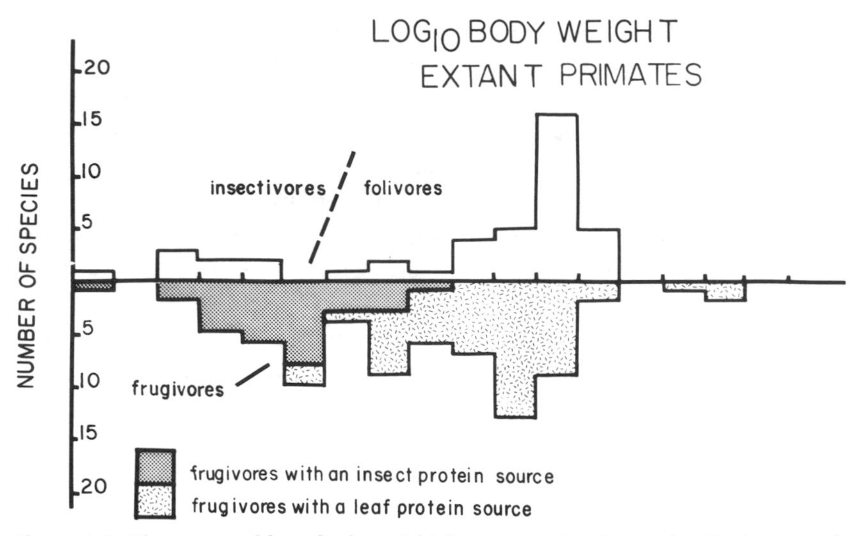

Figure 1.3. Histograms of $\log_{10}$ body weight for extant primate species. Each species is assigned to a dietary category—frugivory, insectivory, or folivory—according to the best behavioral evidence. Frugivorous species are assigned to one of two further categories according to whether they eat large additional amounts of leaves or insects. Each interval in the histogram represents a 0.2 $\log_{10}$ interval. Figure modified from Kay and Simons (1980).

Given the high correlations between length of the second lower molar (M_2) and body weight (Kay and Simons 1980), we may provide a rough predictive guideline for dietetic estimates in extinct species: a mean M_2 length above about 3.2 mm implies a folivorous, frugivorous, or gummivorous diet; a species with M_2's smaller than this probably ate insects, fruits, or gum, but not leaves. Although the molar adaptations to feeding on a diet primarily of insects greatly resemble those of leaf-eaters (see below), this body size Rubicon makes it unlikely that insectivorous species can simply evolve into folivorous ones without going through an intervening frugivorous stage.

Diet and Molar Shearing

The anatomy of the teeth is crucial to reconstructing dietary patterns. A series of studies have established clearly that details of dental anatomy, especially the relative development of the molar crests that cut food, are strongly and functionally correlated with dietary pattern.

The dentitions of mammals are compartmentalized according to several conflicting functional and adaptational demands. Generally the front teeth (the incisors, and to some degree the canines) are modified for obtaining food, and the cheek teeth are designed for chewing it up effectively before swallowing. Among primates, many additional nonfeeding selective demands are placed on the anterior teeth. For example, the incisors in strepsirhines are modified for use in fur-combing in addition to their ingestive roles, which range from bark prying (*Propithecus*) to gumscraping (*Galago elegantulus*). Similarly, among anthropoids, the canines are sexually dimorphic, which reflects the different ways they are used in various social roles (Fleagle et al. 1980), but are also used for cracking open hard objects and for other ingestive purposes. Furthermore, it is often the case among primates (and mammals generally) that, once evolved, a particular incisor morphology becomes useful for a variety of functions that cross-cut dietary patterns. For example, the continuously growing incisors of rodents have become useful in a variety of adaptive roles such as bark-cutting and nut-splitting. Thus, because incisor morphology functions represent a compromise for feeding and nonfeeding adaptive roles, and because incisor use in feeding tends often to be nonspecific for particular diets, incisor morphology can be misleading for dietary reconstruction in fossils. I shall return later to the consideration of incisor morphology as an indicator of ingestive behavior.

Molar morphology generally provides a better guide to dietary adaptation than incisor structure because these teeth have a single adaptive role, the mastication of food before swallowing. Molars are composed of a series of cutting blades and crushing basins. As the teeth occlude, matching crests on the upper and lower teeth fit together, and intervening food is shredded. Matching dental surfaces crush the interposed food as the teeth come together. Once the details of the occlusal patterns are known, we can derive a picture of the relative efficacy of these devices for cutting and crushing food by measuring crest lengths and crushing areas. The physical and chemical properties of food should be expected to influence the relative expression of cutting and crushing features on the teeth, and because of major variation in the physical and chemical characteristics of fruits, leaves, and insects, selection should favor different functional adaptations for species that eat different things.

We should expect the molars of leaf-eating primates to resemble those of insect-eaters, and both to differ from fruit-eaters for two reasons. First, there are physical similarities between chitin (the major constituent of insect exoskeletons) and plant fiber found in leaves. Both take the form of resistant sheets or rods, making them more easily broken down by cutting or slicing devices than by crushing systems. (One can more easily reduce paper to small pieces with a pair of scissors than with a hammer, and similarly, trees are more easily trimmed with a saw than with a mallet.) On the other hand, fruits or nuts, which contain relatively less fiber and are more three-dimensional, may be more effectively deformed and broken into smaller pieces by crushing. So we might expect insectivorous or folivorous species to emphasize shearing and cutting devices on their molars whereas these would be much less developed in frugivores.

A second reason why we might expect functional convergence in the molar systems of insect-eaters and leaf-eaters has to do with chemical similarities between the fibrous portions of both foods. The digestibility of chitin in insect exoskeletons and of cellulose and hemicelluloses in leaves increases as they are more finely cut up, whereas it appears that nonstructural carbohydrates or proteins are digested relatively completely whether chewed or not (Van Soest 1977; but cf. Janis 1976; Levine and Silvis 1980). Therefore, species that eat a lot of structural carbohydrates, whether chitin or plant fiber, will derive more digestive (and energetic) benefit by finely chewing their foods than will those that eat foods that are low in structural carbohydrate.

Various tests confirm the prediction that insectivorous or folivorous species have better developed shearing or cutting features on their molars than closely related frugivorous species. Two examples illustrate this point dramatically.

1. *West African lorises.* Five lorisid species that occur in the large African equatorial rain-forest block are sympatric in Gabon, where

Table 1.1. Diets and Shearing Quotients of West African Lorises

	Dietary Components (%)[a]		$SQ(M_2)$		
	Fruit and Gum	Invertebrates	$\bar{X}$	S.D.	N
Arctocebus calabarensis	13	87	+11.9	4.93	6
Galago demidovii	18	72	+8.4	5.62	8
Galago alleni	81	19	−0.8	3.51	7
Galago elegantulus	81	19	−5.2	4.82	6
Perodicticus potto	89	11	−21.4	4.71	8

[a] All data from Charles-Dominique (1977) are based on average fresh weights of stomach and/or cecum contents.

they were the subject of an ecological study by Charles-Dominique (1977), who analyzed the stomach and intestinal contents of each species. His data are summarized in table 1.1. Although presumably there are biases in this technique of assembling a dietary record (some kinds of foods may be digested relatively rapidly and will be preserved in the digestive tract in lower proportions than their actual occurrence in the diet), this does not affect the relative abundances for between-species comparisons.

Also presented in the table is a measure of the relative importance of the shearing features on the molars of the same species, expressed as a single number, the shearing quotient (SQ). To derive SQ, I first measured the lengths of the lower second molar and six shearing crests found on it in lemurs and lorises. My sample consisted of 28 species of lemurs and lorises. The lengths of the six crests were summed. The log of the mean summed shearing crest length for each species (dependent variable) was expressed as a function of the log of mean M_2 length by a least-squares regression equation. This equation was used to predict the amount of shearing from a known M_2 length for individual specimens of each species. The predicted value was compared with the observed shearing sum and the relative difference expressed as a percentage: SQ = [(observed summed shearing − expected summed shearing) / expected summed shearing] × 100. The larger the SQ, the better the relative development of molar shearing. In table 1.1, the most insectivorous species, *Arctocebus calabarensis* (87%) and *Galago demidovii* (72%) rank first and second in SQ. The least insectivorous species, *Perodicticus potto,* has the lowest SQ. *Galago alleni* and *Galago elegantulus* fall in between those extremes both in insectivory and SQ. Although significant differences in SQ were encountered between the two, a similar dietary separation was not observed by Charles-Dominique (1977). This relationship between the rankings of the percentage of insects in the diet and SQ is remarkably consistent and confirms the prediction that insectivorous primates have better developed molar shearing than their frugivorous or gummivorous close relatives. The correlation might be even higher if more were known about the diet of *Galago alleni;* Charles-Dominique's dietary determination for this species was based on just 12 animals collected at unspecified times over an 8-year period.

2. *Old World monkeys.* The food habits of five monkeys living in western Uganda were summarized by Struhsaker (1978). He collected feeding data by direct observation of habituated animals, scoring the animals' activities at short regular intervals over periods of several days. His data are directly comparable to those of McKey (1979) on West African black colobus, and so all these species will be considered together. Table 1.2 summarizes the feeding observations from Struhsaker and McKey. Also presented is a measure of the relative importance of shearing features on the molars of the same species, a shearing quotient (SQ). This number was derived in a

Table 1.2. Diets and Shearing Quotients of Ugandan and West African Anthropoids

	Diet (%)[a]			$SQ(M_2)$[b]		
	Invertebrates	Nectar, Fruit or Seeds	Leaves	$\bar{X}$	S.D.	N
Cercocebus albigena[c]	11–26	62	12	−12.2	4.05	35
Cercopithecus mitis[c]	20	56	24	−3.4	3.90	32
Cercopithecus ascanius[c]	22	48	30	−8.0	4.79	57
Colobus satanas[d]	5	53	42	+3.7	3.62	11
Colobus badius[c]	3	13	84	+16.0	5.61	57
Colobus guereza[c]	0	15	85	+12.2	5.01	20

[a] Diet is expressed as a percentage of total feeding observations.

[b] Shearing quotient (S.Q.) is calculated as for table 1.1 but based on eight shearing crests for 73 species of Old World monkeys where SE = 2.79 M_2 length 0.982.

[c] Data on diet from Struhsaker (1978).

[d] Data on diet from McKey (1979).

fashion similar to that for the loris example reviewed above with the following modifications. First, eight, rather than six, shearing crests are important on the lower second molars of cercopithecids, and so all were used. Second, the SQs were calculated from measurements of 1,172 specimens in 73 species of extant cercopithecids. Thus, the SQs for cercopithecids should not be compared directly with those for lorises; they are an internally consistent measure of relative shearing crest development *within Cercopithecidae only.* As with the loris data, a larger SQ indicates better development of the shearing crests on the lower second molar, taken to represent the molar dentition as a whole. In table 1.2, the most folivorous species, *Colobus guereza* (85%) and *Colobus badius* (84%) rank second and first in SQ, respectively. The least folivorous species, *Cercocebus albigena* (12%) has the lowest SQ. Given my observation that chitin has physical and chemical properties similar to those of leaves, it may be questioned why I included invertebrate feeding time with time spent eating fruit and not with leaf-eating. I have done so because Struhsaker (1978) notes that *Cercopithecus* monkeys discard the chitinous portions of insects and eat only the soft contents. Invertebrate foods treated in this way are more similar to fruits in their digestibility and physical properties. In terms of folivory, *Cercopithecus ascanius* (30%) and *C. mitis* (24%) fall between the extremes of *Cercocebus* and the aforementioned *Colobus* species and have intermediate SQs.

McKey (1979) reported on the dietary pattern of the black colobus, *Colobus satanas* from the Cameroons. This species eats 42% leaves, only one half the amount eaten by Ugandan *Colobus* species, but more than that eaten by either *Cercopithecus mitis* or *C. ascanius.* The SQ of *C. satanas* is much lower than Ugandan *Colobus* species but still above that seen in Ugandan *Cercopithecus* species. Thus, comparisons of closely related African monkeys strongly support the association between the degree of development of molar shearing and folivory.

Diet and Enamel Thickness

The adaptive significance of another aspect of molar anatomy, the relative thickness of the enamel that invests the molars, has been the subject of much speculation because fossil hominids have relatively very thick enamel. Recently, I quantified molar enamel thickness for Old World monkeys and apes and considered relative enamel thickness in relation to several functional and adaptive parameters of these species (Kay 1981). I found that species with thick enamel reportedly often crack open and eat the contents of extremely hard nuts.

Figure 1.4 is a plot of enamel thickness versus M_2 length in extant catarrhines. The dashed line is the least-squares regression line with enamel thickness as the dependent variable. Plus or minus 20% of expected values are indicated by solid lines parallel to the dashed

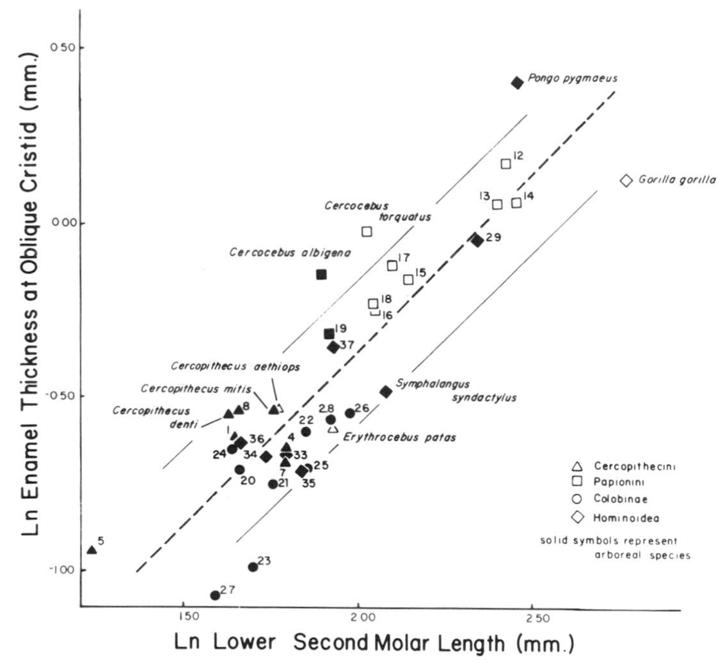

Figure 1.4. Enamel thickness at the M_2 cristid obliqua and M_2 length (in mm). Each symbol represents a species. A few species are named. The remainder are keyed as follows: 1. *Cercopithecus cephus;* 4. *Cercopithecus neglectus;* 5. *Cercopithecus talapoin;* 7. *Cercopithecus lhoesti;* 8. *Cercopithecus ascanius;* 12. *Papio* species; 13. *Mandrillus sphinx;* 14. *Theropithecus gelada;* 15. *Macaca speciosa;* 16. *Macaca fuscata;* 17. *Macaca sylvana;* 18. *Macaca nemestrina;* 19. *Macaca fascicularis;* 20. *Presbytis melalophos;* 21. *Presbytis senex;* 22. *Presbytis pileatus;* 23. *Presbytis obscurus;* 24. *Presbytis frontatus;* 25. *Simias concolor;* 26. *Nasalis larvatus;* 27. *Colobus verus;* 28. *Colobus angolensis;* 29. *Pan troglodytes;* 33. *Hylobates moloch;* 34. *Hylobates lar;* 35. *Hylobates concolor;* 36. *Hylobates klossi;* 37. *Hylobates hoolock.*
Figure is from Kay (1981). See this paper for further details.

line. Arboreal foragers are represented as solid symbols, terrestrial foragers are indicated by open symbols. One of the most obvious features of this plot is that no relationship exists between an arboreal or terrestrial feeding strategy per se and enamel thickness.

A strong negative correlation is found between relative enamel thickness and relative shearing crest development in cercopithecoids (Kay 1981). The association of well-developed molar shearing with thin molar enamel apparently has a functional basis. Thin molar enamel is rapidly perforated by dental wear. Relatively softer, more rapidly wearing dentin appears in enamel windows at the apices of tall, sharp, principal molar cusps. The enamel windows spread rapidly along the principal enamel crests with continued wear. The raised edges of the worn enamel on crest margins form sharp edges that facilitate shredding and slicing during mastication.

A molar system emphasizing high cusps and thin enamel is associated with a leaf-eating diet. In such a diet, maximal food shredding is at a premium because leaves contain large quantities of structural carbohydrate, whose digestion is greatly facilitated by more effective reduction of particle size. At the opposite extreme, species with poorly developed molar shearing crests have relatively thicker enamel. This configuration produces low crown-relief, particularly with advanced stages of wear. Thick enamel is perforated at the apices of the low cusps only after considerable wear. By the time the apical enamel windows have expanded along the principal shearing crests, the molar basins are worn relatively flat. This sort of dental structure may be an adaptive response to selection for the more uniform distribution of very high occlusal forces engendered while masticating hard, tough food objects. Sharp, delicate cusps would be subject to excessive stress concentrations. Molar crowns covered with thick enamel should also have better resistance to wear than those with thin enamel, other factors being equal. On the other hand, poorly developed shearing crests lead to a reduced ability to subdivide food particles maximally, which in turn results in a dropoff in the digestibility of structural carbohydrates. This organization might be expected in species that masticate hard foods but do not use structural carbohydrate as a primary source of energy. A review of feeding behavior in species with thick enamel tends to support these assumptions.

Among extant species, the extreme in enamel thickness is documented in mangabeys (*Cercocebus*), orangutans (*Pongo*), and *Cebus apella*. Each of these thick-enameled species takes advantage of very hard nuts, seeds, and fruits that other arboreal monkeys and apes cannot eat. *Cercocebus albigena* has a diet consisting largely of fruit and berries (Waser 1977). Its powerful, specialized jaws allow it to consume tough fruits such as palm nuts, which most other forest monkeys cannot handle.

MacKinnon (1977) reports that greater size and strength enable orangutans to feed on a variety of hard fruits, and also very large fruits, that gibbons cannot tackle. In the area of his study, such fruits accounted for 35% of all fruit-feeding records for orangutans. When orangutans feed on acorns, loud crunching sounds can be heard for well over 100 yards (MacKinnon 1974).

Cebus monkeys include in their diets the soft contents of palm nuts, resembling small coconuts, which are extremely hard and difficult to open. Struhsaker and Leland (1977) report that the coats of these nuts are so hard that for a human to open them requires a hammer or heavy stone.

Other mammals that have comparatively thick molar enamel apparently feed on very hard, tough foods. For example, peccaries, which have low-cusped teeth, are capable of cracking open hard nuts, and these constitute an important component of their diet. Kiltie (1979) reports that some of these nuts require sustained loads of more than 1000 kg per square inch before cracking.

In summary, extant species with thick-enameled molars generally have reduced molar shearing capabilities and inefficient designs for cutting foods up finely. Since dividing foods finely is at a premium among species that depend on structural carbohydrates for a large proportion of their energy, we may assume that the thick-enameled extinct species were *not* efficient consumers of leaves or other high-fiber foods. A survey of the diets of various relatively thick-enameled primate species reveals that they commonly masticate very hard foods such as seeds and hard nuts, which require large amounts of masticatory force to open but whose contents are easily digested.

Diet and Dental Wear

Recent interest has focused on the use of scanning electron microscopy (SEM) to study dental wear in primates and other mammals (Walker et al. 1978; Ryan 1979; Rose et al. 1981; Covert and Kay 1981; Kay and Covert 1983). In one well-documented instance, enamel microwear was shown to be altered by a seasonal variation in diet. The enamel microwear of two sympatric hyraxes, *Procavia johnstoni* and *Heterohyrax brucei,* were analyzed by Walker et al. (1978) using SEM. The microwear of the two species was similar during the dry season when both browsed on bush and tree leaves. During the wet season, however, *Procavia johnstoni* ate grass, and its microwear was quite distinct from *Heterohyrax brucei,* which remained strictly a browser. The molar enamel of grass-eating *P. johnstoni* was covered with many fine parallel striations, possibly produced by silica in the grasses. The browser's teeth were smooth and virtually lacking in striations.

Covert and Kay (1981; see also Kay and Covert 1983) fed different diets to laboratory opossums to simulate other common sources of dental wear in the natural diets of primates, such as chitin, plant fiber, and grit. The microwear differences between grit-fed and other animals were obvious and follow the lines of the distinction made by Walker et al. (1978) with species that do and do not have silica in their diet. Covert and Kay were, however, unable to distinguish other dietary patterns effectively. Thus, for fossil species, while microwear may be important for discrimination of grazers and/or high-grit terrestrial foragers from browsers (at least insofar as tree leaves are relatively free of silica, whereas grass leaves contain a lot of silica), its utility for other sorts of dietary discriminations remains doubtful. In addition, Walker's study suggests, and Covert and Kay's study confirms, that microscopic wear changes rather rapidly with seasonal diet shifts or modifications in test diets. For fossils, the wear patterns may reflect diet at only one unidentified part of the year and might be misleading for reconstructing overall feeding strategies.

Orbit Size and Activity Patterns

For nocturnal primates, the activity period begins in the twilight of the day; animals move to their diurnal resting sites at the first glimmer of dawn. Such a pattern is ubiquitous among lorises and in *Tarsius.* It is seen frequently in the lemurs but occurs in *Aotus* alone among anthropoids. Other species are diurnally active. Although nocturnal foragers tend to be small, there appears to be no relationship between this activity pattern and a particular diet. Highly folivorous species like *Lepilemur* and *Avahi* as well as insectivorous, frugivorous, and gummivorous ones are nocturnal. Likewise there is a great diversity of dietary patterns among diurnal species. The eyes of nocturnal primates differ from those of their diurnal relatives in several ways, reflecting their need for an eye that can function at low light intensities. Generally, nocturnal species have all-rod retinas as compared with the eyes of diurnal species, which have both rods and cones. Rod receptors can function at low light intensities, whereas cones, found together with rods in the eyes of diurnal species, are sensitive only to higher intensities of light. The sensitivity of the eyes in dim light is enhanced in some primates by a reflecting or diffusing system, the *tapetum lucidum,* a glistening, reflective part of the choroid layer. Finally, nocturnal primates commonly have larger eyes than comparably sized diurnal species, allowing them to have more rods and still preserve visual acuity. Most of these anatomical adaptations to nocturnal and diurnal activity patterns cannot be discerned in extinct species. There is, however, a strong correlation between orbit size, which crudely mirrors the size of the eye, and activity pattern. Figure 1.5 is a log-log plot of orbit size and skull length. Nocturnal primates fall above the line, indicating that they have relatively large eyes, whereas diurnal species fall below it, showing that their eyes are relatively smaller. Kay and Cartmill (1977) used this sort of data to estimate the probable activity patterns of primitive primates. They also discussed the problems and limitations of this technique.

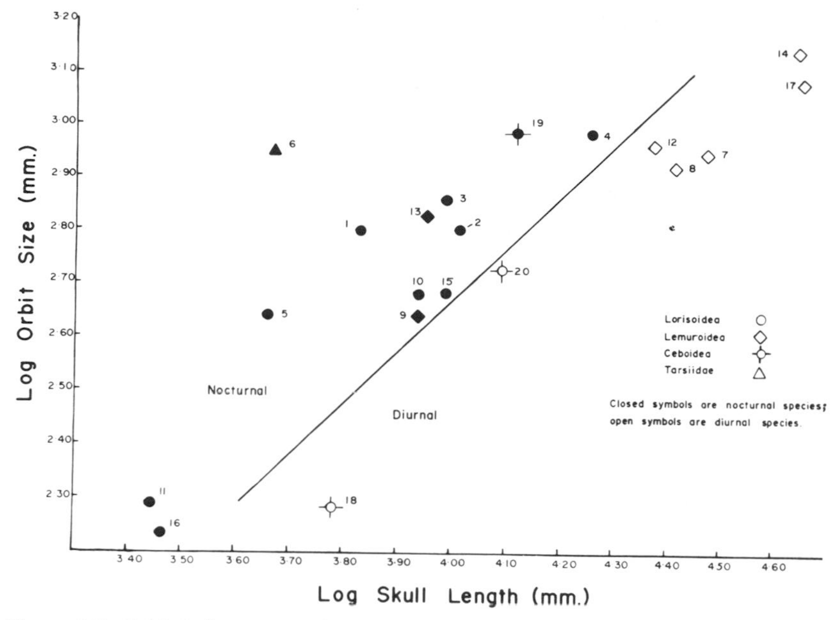

Figure 1.5. Orbital diameter and prosthion-inion length (measurements in mm). Data from Cartmill (1970); Kay and Cartmill (1977). Key to species is as follows: (1) *Loris tardigradus;* (2) *Perodicticus potto;* (3) *Nycticebus coucang;* (4) *Galago crassicaudatus;* (5) *Galago senegalensis;* (6) *Tarsius syrichta;* (7) *Lemur fulvus;* (8) *Lemur catta;* (9) *Lepilemur mustelinus;* (10) *Cheirogaleus major;* (11) *Microcebus murinus;* (12) *Propithecus verreauxi;* (13) *Avahi laniger;* (14) *Indri indri;* (15) *Phaner furcifer;* (16) *Galago demidovii;* (17) *Varecia variegatus;* (18) *Callithrix jacchus;* (19) *Aotus trivirgatus;* (20) *Callicebus moloch.*

Habitat Preference, Vertical Distribution and Anatomy

A widely recognized variable in foraging strategies of primates is the place in the habitat where foods are gathered and eaten. Variations in the use of forest space have been documented among sympatric primate species in Malaya, Uganda, West Africa, and Surinam (MacKinnon and MacKinnon 1978; Struhsaker 1978; Gautier-Hion 1978; Charles-Dominique 1977; Fleagle and Mittermeier 1980).

Several obvious contrasts in locomotor patterns, body size, and dental patterns have been documented for terrestrial versus arboreal foragers. Jolly (1966, 1970) has given details of many distinctions in the postcranial skeleton between arboreal and terrestrial cercopithecoids. Kay and Simons (1980) note differences in body size and molar crown structure between arboreal and terrestrial Old World monkeys. Terrestrial monkeys tend to be larger than arboreal ones, possibly as a response to predator pressure, and terrestrial monkeys tend to have higher molar crowns than their arboreal close relatives, possibly because they have more grit in their diets. In most instances, however, although these adaptations have an obvious association with an arboreal or terrestrial foraging strategy, they are not sufficiently strong to predict habitat use except in a probabilistic sense.

Considerable effort has been made and some success has been achieved in using postcranial anatomy to estimate locomotor patterns of arboreal primates, whether leaping, suspensory, or quadrupedal. Nonetheless, a report by Fleagle and Mittermeier (1980) shows little association between locomotor pattern and food types in the diet. "One can predict neither diet from locomotor behavior nor locomotor behavior from diet" (p. 313). Rather, South American monkeys illustrate how a large number of species with broadly similar diets can be packed into a forest and still avoid competition with one another: animals with grossly similar diets tend to show locomotor and stratification differences and vice versa.

Food Handling and Incisor Structure

Ingestion, the separation of a bite of food from its matrix for further processing, has involved many specializations of the anterior teeth. Ingestive adaptations are often, but not always, closely associated with the dietary categories mentioned above. A good example of this is the relative size of the incisors compared to body size in anthropoids. The relative size of anthropoid incisors is related to the extent of incisal preparation prior to mastication (Hylander 1975). Certain foods, including leaves, berries, grasses, seeds, buds, and flowers, may not ordinarily require any extensive preparation prior to mastication. In contrast, large, tough-skinned fruits may require an appreciable amount of incisor handling. Such extensive use produces large amounts of wear. Enlarged incisors may be the adaptive response to delaying the time when incisors wear out. The relationship between incisor size and food handling has produced a generally strong association between relative incisor size and dietary patterns among Old World anthropoids (Hylander 1975). In most cases the more frugivorous species have relatively larger incisors, implying more incisal preparation, whereas folivorous forms have relatively smaller ones and prepare their foods less before mastication.

There are, however, some unexpected differences in incisor size among species eating the same foods in different ways. Baboons of the genus *Papio* have large incisors, but those in the genus *Theropithecus* do not. The former often pulls grass rhizomes from the ground with its incisors, unlike the latter, which habitually pulls up the same rhizomes with its hands (Dunbar and Dunbar 1974). The species that selects food objects from the ground with its mouth ingests more grit than the one that eats the same food plucked by hand. Thus, increased grit means more dental wear, selecting for larger incisors.

In using evidence of small incisors to infer a dietary pattern it should be remembered that Hylander's model uses body weight as a basis for judging relative incisor size. He notes (personal communication) that there has been an uncritical tendency to use cheek tooth size as a basis for observations on relative incisor size. However, using this standard in place of weight does not separate living species adequately into groups of similar diet, apparently because relative cheek tooth size itself varies in a systematic fashion with respect to a different set of dietary parameters (Kay 1975).

The incisor structure of lemurs and lorises has been modified greatly for nonmasticatory functions. It is used in many species primarily as a comb for grooming the fur. Tooth combs show structural variation related to ingestive adaptations as well. Several authors have discerned that the tooth-comb is elongated in *Phaner furcifer* and *Galago elegantulus* commensurate with their use in scraping tree gum (Petter et al. 1971; Charles-Dominique 1977). Other species have stouter combs composed of fewer teeth, which are used to gouge bark either to eat it, as in *Propithecus,* or to reach insect larvae concealed underneath, as in *Daubentonia.* Similar sorts of adaptive modifications related to ingestion are seen in New World monkeys. Coimbra-Filho and Mittermeier (1977) review adaptations for bark-gouging and gum-collecting in the anterior teeth of callitrichids and other primates.

In summary, the key to the interpretation of the adaptive significance of incisor structure lies in the recognition of ingestion as distinct from mastication. Molar structure responds to selection for the physical and, to some extent, chemical properties of the food being eaten. Incisor structure by contrast may reflect the special requirements for obtaining foods. A one-to-one correspondence of food type and ingestive mechanism should never be assumed. One species may use its incisors to pry up bark it eats, while another may pry up the bark, discard it, and eat the grubs underneath. The key to assessing this adaptive difference would lie in a consideration of the structure both of the incisors and the molars. First one would infer the diet based on the molar structure; then one would evaluate the adaptive role of the incisors within this dietary context.

Discussion

Assessment of the behavior of an extinct species often requires a choice between conflicting analogical models based on different groups of extant species. It is assumed generally (as it is here) that a model should be chosen on the basis of anatomical and functional similarity. For example, the best model for interpreting the anatomy of a fossil ape is that based on living apes (provided it can be demonstrated that there is a close match functionally between the anatomy of the two groups). In most cases, the choice of models is obvious. For example, early Oligocene–Recent apes and parapithecids all show minor variations on a virtually identical molar pattern, and so Kay (1977, 1981) and Kay and Simons (1980) chose living ape molars as the model for assessing the diets of Oligocene and Miocene apes and parapithecids. On the other hand, the molar structure of cercopithecids is quite different from that of other anthropoids, and so an assessment of ape fossils based on the Old World monkey pattern would be inappropriate.

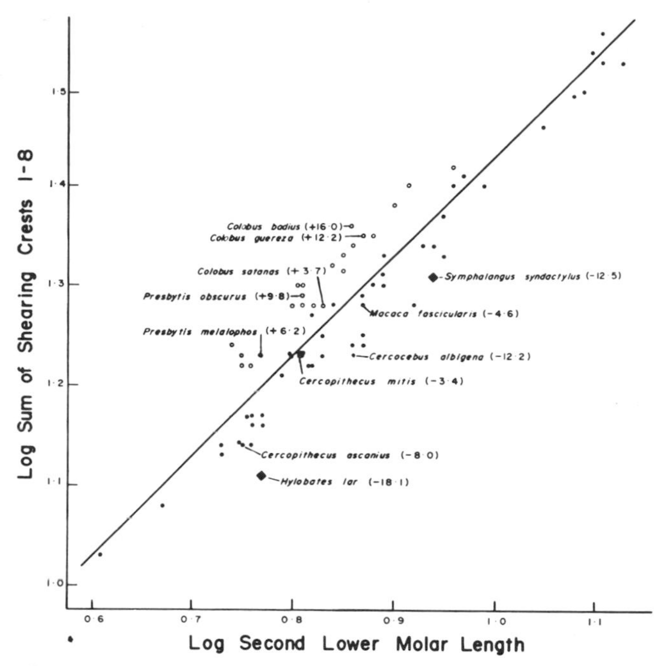

Figure 1.6. Sum of eight M_2 shearing crests and M_2 length (measurements in mm). Data not published elsewhere.

The functional differences between the molars of Old World monkeys and apes are worth considering in an adaptive and evolutionary perspective. Figure 1.6 is a log-log plot of the means of summed M_2 shearing crests and M_2 length for 50 species of cercopithecines, 23 species of colobines, and 2 species of hylobatids. The line represents the least-squares regression equation fit to the 73 cercopithecids. Generally, the colobines (open circles) fall above the regression line and the cercopithecines fall below it, illustrating the better overall development of shearing crests in the former in conjunction with their more folivorous habits. Also, as noted above on the basis of Struhsaker's data, individual species' SQs (given in parentheses in fig. 1.6) are informative vis-à-vis minor details of the diet. Among other species identified are five sympatric anthropoids from Malaya: two species of *Presbytis,* one of *Macaca,* and two hylobatids. The diets of these species are summarized in table 1.3, on the

Table 1.3. Diets and Shearing Quotients of Malayan Anthropoids

	Diet (%)[a]			$SQ(M_2)$[b]		
	Invertebrates	Fruit and Flowers	Leaves	$\overline{X}$	S.D.	N
Macaca fascicularis	17	63	20	−4.6	4.08	47
Presbytis melalophos	3	58	39	+6.2	4.51	19
Presbytis obscura	<1	44	56	+9.8	5.26	16
Hylobates lar	8	61	30	−18.1	—	5
Symphalangus syndactylus	8	44	48	−12.5	—	6

[a] Diet is expressed as a percentage of total feeding time. All feeding percentages are from Raemaekers and Chivers (1980).

[b] All SQs (including those for apes) are computed from an Old World monkey model in the manner described in Table 1.2.

basis of studies of the same groups over a ten-year period by five different observers (Raemaekers and Chivers 1980). The shearing quotients of the three cercopithecoids are consistent with the percentage of leaves in the diet: *Presbytis obscurus,* the most folivorous species, has the best developed shearing crests; *Presbytis melalophos,* which eats less leaves, has a significantly lower SQ. *Macaca fascicularis,* the least folivorous, has the lowest SQ. (Comparisons of Malayan and Ugandan data indicate that apparently none of the folivorous Malayan species eats as much leaves as either *Colobus badius* or *Colobus guereza,* and none has as high an SQ as these African species. Similarly, *Macaca fascicularis* probably eats more leaves than *Cercocebus albigena,* which is consistent with its lower SQ. Compare tables 1.2 and 1.3.)

In comparing cercopithecids with sympatric Malayan hylobatids, however, our model breaks down in part (fig. 1.6; table 1.3). *Symphalangus syndactylus* is more folivorous than *Hylobates lar,* which is consistent with its comparatively larger SQ, calculated on the basis of the cercopithecid model. Hylobatids, however, irrespective of diet, have much less developed shearing crests than sympatric cercopithecids do with apparently similar diets. *Macaca fascicularis* eats lower percentages of fruit than *Hylobates lar* but has much better developed shearing. *Symphalangus* eats more leaves than *Presbytis obscurus* but has a much lower SQ. What is the behavioral or phylogentic significance of hominoids' having less well-developed shearing than cercopithecids with similar diets?

To some degree the anatomical difference between hylobatids and cercopithecids, which supposedly eat the same food in similar proportions, actually reflects problems with the method of segregating food types. Within the fruit category, for example, hylobatids appear to be much more dependent on ripe fruit of a few species and tend to avoid unripe fruits (Raemaekers and Chivers 1980). Cercopithecids tend to be more eclectic in their fruit-eating. They readily exploit many sources of unripe fruit as they become available. Within the leaf-eating category, hylobatids avoid mature leaves and concentrate on a few species of new leaves and buds, whereas cercopithecids eat more mature material and tend to be less choosy. Data on the chemical composition of the foods eaten by sympatric hylobatids and cercopithecids are unavailable, but on the basis of other studies, ripe fruits may contain less structural carbohydrate and more simple carbohydrate than unripe fruits of the same species. New leaves generally contain more water, which presumably makes them softer, and less structural carbohydrate than mature leaves. Thus, within each dietary category, structural carbohydrate may make up a lower proportion of ingested food in hylobatids than in cercopithecids. A strong possibility exists that the hylobatids may be eating less structural carbohydrate than cercopithecids with apparently similar diets as judged by the dietary categories used by behaviorists. If so, one would expect to find less developed molar crests in hylobatids.

A second possibility is raised by the anatomical differences between hylobatids and cercopithecids—it may be that hylobatids extract energy less efficiently from the structural carbohydrate they do eat. We are ignorant of the comparative digestive abilities of *Macaca* and *Hylobates* with respect to structural carbohydrate, but indirect evidence suggests that the former may digest this material more efficiently than the latter. Studies by Walker and Murray (1975) on the particle sizes of foods in the stomachs of cercopithecids and hylobatids strongly indicate that cercopithecids, compared with hylobatids, may chew up their food more effectively before swallowing. Sheine's (1979) study shows for primates that the more finely foods are ground before swallowing, the more nearly completely digested is the structural carbohydrate. Therefore, from the masticatory evidence alone it would appear that cercopithecids can digest structural carbohydrate more effectively than hylobatids can. From this it would appear that *Symphalangus* is a behavioral "folivore" but an anatomical and physiological "frugivore" in terms of the amount of structural carbohydrate it can digest. Similarly, *Macaca fascicularis* is behaviorally a frugivore but anatomically and physiologically more folivorous.

At present there is not enough evidence to decide whether real dietary differences masked by our dietary categories explain the dental differences between hylobatids and cercopithecids, or the dietary categories accurately depict the dietary similarities and the hylobatids are less well adapted for extracting energy from the same foods. I suspect that the answer lies somewhere between the two but closer to the latter. In the same environment, eating roughly similar things, cercopithecids are more abundant than hylobatids in biomass and species abundance. The peculiar brachiating locomotor mode of gibbons may allow them to feed in areas on the periphery of the tree canopies where direct competition with macaques is avoided. In contrast, Miocene gibbon ancestors had more monkey-like quadrupedal locomotion. It is tempting to think that the evolution of competitively superior digestive abilities in the cercopithecids was the selective impetus for the evolution of brachiation among gibbons.

Summary

I have reviewed the ways in which paleontologists may combine behavioral and anatomical information from living species to infer the foraging behaviors of extinct species for which only the anatomy is known. The following information from fossils can be used for such inferences.

1. Body size is useful for eliminating certain dietary possibilities: insectivorous primates are always smaller than folivorous ones. Given the high correlation of tooth size with body size, either insectivorous or folivorous habits can be ruled out on the basis of molar size.

2. The molar structure, particularly the relative size of the shearing crests and the thickness of the enamel, is informative about dietary habits. Folivorous and insectivorous species tend to have better developed molar shearing than their frugivorous or gummivorous close relatives; thick-enameled frugivores eat harder food items than their thinner enameled close relatives.

3. Food wear on the molars is sometimes informative about whether the ingested food contained grit or silica abrasives. In case of the former, terrestriality might be suggested; the latter implies a diet that includes grasses.

4. Nocturnality may be inferred from the presence of relatively large orbits. Relatively small orbits imply diurnal habits.

5. Generally speaking, terrestrial foragers may be distinguished from arboreal foragers on the basis of some aspects of the postcranial skeleton, as well as by the fact that more terrestrial species have relatively larger size and relatively higher molar crowns, at least among Old World monkeys.

6. Incisor size and structure can be informative and food-handling behavior, although not necessarily dietary preferences. Among anthropoids, however, large incisors often go along with a frugivorous diet and small incisors with a folivorous one.

The appropriateness of choosing one or another group of living species on which to base dietary inferences is discussed.

Acknowledgments

I thank those who read and commented on this manuscript at various stages: H. Covert, B. Shea, and K. Glander. Measurements were made on specimens at the American Museum of Natural History; Field Museum of Natural History; British Museum (Natural History); Powell-Cotton Museum (Birchington, Kent); National Museum of Natural History (Smithsonian Institution); Congo Museum (Tervuren, Belgium); Comparative Anatomy Collection and Birds and Mammals Collections, Natural History Museum (Paris); and the private collection of N. Tappen, University of Wisconsin. I thank the keepers of those collections for their help. E. Crumpacker assisted in the statistical analysis of data and preparation of this manuscript.

References

Bearder, S. K., and R. D. Martin. 1980. *Acacia* gum and its use by bushbabies, *Galago senegalensis* (Primates: Lorisidae). *Int. J. Primatol.* 1:103–28.

Boyd, C. E., and C. P. Goodyear. 1971. Nutritive quality of food in ecological systems. *Arch. Hydrobiol.* 69:256–60.

Cartmill, M. 1970. The orbits of aboreal mammals: A reassessment of the aboreal theory of primate evolution. Ph. D. thesis, University of Chicago.

Charles-Dominique, P. 1977. *Ecology and Behavior of Nocturnal Primates.* New York: Columbia University Press.

Chivers, D. J., and J. Herbert, eds. 1978. *Recent Advances in Primatology,* vol. 1, *Behavior.* London: Academic Press.

Clutton-Brock, T., ed. 1977. *Primate Ecology: Studies of Feeding and Ranging Behaviour in Lemurs, Monkeys and Apes.* London: Academic Press.

Clutton-Brock, T. H., and P. H. Harvey. 1977. Species differences in feeding and ranging behavior in primates. In Clutton-Brock, ed. 1977:557–84.

Coimbra-Filho, A. F., and R. A. Mittermeier. 1977. Tree-gouging, exudate-eating and the "short-tusked" condition in *Callithrix* and *Cebuella.* In D. G. Kleiman, ed. *The Biology and Conservation of the Callitrichidae,* pp. 105–15. Washington, D.C.: Smithsonian Institution Press.

Covert, H. H., and R. F. Kay. 1981. Dental microwear and diet: Implications for determining the feeding behaviors of extinct primates, with a comment on the dietary pattern of *Sivapithecus. Amer. J. Phys. Anthrop.* 55:331–336.

Dunbar, R. I. M., and E. P. Dunbar. 1974. Ecological relations and niche separation between sympatric terrestrial primates in Ethiopia. *Folia primatol.* 21:36–60.

Fleagle, J. G., and R. A. Mittermeier. 1980. Locomotor behavior, body size, and comparative ecology of seven Surinam monkeys. *Amer. J. Phys. Anthrop.* 52:301–14.

Fleagle, J. G., R. F. Kay, and E. L. Simons. 1980. Sexual dimorphism in early anthropoids. *Nature* 287:328–30.

Gautier-Hion, A. 1978. Food niches and coexistence in sympatric primates in Gabon. In Chivers and Herbert, eds. 1978: 1:269–86.

Hladik, C. M. 1977. Chimpanzees of Gabon and chimpanzees of Gombe: Some comparative data on diet. In Clutton-Brock, ed. 1977:481–503.

Hladik, C. M., A. Hladik, J. Bousset, P. Valdebouze, G. Viroben, and J. Delort-Laval. 1971. Le regime alimentaire des primates de l'ile de Barro-Colorado (Panama): Resultats des analyses quantitatives. *Folia primatol.* 16:85–122.

Hylander, W. L. 1975. Incisor size and diet in anthropoids with special reference to Cercopithecidae. *Science* 189:1095–98.

Jacobs, L. L. 1981. Miocene lorisid primates from the Pakistan Siwaliks. *Nature* 289:585–87.

Janis, C. 1976. The evolutionary strategy of the Equidae and the origins of rumen and cecal digestion. *Evolution* 30:757–74.

Jolly, C. J. 1966. Evolution of the baboons. In H. Vagtborg, ed. *The Baboon in Medical Research,* vol. 2, pp. 477–57. Austin: University of Texas Press.

—— 1970. The large African monkeys as an adaptive array. In J. R. Napier and P. H. Napier, eds. *Old World Monkeys,* pp. 139–74. New York: Academic Press.

Kay, R. F. 1975. The functional adaptations of primate molar teeth. *Amer. J. Phys. Anthrop.* 43:195–216.

—— 1977. Diets of early Miocene African hominoids. *Nature* 268:628–30.

—— 1981. The nut-crackers—a new theory of the adaptations of the Ramapithecinae. *Amer. J. Phys. Anthrop.* 55:141–51.

Kay. R. F., and M. Cartmill. 1977. Cranial morphology and adaptations of *Palaechthon nacimienti* and other Paramomyidae (Plesiadapoidea, Primates(with a description of a new genus and species. *J. Hum. Evol.* 6:19–53.

Kay, R. F., and W. L. Hylander. 1978. The dental structure of mammalian folivores with special reference to Primates and Phalangeroidea (Marsupialia). In G. G. Montgomery, ed. *The Ecology of Arboreal Folivores,* pp. 173–96. Washington, D.C.: Smithsonian Institution Press.

Kay, R. F., and E. L. Simons. 1980. The ecology of Oligocene African Anthropoidea. *Int. J. Primatol.* 1:21–37.

Kiltie, R. 1979. Nutcrackers on the hoof: Bite force and canine function in rain forest peccaries. Manuscript.

Kleiber, M. 1961. *The Fire of Life*. New York: Wiley.

Levine, A. S., and S. E. Silvis, 1980. Absorption of whole peanuts, peanut oil, and peanut butter. *N. Engl. J. Med.* 303:917–18.

MacKinnon, J. 1974. *In Search of the Red Ape*. New York: Holt, Rinehart and Winston.

—— 1977. A comparative ecology of the Asian apes. *Primates* 18:747–72.

MacKinnon, J. R., and K. S. MacKinnon. 1978. Comparative feeding ecology of six sympatric primates in west Malaysia. In Chivers and Herbert, eds. 1978:305–21.

McKey, D. B. 1979. Plant chemical defenses and the feeding and ranging behavior of colobus monkeys in African rain forests. Ph.D. thesis. Ann Arbor: University of Michigan.

McLeod, M. N., and D. J. Minson. 1969. Sources of variation in the *in vitro* digestibility of tropical grasses. *Brit. Grassland Soc. J.* 24:244–49.

Petter, J. J., A. Schilling, and G. Pariente. 1971. Observations eco-ethologiques sur deux lemuriens Malagaches nocturnes: *Phaner furcifer* et *Microcebus coquereli*. *Terre et Vie* 118:287–327.

Raemaekers, J. J., and D. J. Chivers. 1980. Socioecology of Malayan forest primates. In D. J. Chivers, ed. *Malayan Forest Primates*, pp. 279–316. New York: Plenum.

Rose, K. D., A. Walker, and L. L. Jacobs. 1981. Function of the mandibular tooth comb in living and extinct mammals. *Nature* 289:583–85.

Ryan, A. S. 1979. Wear striation direction on primate teeth: A scanning electron microscope examination. *Amer. J. Phys. Anthrop.* 50:155–68.

Sheine, W. S. 1979. The effect of variations in molar morphology on masticatory effectiveness and digestion of cellulose in prosimian primates. Ph.D. thesis. Durham, N.C.: Duke University.

Smith, R. 1980. Craniofacial morphology and diet of Miocene hominoids. Ph.D. thesis. New Haven, Conn.: Yale University.

Struhsaker, T. T. 1978. Food habits of five monkey species in the Kibale Forest, Uganda. In Chivers and Herbert, eds. 1978:225–48.

Struhsaker, T. T., and L. Leland. 1977. Palm nut smashing by *Cebus a. apella* in Colombia. *Biotropica* 9:124–26.

Van Soest, P. J. 1977. Plant fiber and its role in herbivore nutrition. *The Cornell Vet.* 67:307–26.

Walker, A., H. N. Hoeck, and L. Perez. 1978. Microwear of mammalian teeth as an indicator of diet. *Science* 201:908–10.

Walker, P., and P. Murray. 1975. An assessment of masticatory efficiency in a series of anthropoid primates with special reference to the Colobinae and Cercopithecinae. In R. Tuttle, ed. *Primate Functional Morphology and Evolution,* p. 135–50. The Hague: Mouton.

Waser, P. 1977. Feeding, ranging and group size in the mangabey *Cercocebus albigena.* In Clutton-Brock, ed. 1977:183–222.

2 Motion Economy Within the Canopy: Four Strategies for Mobility

THEODORE I. GRAND

SEEN FROM ABOVE, the tropical forest canopy is a system of overlapping tree crowns that give the illusion of a continuously connected, undulating surface. In reality, only long vines offer structural continuity over any appreciable distance; open space within the canopy is a basic feature of the system. As a result, most nonflying animals move *within* the canopy matrix rather than upon its surface. The main supports of the canopy are variably placed, stable, vertical tree trunks that do not directly interconnect. In the upper regions of each tree radiate smaller, inclined branches and leaf clus-

THEODORE GRAND has been working at the interface of anatomy and natural behavior of primates and other animals for many years. His contributions to knowledge and development of technique in anatomical analysis are unique and fall in the center of organismal biology. The following paper demonstrates Grand's perspective on the relation between substrate and animal function and his characteristically precise description of locomotion. By comparing the problems of movement within the forest canopy to the locomotor solutions of macaques, howler monkeys, spider monkeys, and gibbons, he illustrates the diversity of capacities and skills of primates faced with the ups and downs of arboreal life. His final point, that each animal is a unique compromise among a variety of different and conflicting factors, is important. There has been, as Grand points out, a history of premature generalization about primate locomotion before specific function in each species has been carefully observed. Grand's description and interpretation offer a valuable example of the proper ordering of observation, interpretation, and generalization.

ters that form the primary horizontal intersections between the tree and its neighbors. Branch instability is asymmetrical; because a branch becomes more slender toward its periphery, any weight or strain increases its *downward* deformation, and all canopy dwellers must cope with this mechanical property. The infinitely diverse irregularities, discontinuities, and instabilities of these support surfaces create barriers to direct, point-to-point movements for all nonflying canopy dwellers. Moreover, food, in the form of leaves, fruit, and insects, is the most important motivation for movement and is both dispersed and available only seasonally within this complex space.

Another obstacle to forward movement is size graded in its effects on arboreal animals. Smaller animals of 50 to 1000 g (i.e., opossums, squirrels, marmosets) move with relative ease among the terminals of a given tree and do not shift their locomotor patterns to cross over from one tree to another unless a gap, great in relation to body length, requires a jump or a detour. On the other hand, canopy inhabitants of 20 kg or more (i.e., pangolins, binturongs, orangutans) find their way among only the largest, most stable, and best connected branch supports. They accommodate their travel routes and motor patterns to the mechanical constraints of their large bodies. Thus the scamper-like motor patterns of small animals, despite their evolutionary diversity, are not very different from one another, while the motor patterns of the large forms are so distinctive as to make comparison almost irrelevant.

Of critical interest, however, are the animals in the 3- to 8-kg class (i.e., many rodents and primates, some carnivores), which exhibit an extraordinary range of motor patterns; among the primates, for example, we find suspension, quadrupedal climbing, prehensile-tailed climbing, and jumping. Thus, we have to ask the following: (1) Why, despite the fact that the structural properties of the canopy are the same for these intermediate sized animals, have such diverse motor patterns evolved? (2) How can we compare and contrast these motor patterns? (3) What is the relationship between locomotor and feeding behavior as functional subsets of the motor repertoire itself?

One problem in the description of motor patterns is that field workers, most of whom are interested in social behavior or ecology, miss subtleties of performance. The frequencies of defined, standard gaits (brachiation, bipedalism, quadrupedalism) or differences in brachiation between spider monkey and gibbon are reported (e.g., Eisenberg and Kuehn 1966; Grassé 1955), but the process of transition between one gait and another, and the differences in suspensory mechanism are missed. Further, primate studies are greatly distorted by questions of human evolution. Bipedal walking in the gibbon and the structural similarity of the upper limb to that of man are discussed much more than the adaptive values to the gibbon of bipedal walking or the orthograde trunk. Even Avis' cage experiments (1962) on locomotor performance in monkeys and apes were incorporated into a hypothesis of human origins rather than considered as adaptive patterns valuable to the animals. How can emphasis be focused on examining how animals move and feed within the canopy?

Selecting animals among which to compare locomotor behavior is a tricky business. If the animals are too closely related, their body structures are so similar that no significant contrast may be evident. If the animals are too different, the comparison may be gratuitous or absurd. Thus, the species selected must be far enough apart in anatomy and behavior for distinctive differences to be apparent, yet close enough to be meaningful. Four genera I have studied fit this specification: two old World genera (*Macaca sinica, Hylobates lar*) and two New World genera (*Alouatta caraya, A. seniculus, A. palliata, Ateles geoffroyi*) have approximately the same body size (4 to 8 kg) but differ strikingly both in anatomy and motor behavior. I have dissected animals of all four genera; I have observed three of the genera in the field; and I have studied the movements of all four on film (see Acknowledgments). The descriptions that follow are distillations of frequently employed behaviors that were taken from field notes, photographs, and film records. I shall project the varying movements of these animals onto the same vertical, inclined, and horizontal supports to see how they differ. The descriptions are valid regardless of whether the animals were observed in the wet forest canopy of Barro Colorado Island, the dry zone of Sri Lanka, or the llanos of Venezuela. By contrast, Fleagle and Mittermeier (1980) took advantage of the sympatry of seven species in Surinam to compare locomotor and feeding strategies.

Description begins with common structural arrangements of the support surface followed by the ways in which each of the four genera solve the problem of movement.

Terminal Crossings and Horizontal Displacement of the Body Mass

In figure 2.1, the aligned branches continue in one direction, an intersecting branch angles off at a 30° inclination, and the level drops 1 or 2 feet. The macaque walks quadrupedally across the tops of the branches and simply drops to the new level (fig. 2.1B). The gibbon (fig. 2.1A) approaches the intersection by brachiation but accelerates at the last swing, hops up on top of the distal branch, and moves off with bipedal step and orthograde posture. The howler (not figured) would move quadrupedally across the top of the branches, securing himself posteriorly with his prehensile tail. The spider monkey (fig. 2.1C) approaches the intersection from underneath the first section, swings up onto the distal section, and then moves off quadrupedally. (Note that words like brachiation, bipedalism, or quadrupedalism tend to freeze the flow of motion and the phenomenological meaning of the gait changes themselves may be lost.)

In a longer sequence (fig. 2.2), the gibbon directly cuts through the canopy matrix. As phase I begins, the animal reaches across a gap between terminal branches with his left hand and secures himself first with that hand and then with both feet. His weight is distributed across two trees. He reaches out again with the left hand to take a more advanced position along the same branch, the arm fully extended preparatory to the initial downswing. He lets go with the rear (right) hand; the forward branch bends farther down under his full body weight; he does not let go with his feet. The trunk rotates to the left (around its long axis) and the free right arm takes an advanced position. This first swing is quite controlled. In phase II, the left hand and feet release their grip and the body swings down from the right hand. The animal tucks his feet, reaches up with his left hand, and as his body passes the next branch, he steps upon it with both feet. In phase III, he pushes himself upward with both

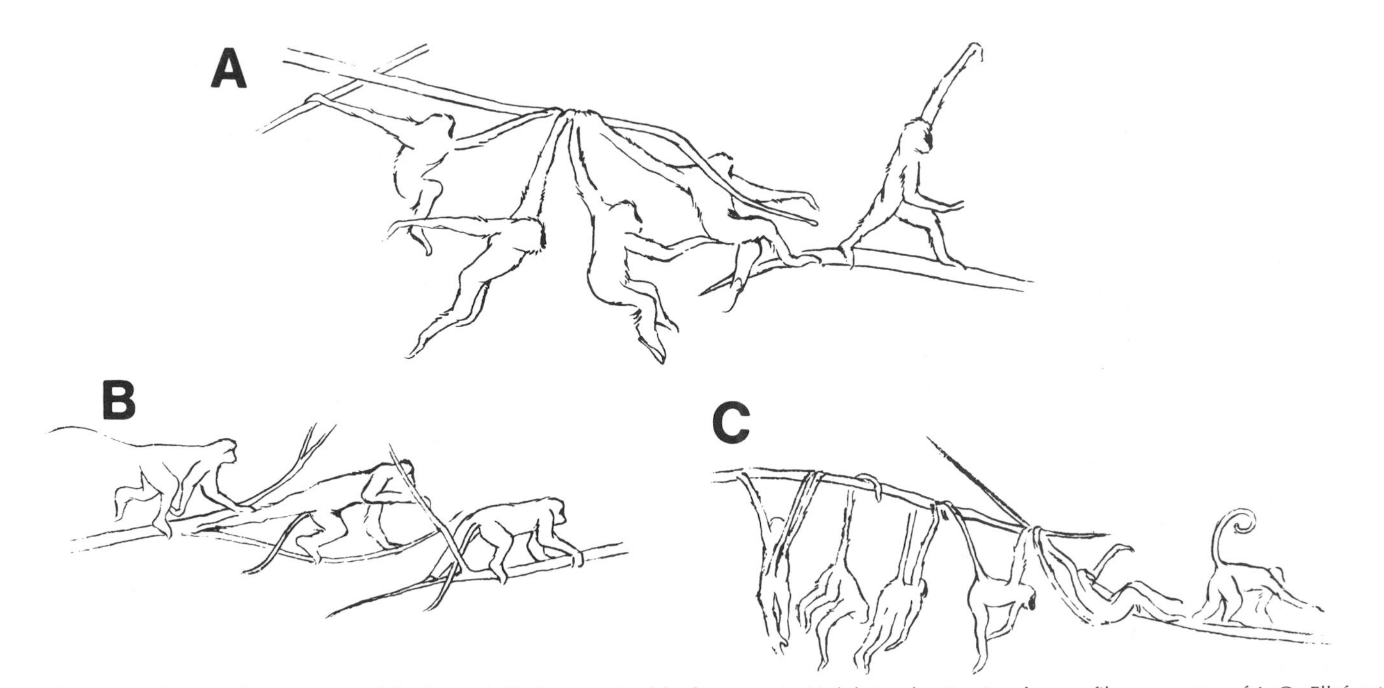

Figure 2.1. Terminal crossings and horizontal displacement of body mass. A. *Hylobates lar* (tracing from a film sequence of J. O. Ellefson); B. *Macaca sinica* (from a film sequence of T. I. Grand and F. S. Shininger); C. *Ateles geoffroyi* (from a sequence of still photos of T. I. Grand).

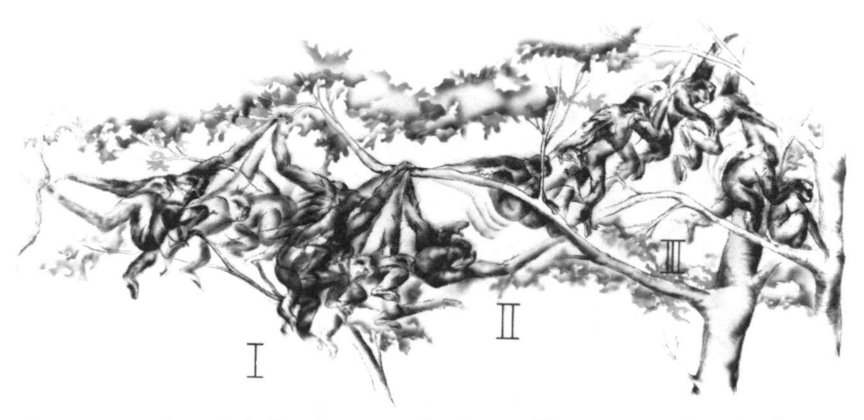

Figure 2.2. Horizontal displacement in *H. lar* (from a film sequence of Naaktgeboren and Slijper).

legs, at the same time releasing with both hands. In the middle of this new trajectory, he reaches a branch with his left hand, swings free, and then drops down to a seated posture.

Terminal Crossing with Descent of the Body Mass

When a macaque crosses a gap or descends, he always stays in an above-the-branch relation (fig. 2.3B). The gibbon (fig. 2.3A) moves forward underneath the first branch, and at the last swing, drops his free hand into contact with the lower target branches. He can reach this far down because of branch deformation and because he is underneath the branch rather than on top of it. He halts the body's sway with the hand and both feet. As he becomes stable, his lower, forward arm reaches for a more interior position in the target tree and the animal pulls himself forward to take this advanced position. Once more, the arm pulls the body forward. The body may rotate about the fixed (elevated) wrist and fingers. When the feet and hand are securely in contact, the elevated hand releases, and in one forward motion, the animal reaches to take a grip on the target cluster of leaves and branches. (Reach and forward grasp of the hands and

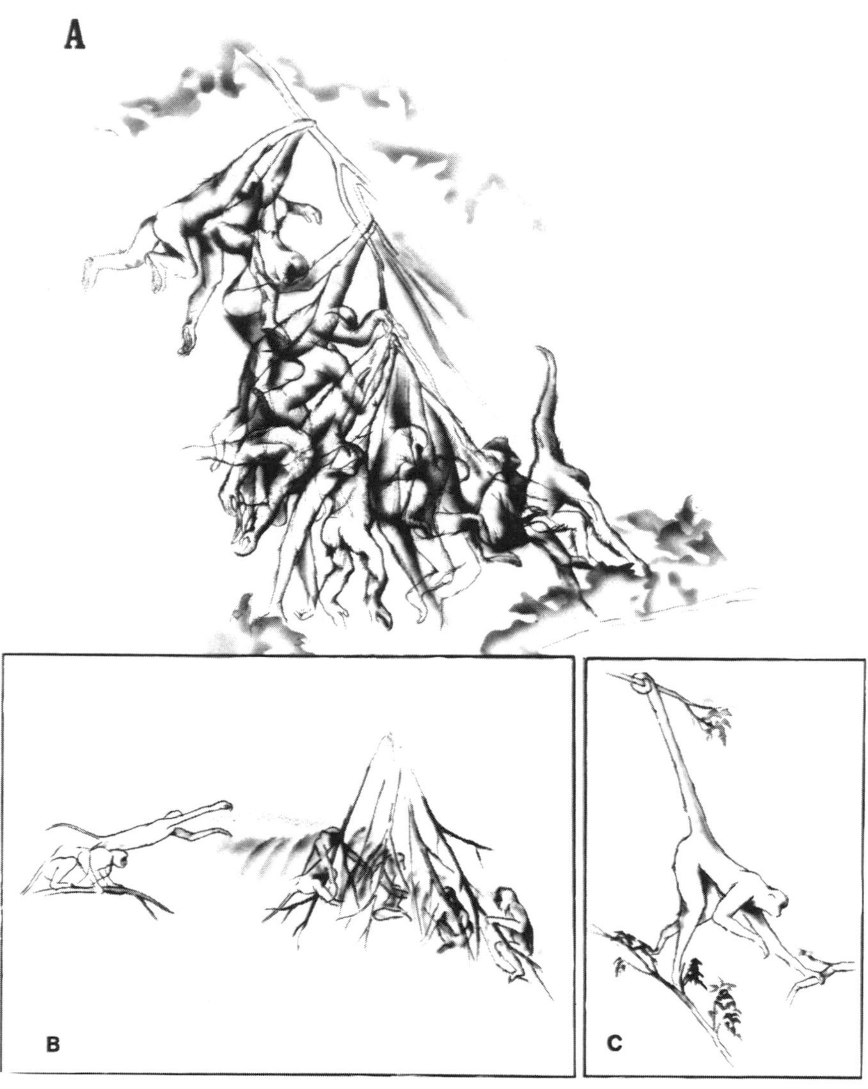

Figure 2.3. Terminal crossings and descent. A. *H. lar* (tracing from a film sequence of J. O. Ellefson); B. *M. sinica* (from a film sequence of T. I. Grand and F. S. Shininger); C. *A. geoffroyi* (from still photos of T. I. Grand).

feet depend on the degree of downward deformation of the upper branch and the vertical suspensory posture of the body on the hand.)

Howler and spider monkeys have similar solutions to descent (fig. 2.3C), but unlike the gibbon, the tail is the suspensory member. As the fingers and toes become more secure, the tail loosens its grip and disengages completely from the upper support. The animal walks forward on all four limbs into the receiving thicket of small branches and leaves.

In a long, relatively elaborate descent sequence (fig. 2.4A), the gibbon uses the widely spaced forks of the trunk, instead of following one fork to the bottom. His approach jump increases his momentum and speed of descent, while at the same time, the guiding, controlling motions of his arms show how fully and fluidly he can use angular motion and his potential energy in the fall.

In phase I, he swings downward toward the center of the tree, gathers himself at the fork with his feet, and tucks his body for the initial jump. His head and arms are turned toward the target early in the jump; then his legs and feet are turned so that he faces forward. (This turning of the body on its longitudinal axis during free flight is commonly seen in vertical clingers and leapers [Napier and Walker 1967.]) The eyes and head are oriented to the target; after pushing off with the feet, the lower body turns toward the support. In phase II, as the hands and feet contact the inclined target, the body is decelerated, but the acquired momentum carries the gibbon counterclockwise around the branch. His feet swing free, his right hand is released, while the secure left hand pivots the body's continued rotation. The right arm is flexed and set for the next phase of descent and the feet contact the branch. The left hand remains grasping the branch but slides downward. The feet are lifted, the body drops. While the feet contact the trunk once more, the left arm still guides the body down. The body rotates on its longitudinal axis around the arm until both arms are in contact. In phase III, the animal backs down the next section of the incline, steps backward with both arms spread-eagled above him at the major fork. Here he rotates about the left arm and reaches around the tree trunk with the right. Now upright, the body is suspended from the left arm, the right encircling the trunk once again. As he slides downward almost

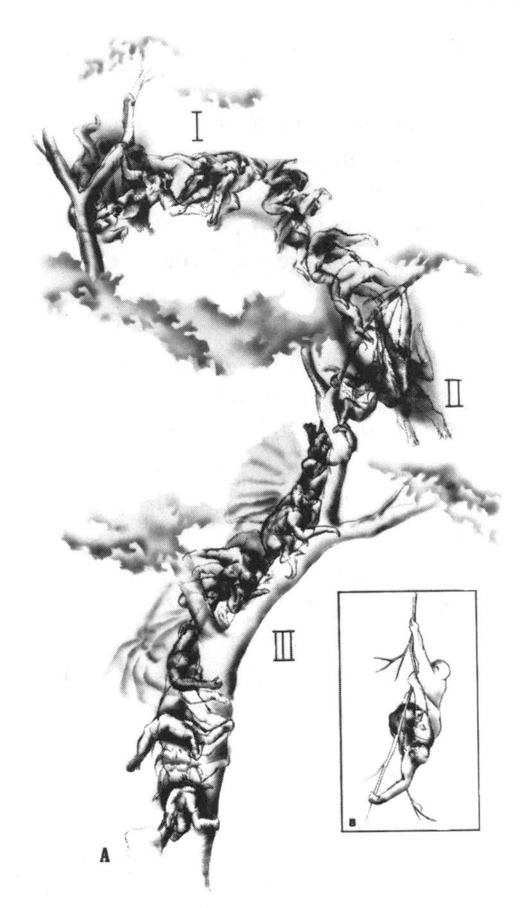

Figure 2.4. Descent of *H. lar* (from a film sequence of J. O. Ellefson).

free, his arms exert counterpressure to guide him inward so that he will not tilt backward. More than seven body lengths are traversed in about 6 seconds.

Another specific descent of about five body lengths took less than 5 seconds (as estimated by the length of the film clip). In this case, the animal remained upright, almost dropping vertically with the pull of gravity. He regulated his descent somewhat like a man coming down a gymnast's rope, with right and left hands alternately grasping a branch support. By contrast, in figure 2.4B, the macaque, because of the limited flexibility of his upper body, rotates its entire body from the head-up to head-down position.

Ascent

In his ascent to a new level, the gibbon sometimes exhibits a novel combination of motor skills. He maximizes the development of energy in his downswing (Fleagle 1974) by lengthening the radius of gyration and accelerates in his upswing by tucking his feet and flexing his pivoting elbow. At the peak of trajectory, he reaches upward, perhaps also steps up on the branch, and then gains still further height by another combination of brachiation and stepping. This is clearly seen in the films of Naaktgeboren and Slijper (1965). Ascent appears to be energetically efficient but requires some linear travel distance as well.

Reaction to an Environmental Accident

> The buff female . . . swung quickly through an open tree and out on a limb. As she was preparing to spring into an adjacent tree, the limb broke, leaving a stub which was about 6 inches long. As the limb broke and fell, the gibbon recovered by turning almost in mid-air and catching the remaining stub of the branch. With extreme rapidity she swung around under and then on top of the limb and then, with only a slight loss of time and momentum, jumped outward and downward 30 feet to an adjacent tree top (Carpenter 1964:186).

Discussion and Conclusions

In attempting to understand motor skill differences among four genera of primates, I have rediscovered the injunction of Auguste Comte. Just as he emphasized the interdependence of structure, function, and milieu (Jacob 1976), so have I integrated the motor behaviors of these primates and their musculoskeletal structures with the common mechanical properties of the forest canopy. I wanted to portray *each* animal within the same space, in order to contrast as strongly as possible their different motor strategies. Too intense (or myopic) a focus on gaits or postures without regard for the support surfaces would have upset the balance of these interdependent variables.

The use of the term *brachiation,* for example, isolates and defines the movements of gibbons and spider monkeys but does not focus on the skill differences between them. Application of the word discourages deeper analysis of the motion being observed; emphasis on pronograde *versus* orthograde, brachiation *versus* quadrupedalism, feeding *in contrast to* locomotion would have taken my conclusions along well-known paths. If, however, we think more in terms of the mechanics and functional convergences of suspension, then we can understand the elements of motion more in terms of mobility of each limb joint. The baseline for any analysis is: "What is the shoulder doing? How far into extension is the elbow? How much does the wrist deviate to the ulnar side?" Both animals use ballistic swing and lower themselves using the arm (the spider monkey also uses its tail) to regulate the pull of gravity and to accommodate to downward branch deformations. The gibbon shifts from brachiating to bipedal walking, the spider monkey from brachiating to quadrupedal walking, to move more directly through canopy space. From one- or two-point suspension, the oscillations of the spider monkey and the gibbon are damped and controlled by gravity. In pendular motion, more than 70% of body weight (truncal mass and the free limbs) progresses by a series of arcs with peaks of acceleration (Jenkins et al. 1978). This weight constitutes an energy reservoir since the swing is induced by gravity; forward motion *does not require* constant energy input. This potential energy increases step length, but the animal may use other techniques to achieve forward motion. The gibbon and spider monkey may flex an elbow of the suspending limb and tuck the legs at the bottom of the swing (Fleagle 1974), shortening the radius of gyration about the suspended hand and imparting to the body greater angular acceleration; or, by using suspension from the two hands, the gibbon creates a natural twisting of the forward limb joints so that as the posterior hand is released and the body descends along its arc, the twisted joints release energy as in the torsion pendulum clock (fig. 2.5).

Whereas macaques appear to be uncomfortable hanging from one or both forelimbs, as suggested by their frequent change of position, gibbons and spider monkeys change position over longer periods of time and transfer suspended weight easily between upper limbs and tail. The macaques and howler monkeys are not as flexible at shoulder-elbow-wrist joints, and when the body turns downward from a suspending limb, the animal tends to let go.

In addition to the tensile strains of body weight, each limb joint is subjected to torsional strain because the joints do not fully extend. The animal rotates his body in order to accommodate to the discomfort of these unusual trunk-support surface relationships. By contrast, descent movements in gibbon and spider monkeys are enhanced by upper limb and tail flexibility; weight shifts and regulated lowering of the body are achieved more economically. During hang feeding or suspensory progression, these animals use their feet to support themselves on randomly available lower supports. Hanging puts them closer to these lower branches because of the natural downward deformation of the slender terminals. (When a macaque deforms a branch downward, his hands and feet are already occupied.) The body of gibbon or spider monkeys hangs straight down from hands or tail; the bones and ligaments may resolve these pure tensile strains regardless of muscular input (Kummer 1970). Wrist-elbow-shoulder mobility in the gibbon, which is reflected in rotations of the body from the suspended hand, has a positive selective value in brachiation, terminal branch feeding, terminal crossing, and vertical descent. The gibbons probably have greater upper body flexibility than the spider monkeys because the former depend on one- or two-point suspension, while the latter can use two or three points (Lewis et al. 1970; Lewis 1971, 1972).

The primary suspensory equipment of the gibbon is its long muscular arms, which constitute more than 12% of body weight and its hands with their curved phalanges and shortened digital flexors. In both howler and spider monkeys the tail represents about 6% of weight (Grand 1977). This prehensile appendage, though not propulsive during quadrupedal walking, is essential to under-the-branch

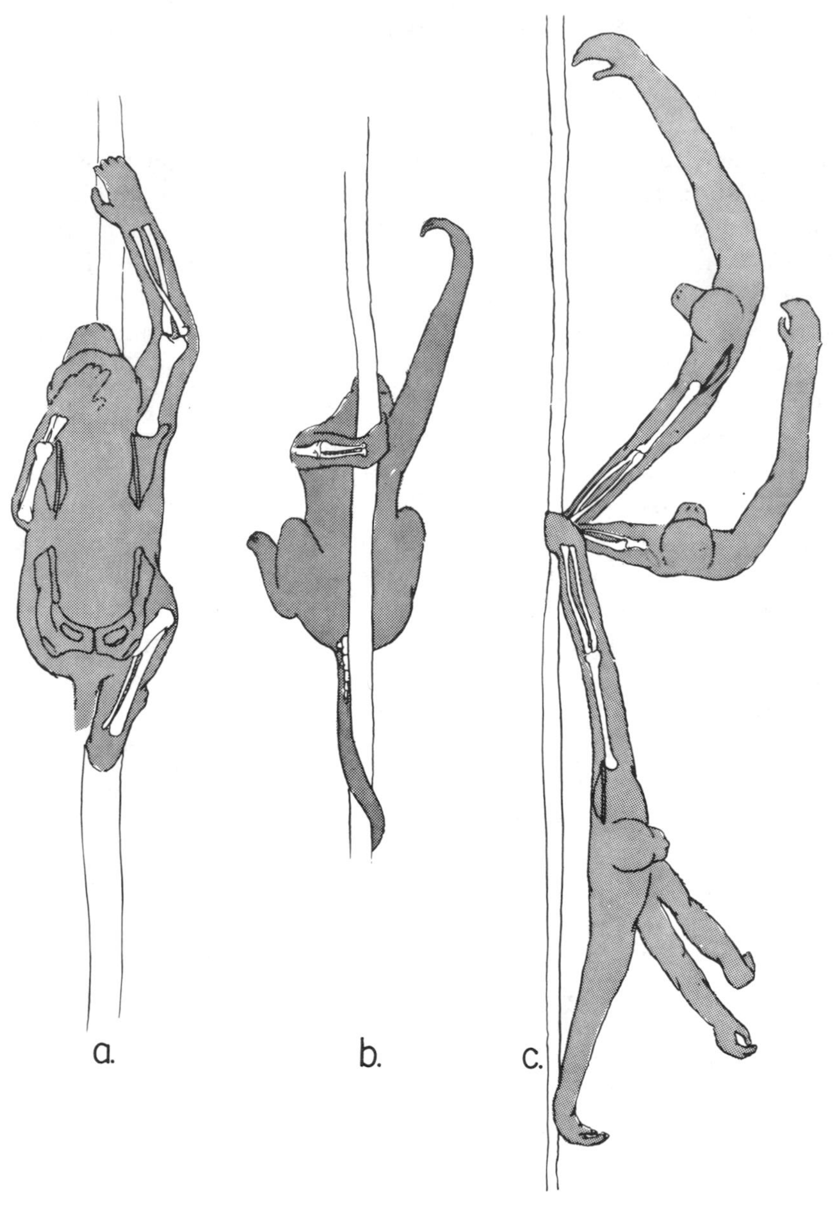

Figure 2.5. Above- and below-the-branch support, suspension, and torsion of the limb joints. A. *M. sinica;* B. *A. geoffroyi;* C. *H. lar.*

feeding, to regulating descent, and to stabilizing the body during horizontal transport and terminal crossings. The tail, as heavy as one hindlimb, possesses a longer reach and greater range of motion than the limb. And because of powerful, shortened flexor muscles, the tail naturally curls into a suspensory hook at its tip. (The tail of *M. sinica* is half the relative weight, far less muscular and, as in other conventional quadrupedal monkeys, functions as a simple counterweight.) That so much weight, musculature, and central coordinative control are invested in the tail of the spider and howler monkeys argues for its pervasive adaptive importance.

The overall displacement of the body in space provides an interesting means of assessing motion efficiency. The macaque, restricted by forelimb inflexibility to a few trunk-support surface relationships, must follow the tops of the support surfaces through all their vagaries of inclination. In descent, he compensates for his inflexibility by continual truncal rotations (Grand 1972). The gibbon and spider monkey proceed more directly through space and with apparent ease through regions that are obstacles to the macque. Both of them "cheat" and take shortcuts by reaching around an incline (figs. 2.1A and 2.1C), by changing gait, or by using the energy stored up in the swing. By means of the force of gravity, the gibbon and spider monkey drop straight downward without truncal rotations or side-to-side motions. The howler's technique during descent is the same as that of the spider monkey.

In ascent motions the macaque levers his body upward by means of his hindlimbs and pulls upward with his arms. Frequently, the animal moves into and then outward from the core of the tree searching for continuous paths of ascent or descent, whereas the gibbon and spider monkey move more freely at the tree's periphery. The gibbon maximizes the potential energy during the swing by coordinating his reach with the upper limb or steps with the leg to ascend.

In sum, greater control of space by spider monkey and gibbon permits more direct movements, greater choice of travel route, and increased access to food. If the macaque is turned underneath a target branch at the end of a jump, he must cling to it till it stops oscillating. Then he clambers topside to continue his forward progress. By contrast, if the gibbon or spider monkey is rotated underneath the support, each continues forward almost immediately beneath the support. The animals halt branch sway with arms, legs, or tail—elements of the motor repertoire quite unavailable to a conventional quadruped like a macaque. In feeding, where the macaque moves along the top of the branch, the gibbon and spider monkey move beneath, able to place their weight farther back on the branch, closer to the axis of branch rotation. This latter technique reduces branch deformation and also puts the food right in front of the animal (Grand 1972). By the interpretation of such differences, macaques are not as economical in movement as gibbons, nor are howlers as skilled as spider monkeys. Thus, the adaptive significance of suspension is the refined accommodation to the downward deformation of the branch surfaces.

Once primates began to restrict themselves to life in the canopy (Jenkins 1974), the locomotor consequences were extraordinary. Movement from one point to another in three-dimensional space is a necessary precondition to feeding, to social coordination, and to antipredator behavior; and locomotor behavior became part of this feedback relationship of interdependent variables, which has never ended. Increasing size, however, has had special consequences. Whereas 50- to 1000-g animals could retain a basically scampering mode of locomotion (Grand, in press), animals more than 2 or 3 kg had to evolve novel motor patterns by which to resolve their increased difficulties in forward movement. The larger the animal the more numerous the obstacles to canopy life.

The progressive control of space can be argued from the New and Old World radiations. Among the New World forms, the squirrel monkeys are quadrupeds that use their tails as counterweights; the cebus use their heavy muscular tails occasionally in descent where additional support is advantageous (Grand, personal observation; Robinson, personal communication). Howler monkeys are above-the-branch quadrupeds, but their fully prehensile tails are better suited to the demands of quadrupedal climbing and terminal branch feeding than the cebus. Spider monkeys represent the peak of skill in that they have both a prehensile tail and an elongated, mobile upper limb (fig. 2.6C). Among the Old World monkeys and apes the ma-

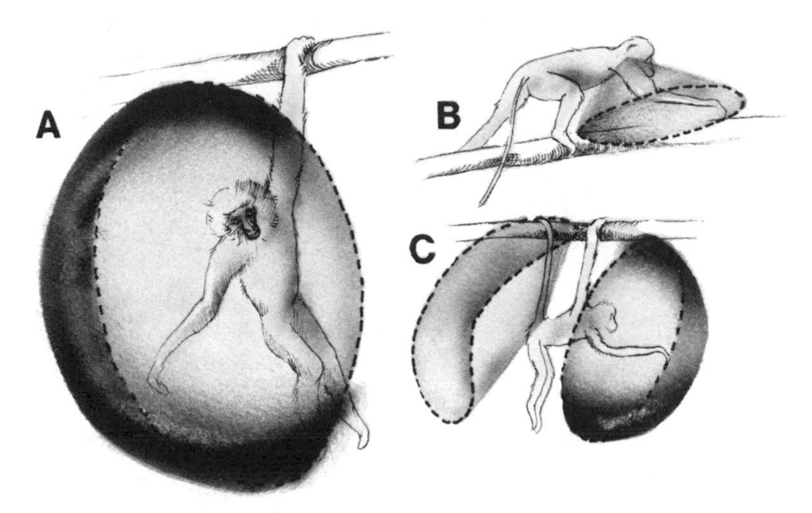

Figure 2.6. Control of space and the kinesphere. A. *H. lar;* B. *M. sinica;* C. *A. geoffroyi.*

caques are above-the-branch quadrupeds (fig. 2.6B), and the number of guenon and colobine species that are similar argues for the common origin of this motor pattern. By contrast, the suspensory gibbons represent a revolutionary new stage in skill and economy of motion through their ability to reach about themselves in almost a hemisphere of space (fig. 2.6A).

At the same time, the anatomical investment in the tissues of posture and movement remains enormous and bespeaks the impact of the motor repertoire on everyday life. More than 70% of the primate body is bone, muscle, and skin, and a substantial metabolic expenditure is committed to growth and maintenance of these tissues, even before motor activities are brought into the accounting. Since feeding and moving are interdependent activities, the same motor elements in the gibbon are incorporated into locomotor economy and terminal branch feeding: control of space, truncal orthogrady, brachiation, and terminal branch suspension. Selection for increased feeding efficiency would remodel the body, secondarily modifying the locomotor repertoire; selection for greater locomotor efficiency would remodel the body, secondarily modifying feeding skill.

Tuttle's four-stage reconstruction of hylobatid evolution (1972) interlocks these functional motions. Since the precise evaluation of the economy of these activities in our four species is not yet possible, the impact of economy of feeding and movement cannot be judged. Let us say that motor skill gives the individual animal a "striking position" in the struggle for a resource. It is the mechanism for the acquisition of a resource; in some way increased skill cuts energy expenditure. Thus, economy has many guises and the evolution of the motor repertoire has many explanations. Selection may be for endurance, for distance, for agility, for speed and acceleration, for rapid change of direction, or for increased "foraging radius" (Pennycuick 1979). Because the cost of transport and its benefit to the organism vary with the circumstance, no one answer in terms of anatomy and motor pattern is sufficient; each animal is the special compromise and product of these factors and forces.

Acknowledgments

The subtitle is borrowed from Wilfred Owen's *Strategy for Mobility* (1964), a book of great and lasting influence for me but one for which a single textual citation would be inadequate. John Ellefson took the films of *Hylobates lar* while on a field trip to Malaysia and then lent them to me so that I could study them repeatedly over the years. Stuart Shininger took the films of *Macaca sinica* in Sri Lanka and Harry J. Wohlsein Jr. took the films of *Alouatta seniculus* in Venezuela; both photographers worked with me under the auspices of the Smithsonian Institution. Joel Ito did all the illustrations, building up several of them from frame-by-frame tracings of the films. My field work has frequently been supported by the Smithsonian Institution and the Wenner-Gren Foundation, and I owe to Lita Osmundsen and John Eisenberg a particular debt for their continued interest and encouragement. Critical comments on drafts of the manuscript were made by John Cant, Peter Rodman, and Alis Temerin. Barbara Questad tightened, polished, and typed the final draft.

References

Avis, V. 1962. Brachiation: The crucial issue for man's ancestry. *S. W. J. Anthrop.* 18:119–48.

Carpenter, C. R. 1964. *Naturalistic Behavior of Non-Human Primates.* University Park, Pa.: Pennsylvania State University Press.

Eisenberg, J. F., and R. E. Kuehn. 1966. The behavior of *Ateles geoffroyi* and related species. *Smithsonian Miscellaneous Collection* 151:1–63.

Fleagle, J. 1974. Dynamics of a brachiating siamang (*Hylobates [Symphalangus] syndactylus*). *Nature* 248:259–60.

Fleagle, J., and R. Mittermeier. 1980. Locomotor behavior, body size, and comparative ecology of seven Surinam monkeys. *Amer. J. Phys. Anthrop.* 52:301–14.

Grand, T. I. 1972. A mechanical interpretation of terminal branch feeding. *J. Mammal.* 53:198–201.

—— 1977. Body weight: Its relation to tissue composition, segment distribution, and motor function. I. Interspecific comparisons. *Amer. J. Phys. Anthrop.* 47:211–40.

—— (in press). Body weight: its relationship to tissue composition, segmental distribution of mass, and motor function. III. The Didelphidae of French Guyana. *Australian J. Zool.*

Grassé, P.-P. 1955. *Traité de zoologie. Anatomie, systématique, biologie.* 17 (fasc. 2). Paris: Masson et Cie.

Jacob, F. 1976. *The Logic of Life. A History of Heredity.* New York: Random House.

Jenkins, F. A., Jr. 1974. Tree shrew locomotion and the origins of primate arborealism. In F. A. Jenkins, Jr., ed. *Primate Locomotion,* pp. 85–115. New York: Academic Press.

Jenkins, F. A. Jr., P. J. Dombrowski, and E. P. Gordon. 1978. Analysis of the shoulder in brachiating spider monkeys. *Amer. J. Phys. Anthrop.* 48:65–76.

Kummer, B. 1970. Die Beanspruchung des Armskeletts beim Hangeln. Ein Beitrag zum Brachiatorenproblem. *Anthrop. Anz.* 32:74–82.

Lewis, O. J. 1971. The contrasting morphology found in the wrist joints of semibrachiating monkeys and brachiating apes. *Folia primatol.* 16:248–56.

—— 1972. Osteological features characterizing the wrists of monkeys and apes, with a reconstruction of this region in *Dryopithecus (Procunsul) africanus. Amer. J. Phys. Anthrop.* 36:45–58.

Lewis, O. J., R. J. Hamshere, and T. M. Buckill. 1970. The anatomy of the wrist joint. *J. Anat.* (London) 106:539–52.

Naaktgeboren, C., and E. J. Slijper. 1965. *Hylobates lar.* Fortbewegung im Geäst. (Film.) *Encyclopaedia Cinematographica.* E 1107.

Napier, J. R., and A. C. Walker. 1967. Vertical clinging and leaping: A newly recognized category of locomotor behavior in primates. *Folia primatol.* 6:204–19.

Owen, W. 1964. *Strategy for Mobility.* Washington, D.C.: Brookings Institution.

Pennycuick, C. J. 1979. Energy cost of locomotion and the concept of "foraging radius." In A.R.E. Sinclair and M. Norton-Griffiths, eds. *Serengeti: Dynamics of an Ecosystem,* pp. 164–84. Chicago: University of Chicago Press.

Tuttle, R. H. 1972. Functional and evolutionary biology of hylobatid hands and feet. In D. Rumbaugh, ed. *Gibbon and Siamang,* 1:136–206. Basel: S. Karger.

3 Foraging, Habitat Structure, and Locomotion in Two Species of *Galago*

R. H. Crompton

THE LOCOMOTOR BEHAVIOR of an arboreal animal may be thought of as its adaptive system for obtaining access to food resources distributed nonrandomly in three-dimensional space and equally for preventing access to itself as food for potential predators. To understand locomotor behavior, then, we should consider the distribution of food sources and predators and the qualities of supports available for locomotion within the space. We should also consider some basic characteristics of the animal, notably its size and its diet; the diet is partly determined by its body size. Additionally, we should consider some features of food, such as its quality, package size, and concentration, and of the predators, notably their locomotor capabilities.

Napier suggested some possible relationships of diet, habitat, and locomotion as long ago as 1966, indicating that the continuity of supports is a key factor, and Ripley (1967) laid out the elements of a full naturalistic study of locomotion not long afterward. It took until

ROBIN CROMPTON has observed two species of prosimians, *Galago senegalensis* and *G. crassicaudatus,* in *Acacia* forests of South Africa. His comparative analysis of quantitative differences in locomotion, postures, and support use and the changes in these with seasonal changes reveals clearly the relationships among dietary preferences, habitat use, locomotor patterns, and animal body size.

the late 1970s, however, for anatomists to attempt field study of locomotor adaptation. Fleagle (1977) found that quantitative differences in limb anatomy could be predicted broadly from quantitative differences in behavior (movement types, support use, etc.) in sympatric *Presbytis* species but that diet was by no means a predictor of locomotion. Charles-Dominique's (1977) study of sympatric lorisids showed closer ties between diet and locomotion, but Fleagle's conclusion has been echoed by several others (e.g., Rose 1979; Morbeck 1979).

That diet and locomotion are linked is certain, however. The issue is the kind of relationship between them. To be more precise, there will be multiple kinds of relationships, for locomotion must derive from multiple direct influences, such as the needs of manipulation and feeding posture, as well as from the less direct influences of habitat structure and the invisible influences of foraging strategy, energetic constraints, and interspecific competitive relations in the community. I shall examine some of the more direct influences elsewhere (Crompton, in preparation), and Fleagle (1979) has reviewed them recently in some detail.

It is the less direct influences of behavior and habitat on locomotion that this paper deals with. In particular, I examine how locomotion can ensure access to food of changing seasonal distribution, while not subjecting the animal to unacceptably high risks of predation. The role of seasonality in primate locomotor adaptation has received very little attention, but it is vitally important to any consideration of early primate evolution, since Eocene climates of the northern continents were apparently quite strongly seasonal (see, for example, McKenna 1980) and nearly all known Eocene primates came from northern latitudes. Further, I discuss the role of body size in determining diet, and the consequences of body size for foraging strategy, habitat utilization, and locomotion.

I have studied the two most successful lorisid species, *Galago senegalensis* and *G. crassicaudatus,* in two previously established study sites in the intensely seasonal *bushveldt* of the northern Transvaal of South Africa. The advantages of studying congeners are that some control for evolutionary history is provided and that, with reservations, current differences in behavior and morphology may be taken as differential adaptations to current conditions. The galagines may be a good model for the ancestral stock of primates of modern aspect (Charles-Dominique and Martin 1970), and they show a great diversity in body size, locomotor morphology, and ecology. Charles-Dominique (1977) had provided some data on galagines before my work, and studies of the social behavior and ecology of the Transvaal galagines by Bearder, Clarke, and Harcourt provided an excellent background for my study of these two species.

Background

Subjects

Tables 3.1 and 3.2 summarize basic information about the two species. There is a sixfold difference in body weight between *G. senegalensis,* the smaller, and *G. crassicaudatus.* Diets are qualitatively similar but quantitatively different, with greater emphasis on gums (and fruits at other sites) in *G. crassicaudatus* and on arthropods in *G. senegalensis.* Gums become more important in winter, especially in a more severe winter, as evident from increase of both nightly duration of gum-licking and time spent at individual sources in the typical severe winter (Bearder, personal communications). The species are less selective of food size in winter, probably as a reflection of scarcity. On cold nights even *G. senegalensis* may eat nothing but gum. Food shortage in cold winters often leads to severe weight loss (Bearder and Martin, 1980), and there can be little doubt that a "bottleneck" effect operates during winter. Fruit may become a major dietary element in other areas, but because its digestive qualities are similar to those of gum (see discussion), the arguments in this paper are little affected by the lack of fruit in the diet of *G. crassicaudatus* at my study site.

Study Sites

The observations on *G. senegalensis* were made on "Mosdene," a farm and nature reserve owned by E. A. and R. G. Galpin. It is

Table 3.1. Basic Biogeography and Ceology of *Galago senegalensis* and *G. crassicaudatus*

	Distribution, Biogeography	Dietary Constituents	Competitors	Known Predators	Potential Predators	Approximate Mean Adult Male Body weight, $\bar{X}$
G. senegalensis	Wood savannah south of Sahara, avoiding equatorial forest but present [a] on Kenyan coast. Extend into semidesert regions of Botswana, Somalia Namibia. [b,c] Peak Southern African densities in *Acacia thornveld* [d]	Arthropods,[e] gums.[e] No fruit [b] (fecal samples) but do not take fruit in Kenya [a,f]	*Thallomys* take gum at Mosdene.[e] Numerous possible competitors for arthropods	Hawks—the gymnogyre [d]	Various snakes, jackals (alarm calls),[e] possibly genets	195 g
G. crassicaudatus	Woodland, especially riparian woodland south of Ubangi River.	Arthropods,[e] gums,[e] possibly *Zizyphus*	*Thallomys*, baboons, vervets present and known to take gum	Eagle owls (Van Pienaar)[g]	Few possible for adults—possibly leopard,	1300 g

	Distribution contacts but does not overlap that of *Perodicticus* in Kenyan equatorial forest. Avoid open or xeric regions[b,c]	berries. Do take fruit elsewhere	elsewhere. Birds, bats take fruit. Numerous possible competitors for arthropods.		various snakes, genets, monitors for infants. These do not produce alarm calls from adults[e]

[a] May be *G. zanzibaricus*.
[b] Hill (1953).
[c] Kingdon (1971).
[d] Bearder (1972) and personal communications.
[e] Personal observation.
[f] Harcourt (1980) and personal communication.
[g] Personal communication to A. Clark.

Table 3.2. Diets and Seasonal Changes in Diets[a]

	Summer	Winter	Overall
G. senegalensis			*G.s.*<*G.c.*
Gums		Longer time spent licking each night but not longer periods at individual gumlicks.[b] Longer at individual gumlicks.[c]	
Arthropods			*G.s.*>*G.c.*
G. crassicaudatus			
Gums	49%	82%. Significantly longer spent licking each night and longer at individual gumlicks [b]	62%
Arthropods	Take individuals of larger size class than *G.s.* [b]	Drop in numbers taken, take individuals of same size class as *G.s.* [b]	5%
Fruits	33%	2.3%	21%
Nectar			8%

[a] From work by Harcourt (1980) and Bearder (1972) (and personal communications.) Percentages are Bearder's counts of "different types of food taken" from studies in the NE Transvaal, in Zululand, and Natal. The study population of *G.c.* did not take fruit (with the possible and rare exception of *Zizyphus* berries). Harcourt's and Bearder's studies of *G. senegalensis* were made under very different weather conditions at "Mosdene"—Bearder's in a cold winter, Harcourt's in a very mild one.

[b] Harcourt (1980).

[c] Bearder (1972).

located in the northern Transvaal of South Africa (28° 47′ E, 24° 25′ S) at 1085 m above sea level, and the climate is highly seasonal. Mean annual rainfall is 610 mm, but flood and drought years are common (maximum of 810 mm in 1953, minimum of 276 mm in 1966). Summer maximum temperatures of 38° C are common, and minima in winter reach −5°C (personal communication from E. A. Galpin to Bearder). The diurnal temperature range may be 25°C or more in summer.

Observations of *G. crassicaudatus* were made on "Wallacedale," owned by the Thompson-Vise family, some 250 km farther north of Mosdene near Louis Trichardt at 985 m above sea level (23° 57′ E, 29° 56′ S). The site lies within a small valley in the Soutpansberg mountains and is composed of a main strip of trees some 50 m x 600 m on either side of a stream, with scrubby bush to the east and a pasture and road to the west. Mean annual maximum temperature at Louis Trichardt is 32°C, and mean annual minimum temperature is 8°C with a range down to −8°C (Weather Bureau of the Transvaal, Statistics 1954–1964). The farm probably receives more than 450 mm annual rainfall (recorded for Louis Trichardt) since it lies in the Soutpansberg.

Line transects were made at each site at three or four locations, and the species, height, breast height diameter, and number of primary trunks were recorded for all trees, shrubs, and saplings over 1 cm at base that fell within 1 meter on either side of the transect line. Estimates of canopy cover were made from aerial photographs of each site. The main site at Mosdene is a fairly open wooded area of some 10,000 m^2 lying next to an open plain, with 15% canopy cover; 70% of trees were *Acacia karoo,* and 7% were other *Acacia* species, predominantly *A. tortilis* with some *A. nilotica.* Other genera present at Mosdene were *Maytenus, Dichrostachyes, Peltiphorum,* and *Terminalia.* Other parts of Mosdene, such as Harcourt's (1980) site, may be dense and have a greater variety of trees, but the primary site of my work was typical of much of the open bush.

Canopy cover at the site on Wallacedale was 32%; 64% of trees and saplings were *Combretum erythrophyllum,* and 17% were *Acacia karoo.* Smaller numbers of other *Acacia* species were present, and other genera included *Peltiphorum* and *Zizyphus.* Because the *A. karoo* are usually large trees and the *C. erythrophyllum* are more often saplings, *A. karoo* dominated the canopy.

Methods

Observations were made by using a miner's lamp covered with red celluloid from just before nightfall for the next three or four hours, depending on weather conditions. In heavy rainfall, visibility was inadequate. Individual animals were located and followed for as long as possible. Charles-Dominique's (1977) method of recording only the first sighting was not used, because it is likely that the animals sighted the observer well before they were discovered. If possible, data were taken on two animals simultaneously. Observations were spoken onto a cassette recorder for transcription the next day.

The units of observation were a single *displacement* (change of position), recorded whenever a displacement occurred, or a postural observation, recorded once every 2 minutes up to a maximum of ten observations or until a displacement or postural change occurred and a new record began.

After an initial trial, whose results were discarded, winter counts were made from June 1 to 10 on *G. senegalensis* and from June 17 to July 2 on *G. crassicaudatus.* Recounts were made during a dry period, when the spring rains were delayed, for *G. crassicaudatus* from October 23 to November 1 and for *G. senegalensis* from November 16 to 26. Summer counts were made after a period of heavy rains for *G. senegalensis* from January 13 to 24 and for *G. crassicaudatus* from February 20 to March 2. An additional count was made from February 10 to 17 at a second site with denser vegetation for *G. senegalensis.* The final set of data consisted of 17,899 records. Complete statistical documentation of results discussed below is provided by Crompton (1980, 1983).

Problems in Method. In retrospect I find there were some deficiencies in my approach that became apparent only after the analysis had developed. The study began as purely an analysis of locomotor behavior. I did not expect to carry out a detailed ecological study, because this was already being done by Harcourt and Bearder. Further, the importance of ecological factors such as food resource distribution was not apparent to me at the time of the study. Consequently some estimates of resource distribution, frequency of predation, and support availability that would strengthen this analysis can be made only inferentially from less than satisfactory sources. Harcourt and Bearder attempted to assess the availability of arthropods at or near ground level by using ultraviolet light, but they were not able to assess availability of arthropods in the canopies of large trees. I attempted to determine the availability of supports of different diameters and orientations by counting them in trees of modal height in the samples taken along transects at each site. I had no way of taking into account branch length, and since analysis has revealed that trees were not used according to the frequency of their size classes, the measure is of dubious value. Attempts to reconstruct the frequencies of supports in the most commonly used trees from photographs were thwarted by bias due to differential visibility of branches of different sizes and the intertwining of supports. Finally, I have not measured predator pressure. While measures may exist, they are difficult to apply to larger vertebrates, and it is generally true that predation is quite difficult to observe except in very long-term studies or in studies of the predators themselves.

Additional methodological difficulty stems from the issue of subjectivity of categories such as movement, which are not naturally scaled. As Rose (1979) has observed, however, the variables do tend to cluster at various points even though they fall along a continuum. Standardization of categories between studies as suggested by Prost (1967) is an admirable goal, but it is impractical at present given the diverse conditions of the various studies. Precise definition of categories used in each study may be an adequate substitute. I have attempted to define carefully the categories used in this analysis.

Despite the limitations I have described, I believe this study was successful in its original goal of putting consideration of locomotion in galagos on an objective, quantitative basis, and it may, I hope, document some clear links between diet and locomotion.

Distribution and Availability of Food Items

For the reasons just given, I can describe the nature of food dispersion and availability only subjectively. Arthropods should be more abundant in warm, humid conditions. In winter, then, though they may be scarce overall, arthropods may still be found in leaf litter and dead trees on the ground, where warm, humid conditions remain. Because of the increase in warmth and humidity, we expect a more general distribution of arthropods in the summer. On the other hand, flowers and gumlicks may attract flying arthropods to localized places.

Gums are taken by *G. crassicaudatus* themselves from *Acacia karoo* at Wallacedale, and I also observed them to feed on droplets from *Combretum erythrophyllum* occasionally. At Mosdene, Bearder and Martin (1980) found that gums were taken by *G. senegalensis* from three species of *Acacia: A. karoo,* which bears 84% of gum sites; *A. nilotica;* and *A. tortilis.* Gums from *A. karoo* are structural carbohydrates (Lehninger 1972), which are complex polysaccharides made up of arabinose, galactose, rhamnose, and aldobiuronic acids (Bearder and Martin, 1980) with about 1% by weight of protein and 2% of minerals. As structural carbohydrates, the gums require fermentation by gut microflora for digestion. The availability of gums varies seasonally, depending on the amount of rainfall (gums are a tacky fluid in wet weather but become a crystalline solid in dry weather), which peaks in summer and on the number of gum sites, which peaks in winter (Bearder and Martin, 1980). Gum sites seal holes made by wood-boring larvae, and they are concentrated on the undersides of major branches and on the trunk, thus particularly near the main branchpoints of trees. They vary in size from droplets to accumulations of as much as a liter in a major hole.

Variation in rainfall and temperature between seasons leads to a consistent modification of dispersion of arthropods and gum, which are the major food sources of *G. senegalensis* and *G. crassicaudatus* respectively. We should expect to see changes in behavior between seasons that reflect this annual modification.

Tree Zones and Locomotion

Despite the fact that the two study sites differed somewhat, the overlap in habitat structure is sufficient that we may divide the *Acacia* bush into four arboreal zones based on the frequency, type, and continuity of supports (fig. 3.1.).

The first, *low zone* may be crossed by traversing laterally on the ground, by leaping between vertical tree trunks and high-angled main branches, or by climbing into the lower part of the crowns of smaller trees to cross shorter breaks between tree crowns. While movement in this zone is unobstructed, and a large area can be explored rapidly, only the area immediately surrounding the tree trunks and the ground itself is accessible to an arboreal animal for arthropod foraging. Predation by ground-living or avian predators should be a high risk here because of the lack of cover.

The second, *intermediate zone* does not offer the same speed of lateral exploration, but the bush and small tree crowns allow a considerable space to be explored for insects. The first substantial gumlicks

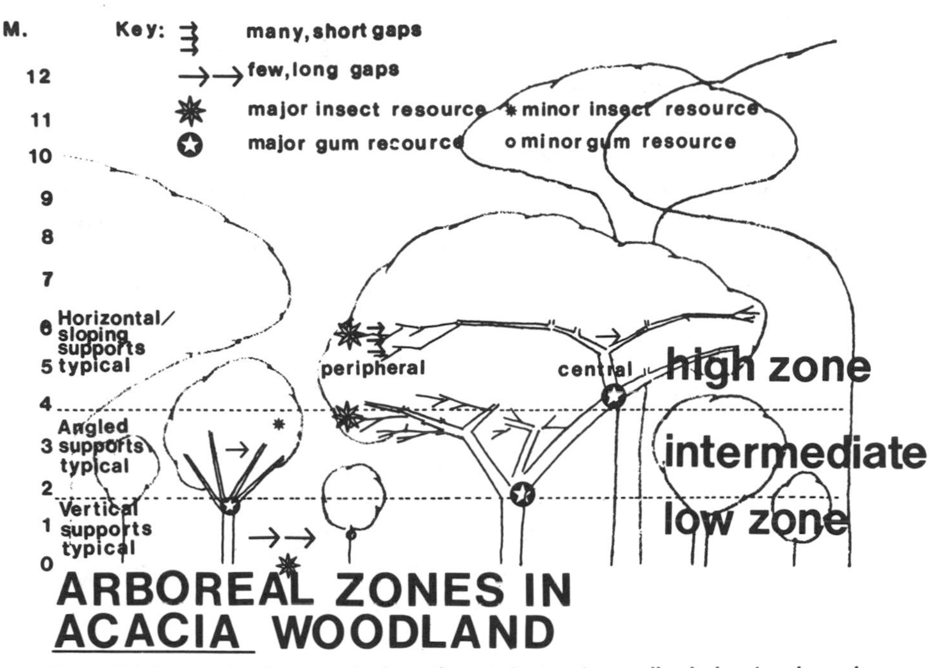

Figure 3.1. Schematic diagram of arboreal zones in *Acacia* woodland, showing the major characteristics of resource distribution and support structure in each zone.

are to be found on the major branches and trunks of trees in this zone. Lateral movement is possible by walking, running, or climbing between the central part of a tree and the terminal branches and then leaping or making slow crosses across the short gaps between the foliage at the periphery of the canopy of the neighboring trees.

The *central high zone* offers rapid lateral movement to and from the gumlicks concentrated on the trunks and major branches of the larger trees. Paths along the main low-angled branches, crossing carefully between peripheral foliage onto similar routes in the next tree or making short leaps, allow substantial point-to-point distances to be traversed with little effort. There is, however, little access to insects in this zone.

The *high peripheral zone,* in the canopy of larger trees, offers maximum access to space for insect foraging, but gumlicks are rare and usually of droplet size. The many short discontinuities make movement slow. Supports in this zone are generally small and low-angled.

We have seen that the two species of *Galago* differ in diet and that elements of the diet are distributed differentially within and between the "arboreal zones" of *Acacia* woodland. The zones differ in abundance, distribution, and location of food resources; in "openness" (and hence the risk of predation); and in orientation, diameter, and spatial separation of the supports available for locomotion in search of food or in avoidance of predators. We can therefore expect quantitative differences in arboreal zones utilized, support use, and locomotion between the two galagos corresponding to their search for different diets. The following section presents the results of my analysis of patterns of locomotion and substrate use and of differences in those patterns. In the discussion I intend to show first that intraspecific differences in locomotor behavior related to dietary change indicate that the basic differences in locomotion between species are alternate strategies that have evolved to cope with seasonally unstable food supplies, and second I show how the alternate strategies may be "predictable" consequences of body size.

Results

Interspecific Comparisons

Height (fig. 3.2). *G. senegalensis* uses all height classes below 4.3 m more than *G. crassicaudatus,* but all heights above 4.3 m less (except the tenth meter above the ground). The mean height of observations in *G. senegalensis* is 2.97 m, but for *G. crassicaudatus* it is 6.46 m. Only *G. senegalensis* is frequently found on the ground. These results must be interpreted with caution since at a second site at Mosdene I found *G. senegalensis* at a mean height of over 5 m, and A. Walker (personal communication) has observed this species in Kenya at over 20 m in *Acacia xanthophloea.* It is certainly true for the South African bushbabies, however, that *G. senegalensis* is frequently found crossing open ground or foraging for arthropods on the ground, whereas *G. crassicaudatus* rarely ventures to the ground. Observations by myself and others indicate that *G. crassicaudatus* avoids the ground unless forced

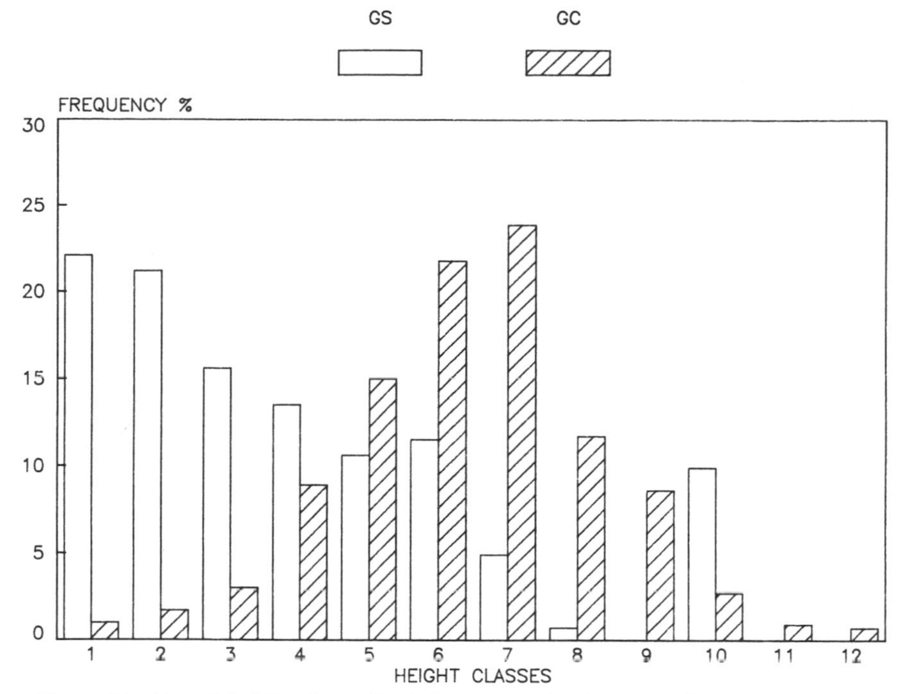

Figure 3.2. Use of height classes by *Galago senegalensis* (GS) and *G. crassicaudatus* (GC). Height is that of the initial support only. 1 = 0.0–1.0 m, 2 = 1.1–2.2 m, 3 = 2.2–3.2 m, 4 = 3.3–4.3 m, 5 = 4.4–5.4 m, 6 = 5.5–6.5 m, 7 = 6.6–7.6 m, 8 = 7.7–8.7 m, 9 = 8.8–9.8 m, 10 = 9.9–10.9 m, 11 = 11.0–12 m, 12 = over 12 m, N = 9,096 (GS, 8,600 (GC).

to make a ground crossing or when foraging for particularly rich resources such as fallen fruit, large (5–10 cm) millipedes, termites, or flying ants. In other words, the mean height of observation is not as important as the difference in use of vertical strata; *G. senegalensis* habitually forages on or near the ground while *G. crassicaudatus* habitually avoids it.

Since the animals were not being studied in sympatry, these results could be purely consequences of different habitat structures at the two sites. Data on availability versus use of the different tree size classes tend, however, to support the conclusions made. Figures 3.3 and 3.4 together show that in neither case are tree size classes used according to their abundance. In both cases the *tree under 12 m* class

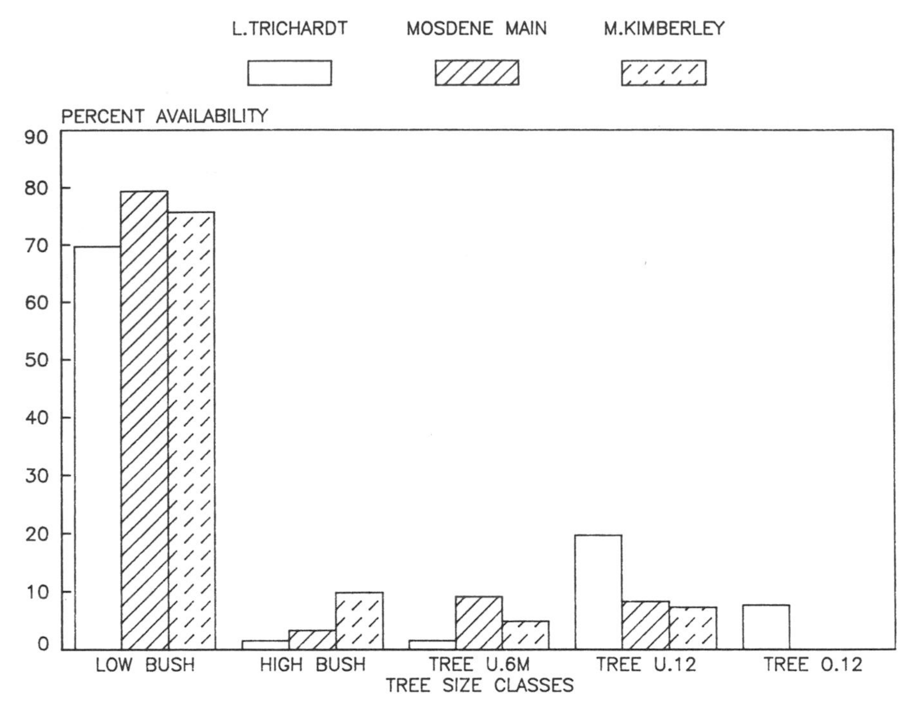

Figure 3.3. Availability of trees of various size classes at each study site. L. Trichardt = "Wallacedale," the *G. crassicaudatus* study site; Mosdene Main = main *G. senegalensis* study site; Mosdene Kimberley = second *G. senegalensis* study site; low bush = under 6 m, diameter at breast height (dbh) less than 10 cm; high bush = over 6 m, dbh less than 10 cm; tree U. 6 M. = under 6 m, dbh 10 cm or more; tree U. 12 = 6–12 m, dbh 10 cm or more; tree 0.12 = over 12 m, dbh 10 cm or more.

is used preferentially, although to a lesser extent in *G. senegalensis* than in *G. crassicaudatus.* The smaller species makes greater use of the smaller tree classes. Further, in trees of the same size class *G. senegalensis* is found consistently lower. In *trees under 12 m,* 46.7% of observations of the large species, *G. crassicaudatus,* were made between 3.3 and 6.5 m but 65% of those of *G. senegalensis.* From 3.3 m downward, 4.4% of observations of *G. crassicaudatus* contrast with 19.3% of observations of the smaller species. For the larger species 47.9% of observations were made above 6.5 m but only 19.6% of the observations of the smaller species.

Some of the height difference is certainly accounted for by the

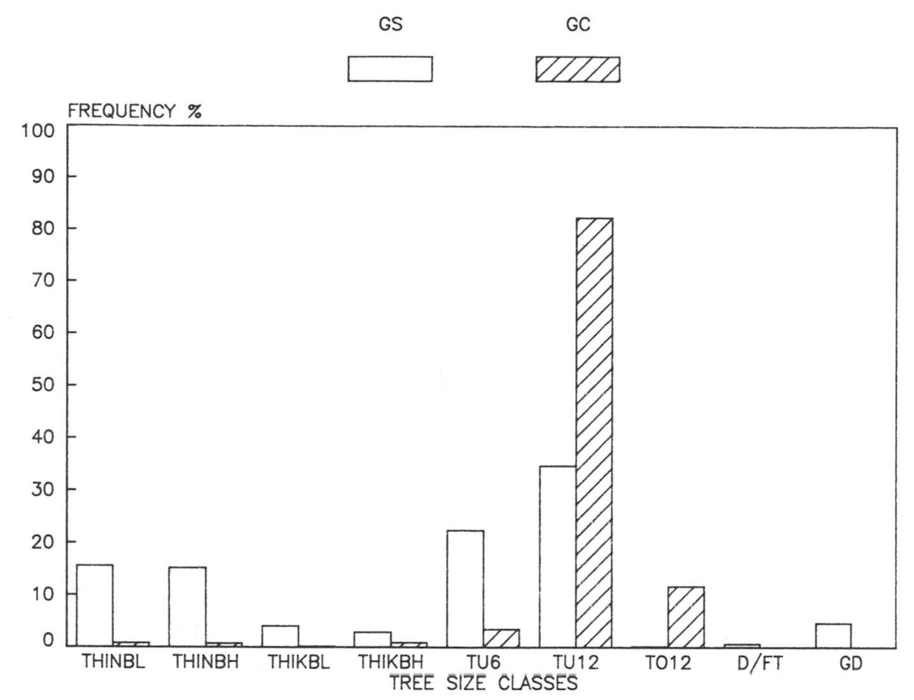

Figure 3.4. Use of different tree size classes in *Galago senegalensis* (GS) and *G. crassicaudatus* (GC). THINBL = thin low bush; THINBH = thin high bush; THINBL = thick low bush; THIKBH = thick high bush; TU6 = tree under 6 m; TU12 = tree under 12 m; TO12 = tree over 12 m; GD = ground; D/FT = dead or fallen tree; Thick and thin are subjective estimates of density; otherwise definitions as above. N = 9,096 (GS) 8,600 (GC).

absence of a high, continuous canopy at Mosdene. A sympatric study would be necessary to factor out the effects of a different habitat structure completely.

Activity (*fig. 3.5*). The differences in activity profiles may be explained as consequences of major differences in diet. *Foraging* (searching for food) is much more common in *G. senegalensis,* but *travel* (movement without searching), *feeding* (actual consumption of food), *rest* (total inactivity), and *grooming* (auto grooming or allogrooming) are more common in *G. crassicaudatus.* There is a bias of observability toward *foraging* on insects but toward *feeding* on gum. In other words, the frequencies of the two activities reflect the compo-

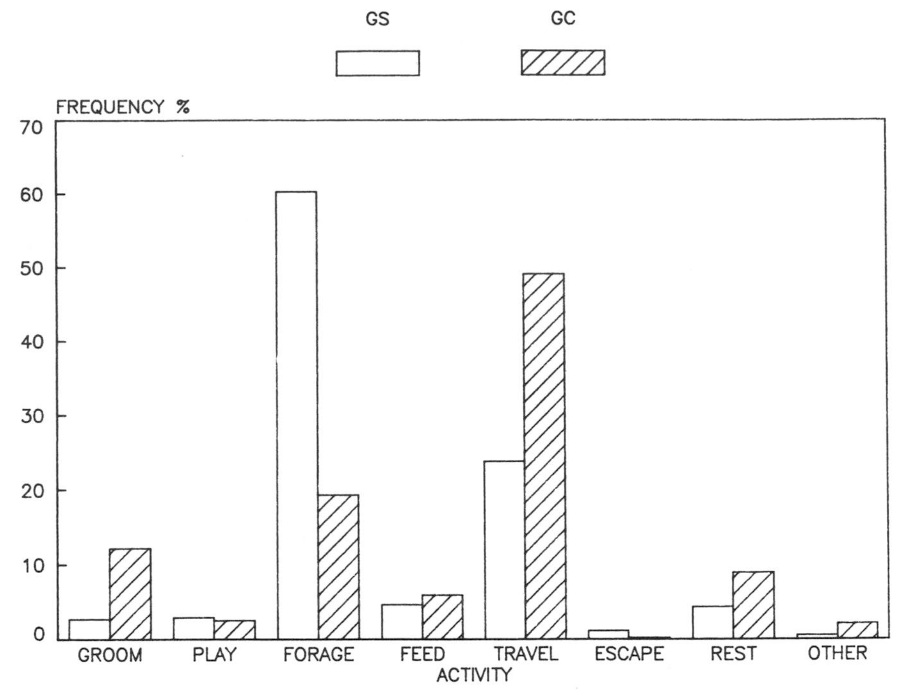

Figure 3.5. Activities of *Galago senegalensis* (GS) and *G. crassicaudatus* (GC). Definitions in text. N = 9,096 (GS), 8,600 (GC).

sition of the diet in each species. Thus *G. senegalensis* behavior is typically continuous, bird-like foraging, while *G. crassicaudatus* behavior is characterized by slow movement between gum sites, punctuated with feeding bouts and periods of inactivity.

Posture and Feeding Behavior. *G. senegalensis* uses clinging postures (in which the body is not horizontal and both fore and hind limbs are used for support) in more than a third of cases, while *G. crassicaudatus* uses them less than half as frequently. About 70% of *G. crassicaudatus'* postures are *standing* (body horizontal, above substrate, four limbs used in support) but half of *G. senegalensis'*. These results are not a direct consequence of dietary type: although, during feeding on arthropods, *standing* postures are commoner than *clinging*, *G. senegalensis* always *cling* more than *G. crassicaudatus*, whatever the food type. *Head-down clinging*, associated with gum-feeding on tree trunks, supports 29.7% of *G. senegalensis gum feeding* but only 18.9% of that of *G. crassicaudatus*. Two small differences in the frequency of rarer behaviors—a greater frequency of *hanging* by the back legs in *G. senegalensis* and of *hanging* by all four limbs in *G. senegalensis*—do reflect dietary differences and feeding style fairly directly. *G. senegalensis,* but not *G. crassicaudatus,* quite commonly catches flying arthropods by extending the body out like a spring from the fixed hind limbs (*cantilevering,* see below, typical of several small Lorisidae and Cheirogaleinae) and making a stereotyped two-hand catch. *G. crassicaudatus* more commonly takes gum from underneath major, low-angled branches. Both species usually take gum at a mean height below the overall mean (with one exception, for one study in *G. senegalensis* where the overall mean was unusually low and not significantly different from the gum-feeding mean). Arthropods are taken at a mean height not significantly different from the overall mean in both species.

Orientation of Supports (fig. 3.6). In considering the use made of supports of different orientations and sizes, the problem of availability crops up again. Although no adequate measure was devised, some indirect evidence allows interpretation of the data on support usage. Supports of different orientation were used in a regular order of height in each study. Lowest mean height occurred in *vertical* supports (80°–90°) then *angled* supports (45°–80°). *Sloping* supports (10°–40°) came next, then finally *horizontal* supports (0°–10°). In two of the nine studies, the mean height for *horizontal* and *sloping* supports was the same. The pattern of gradually decreasing slope was followed by the branches of the typical Acacia tree (both species used *Acacia* in about 80% of observations made): in other words, the main component in determining the angle of support was the height of the support. Similarly, it is common knowledge that branches in a tree get smaller the higher and more peripheral they are.

Vertical supports are much more frequently used by the smaller species (see fig. 3.6) and *angled* supports are also more typical of the small species. Following the argument above, we see that the difference is consistent with the difference in mean height of observation. Part of the difference is accounted for by the fact that 70% of *G.*

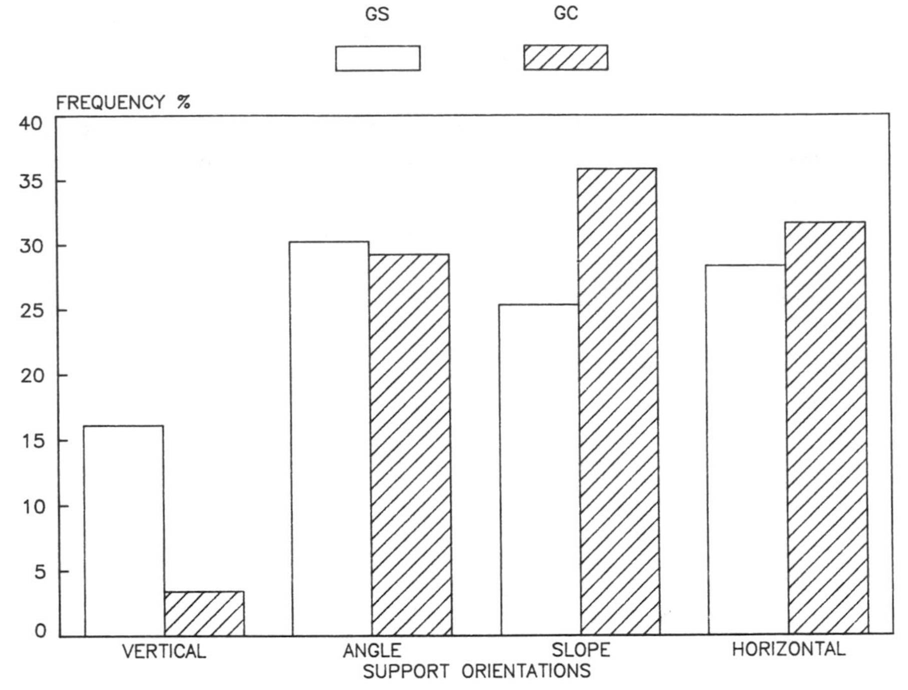

Figure 3.6. Frequencies of different initial support orientations in *Galago senegalensis* (GS) and *G. crassicaudatus* (GC). Definitions in text. N = 9,096 (GS), 8,600 (GC).

senegalensis' gum-licking is on *vertical* supports—partly a reflection of tree class use—*G. senegalensis* uses *vertical* supports at twice the overall frequency when in *thin, low bushes* (single-trunked saplings less than 5 m high with a breast height diameter less than 10 cm). *Vertical* supports are used more frequently in escape and when *leaping* longer distances, and the speed of lateral movement possible between the vertical trunks of the low zone has already been noted.

Diameters of Supports (*fig. 3.7*). Results are not as clear when we consider differences in the diameters of supports used. There is a difference in the frequency of use of supports of different diameters between *G. senegalensis* and *G. crassicaudatus,* with *G. senegalensis* using single supports of *small branch* size (1–5 cm) and less more frequently than *G. crassicaudatus. G. crassicaudatus,* however, uses *foliage* (multi-

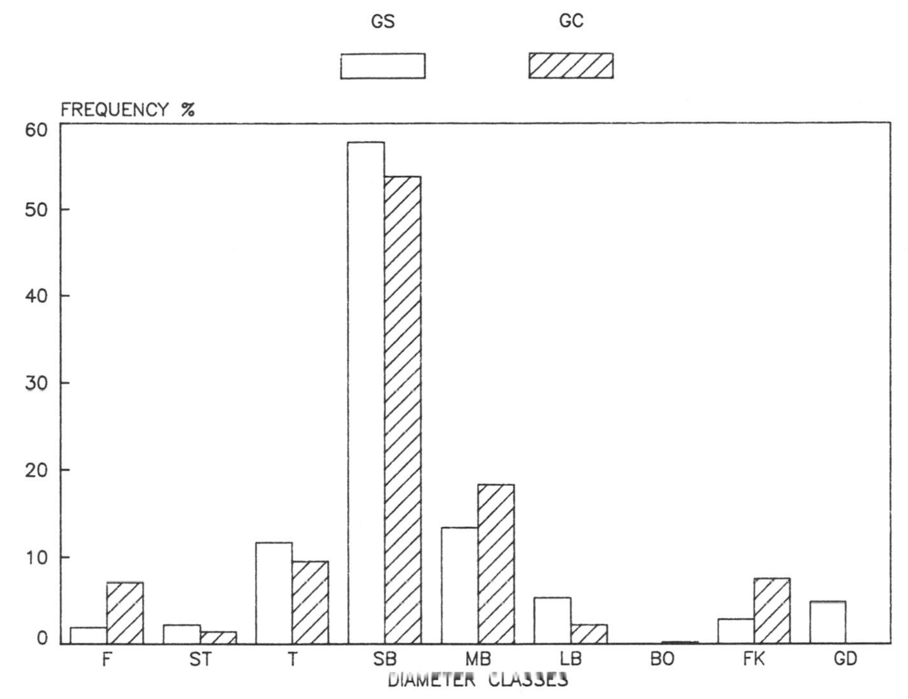

Figure 3.7. Initial support diameters in *Galago senegalensis* (GS) and *G. crassicaudatus* (GC). Fork = multiple supports over 1 cm; bough = over 20 cm; F = foliage; ST = small twig; T = twig; SB = small branch; MB = medium branch; LB = large branch; BO = bough; FK = fork; GD = ground; other definitions in text. N = 9,096 (GS), 8,600 (GC).

ple supports less than 1 cm diameter) more often than the smaller species, which itself uses the *large branch* category (10–20 cm) more frequently. Probably these two exceptions reflect the greater number of tree-to-tree crossings made by *G. crassicaudatus* at canopy level and more frequent use of low vertical tree trunks by *G. senegalensis.*

Movement (*fig. 3.8*). Galagine locomotion consists of four major categories of movement subdivided into smaller categories for analysis: (1) Striding quadrupedalism with a floating phase is *running* and without a floating phase is *walking.* (2) Quadrupedalism in which the forelimbs play an important part in propelling the body is *climbing,* which has the two variants: *foliage crossing,* or slow crossing between

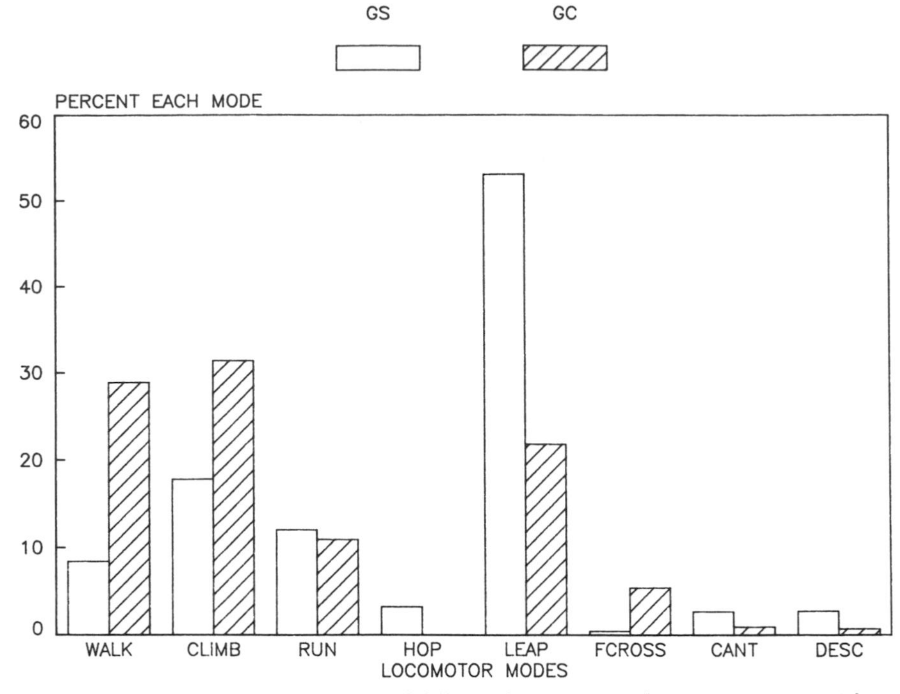

Figure 3.8. Comparison of frequency of different locomotor modes (movement types) in *Galago senegalensis* (GS) and *G. crassicaudatus* (GC). Definitions in text. N = 6,170 (GS), 5718 (GC).

supports in which the supports are drawn to the body *before* weight is transferred, and *descending,* or climbing with rotation of the subtarsal and midtarsal joints to "reverse" foot direction. (3) Two types of saltation, *leaping* (with forelimb participation) and *hopping* (without forelimb participation). (4) A movement in which the feet grasp the substrate firmly and the rest of the body is propelled away from and back to the anchor point like a spring, which I call *cantilevering.* More than half of *G. senegalensis'* displacements are *leaps;* in fact, more than two-thirds of every kilometer is covered by *leaping* (Crompton 1980). Leaps in *G. senegalensis* become longer the higher the angle and the greater the diameter of the initial support, and *leaps* are made more often in the smaller tree size classes (Crompton 1980). In contrast *walking* and *climbing* are the most frequent movement types in the larger species, and *running* accounts for a greater proportion of distance covered by *G. crassicaudatus* than *leaping* does. The frequency of *running* is, however, about equal in both species. When escaping, *G. senegalensis* moves below the overall mean height and uses *leaps* and *verticals* (fig. 3.6) more frequently than usual. In contrast, *G. crassicaudatus* moves *above* the overall mean height when in danger. *G. senegalensis hops* farther than it *walks,* whereas G. *crassicaudatus* does not *hop* at all in this population under normal conditions. The importance of *verticals* in *G. senegalensis* locomotion is underlined by the higher frequency of *descending,* the specialized headfirst movement used commonly on large, high-angled supports. Again, *G. senegalensis cantilevers* more; this makes sense given *G. senegalensis'* dependence on arthropods and since cantilevering is a specialized insect-catching maneuver. In *G. crassicaudatus, foliage crossing* is more common and appears to allow slow, careful transfers between the peripheral branches in the high canopy.

Seasonal Changes

Table 3.2 shows that there are important changes in diet in both species between the winter and summer. Harcourt's work shows in particular that distinctions in arthropod prey size between the species under humid, optimal summer conditions collapse in winter. Harcourt's data were taken in a particularly mild winter, and conditions during the winter of the present study were certainly more extreme. Harcourt's and Bearder's studies show how *G. senegalensis* turns to gum as a mainstay of its diet in winter. In addition to a count at both sites under dry, cold winter conditions and counts during warm and humid summer weather, a third count was made in late spring that revealed the biggest changes in behavior. At this time the weather was still cool at night, and more particularly, the spring rains were seriously delayed. Gumlicks were dry and crystalline and flying insects not yet abundant. Space precludes extensive presentation of data here, but full documentation is available elsewhere (Crompton 1980 and in preparation). Some of the major differences are summarized in figure 3.9.

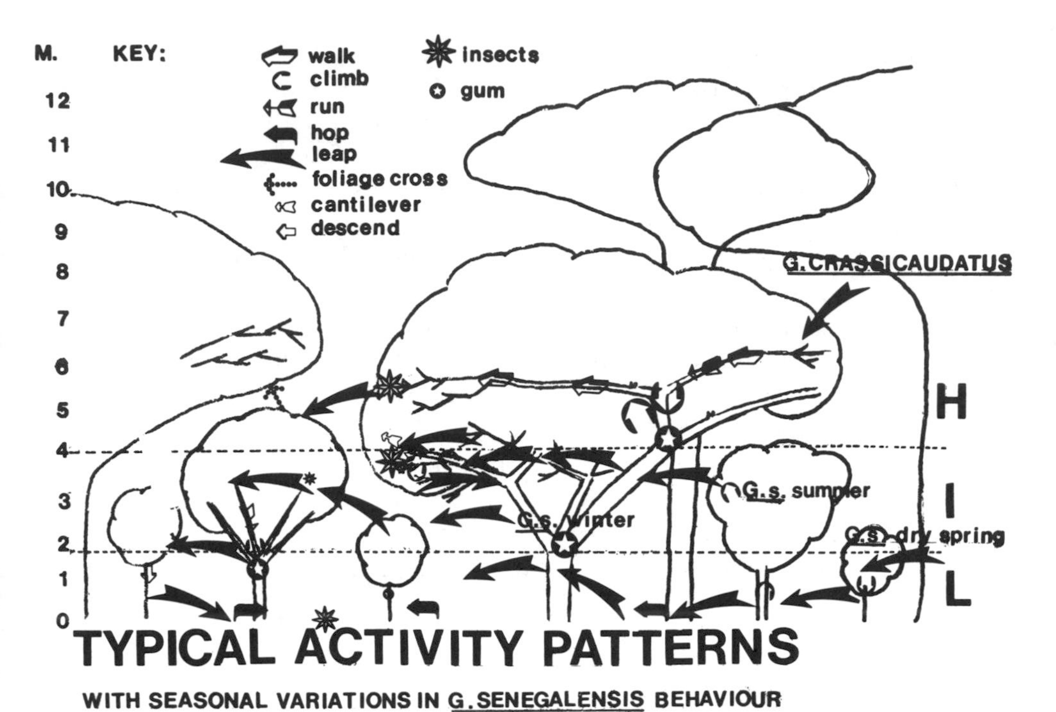

Figure 3.9. Schematic diagram of typical activity patterns in *Galago senegalensis* and *G. crassicaudatus*.

G. senegalensis, Summer. In summer, high temperatures and heavy rainfall bring abundant gum and arthropod food. *Foraging* behavior takes up fully 76.6% of activity but *feeding* only 1.5%. As I have noted, these behaviors are biased toward arthropod-foraging and gum-feeding because of observability. In the summer the mean height of observation is the greatest, at 3.2 m, ground use drops to 0.6%, and the *tree under 12 m* class becomes the one most commonly used (fig. 3.10). Supports used are generally smaller than in winter. Sloping supports are more commonly used, but the use of *horizontal* supports changes little, reflecting the balancing effect of the decline of use of the ground (fig. 3.11).

Clearly, the bushbabies are using the peripheral canopy of larger trees (the *high peripheral zone*), where access to space is maximized and predation danger seems to be minimized.

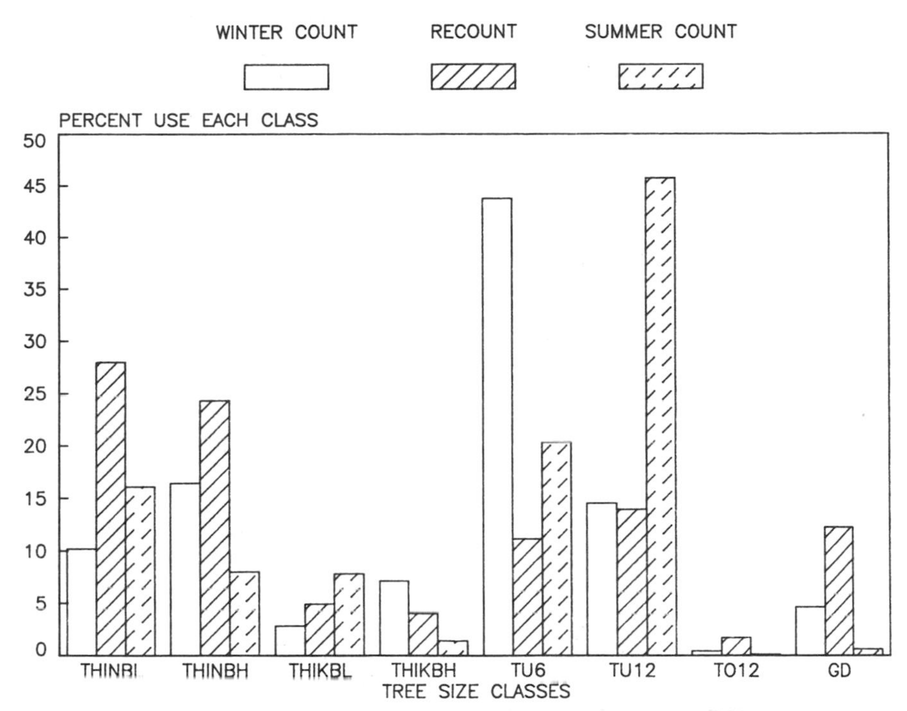

Figure 3.10. Seasonal changes in use of various tree size classes in *Galago senegalensis*. Winter count = count in cold, dry season. Recount = count in cool, very dry late spring. Summer count = count in hot, humid season. Other definitions as in figures 3.3 and 3.4. N = 1,515 (winter), 2,716 (recount), 2,132 (summer).

The frequency of *leaping* undergoes no seasonal change, but the frequency of *climbing,* associated with *foraging* observations in both species, increases. The length of individual displacements decreases as the short, exploratory movements of arthropod-foraging replace the point-to-point *travel* associated with gum-feeding. *Verticals* are used rarely at this height, and so *descending* is reduced. *Hopping* declines with the reduction in use of the ground. *Foliage crossing* also declines, probably because at Mosdene the canopy is discontinuous at this height.

G. senegalensis, Winter. In the Transvaal winter *G. senegalensis* becomes dependent on gum, although it still takes a considerable quantity of arthropods. *Foraging* is less than one-third as frequent in

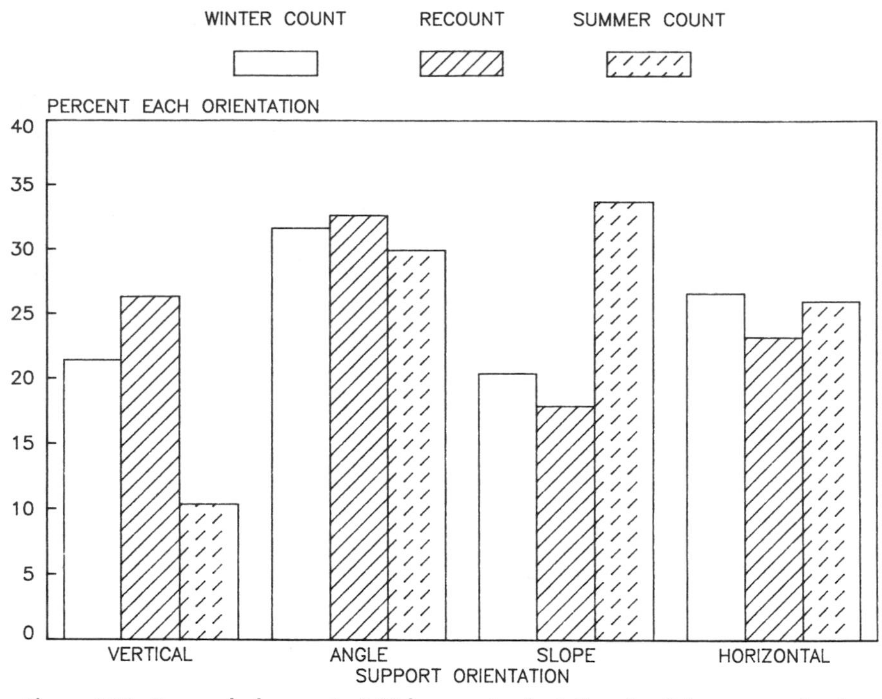

Figure 3.11. Seasonal changes in initial support orientations in *Galago senegalensis*. Winter count = count in cold, dry season. Recount = count in cool, very dry late spring. Summer count = count in hot, humid season. Other definitions in text. N = 1,515 (winter), 2,716 (recount), 2,132 (summer).

this season as in the summer, and *feeding* rises to 11.3% of observations. The decline in the numbers of flying arthropods accompanies a rise in the number of arthropods taken on the ground, now 21.4%. Ground use thus increases to 5%, and *G. senegalensis* is found at a mean height of 2.6 m. The most commonly used tree size class is now the *tree under 6 m,* which is used in 44% of observations. A broken low canopy is formed by these lower trees, which allows lateral movement between the gum sites on their trunks and major branches. Thus there is a rise in the frequency of *foliage crossing*. The lateral movement between gum sites is also reflected in the high frequency of observations of *travel* (at its highest frequency) and in the long average displacement length or bout length. The use of *vertical* supports doubles with the increased use of gum sites and the decrease in mean height. With this rise come a rise in *clinging* and *descending* compared with the summer and a general increase in the size of supports used. The use of *horizontal* supports is greatest at this season: this may reflect both an increased use of the ground and the high frequency of *foliage crossing,* which, being usually between terminal branches, is most frequently on *horizontal* supports. *Hopping* increases in frequency with the increased use of the ground. *Running* becomes more frequent since it is associated with *travel* observations. In general, the features of habitat use described above are consistent with the characteristics of the *intermediate zone.*

G. senegalensis, Dry Spring Recount. During this period, delays in the spring rains threatened drought. Gum was dry and crystalline, and therefore difficult to obtain, and it was in short supply. Now *G. senegalensis* is observed most often in the thin-bush-class, single-trunked saplings. The ground is used as often as either the *under 6 m* or *under 12 m* class—three times its use in winter and fifteen times its use in summer. The mean height of observation drops to its lowest—only 1.3 m. The reason for this is a rise in the use of the ground for *feeding* on arthropods—now 30.6% of observations—and a decline in the importance of gum sites on trees, which were often blocked by crystalline dry gum. As might be expected, *vertical* and *angled* supports are used most commonly in the recount, as *G. senegalensis* is now moving below the tree crowns. *Horizontals* are used most rarely in this season, despite the increase in use of the ground, probably because of a decline in the use of the periphery of trees and of *foliage crossing.* Support diameters tend to be large, as accords with the frequent use of supports of high angle. Leaping between tree trunks and the ground is reflected by the large contribution of leaping to a kilometer of travel. Descending becomes most common at this season, which again follows from the frequent use of vertical supports. Climbing increases in frequency somewhat over the winter, with an increase in arthropod foraging. With declining importance of gum comes a decrease in the frequency of *travel* observations and so a decrease in the frequency of *running.* Behavior at this season clearly corresponds to use of the low arboreal zone.

G. crassicaudatus, Summer. The changes in *G. crassicaudatus'* behavior are much less marked than for the *G. senegalensis.* In summer, as with *G. senegalensis,* there is an increase in *foraging,* and with this a decline in bout lengths. The less frequent types of movement decrease further in frequency. Supports used are usually small, with the exception of the *large branch* category. The reason for this exception is probably that *G. crassicaudatus* made greater use of *Combretum erythrophyllum,* a tree that bears characteristically long, large, horizontal lateral branches that are used by the animals as "arboreal highways." Horizontal supports are indeed used more commonly than in winter, but the increased use of small, low-angled branches in the periphery of tree crowns is also responsible for this increase in frequency. There is a small *decrease* in mean height of observation, the opposite change to that in *G. senegalensis,* reflecting more activity in the lower part of the peripheral canopy. There was no change in the category most often used, which remains the *tree under 12 m.* Some of the activity in the low peripheral canopy consisted of investigation and foraging at termite mounds from the safety of terminal branches of large trees.

G. crassicaudatus, Winter. In winter, activity centers around gumlicks, and there is only a slight change in emphasis from the overall description of *G. crassicaudatus'* behavior. *Vertical* and *sloping* supports are used most commonly at this season but *angled* supports relatively uncommonly. The increase in preference for *vertical* supports probably results from the decrease in height of gum-feeding. In other seasons, the mean height is 5.9 m, but in winter it is 5.4 m, which may reflect a greater use of trunk rather than major branch or branchpoint gumlicks. Associated with this, *descending* and *head-down clinging* reach their highest frequency. Travel is at its highest frequency, since most activity centers around movement between major gumlicks, and as with *G. senegalensis* under similar conditions, bout lengths are long and the frequency of *foliage crossing* high. Increased *leaping* probably represents crossings between peripheral branches of trees. *Clinging* is infrequent overall. *Standing* is common compared to the summer, but *clinging* is commoner in *feeding* observations than in the summer. This indicates that the high frequency of *standing* reflects that of *resting* because long periods of inactivity alternate with traveling between gum sites and gum-feeding in winter, possibly to facilitate digestion.

G. crassicaudatus, Dry Spring Recount. In this dry late spring period, with a shortage of flowing gum sites, *G. crassicaudatus*' behavior undergoes some major changes. Arthropods become a more important element of the diet than during the winter, and as a consequence *foraging* becomes most frequent and *travel* most infrequent in this season. The latter decline also reflects the fact that the bushbabies concentrate their movements in a single group of trees, known as the "Gumlick" which was the most reliable collection of gum trees. These trees are tall, with broad, flat crowns that form an extensive, continuous canopy. This concentration of activity caused the *tree over 12 m* class to become the most commonly used category, and there is a small increase in the mean height of observations. Associated with the decline in *travel,* the frequencies of *leaping* and *running* drop to their lowest levels. *Climbing,* on the other hand, reaches its peak, and support use generally resembles that in summer, except that *angled* supports are used more commonly. The latter result may be a consequence of increased overall mean height. In this period, increased *foraging* for arthropods in the periphery of the trees is not accompanied by a movement downward toward the ground, as in the summer, nor are species other than *Acacia* used regularly. Rather, activity concentrates in a few *Acacias* with particularly reliable gum flows. As the number of *feeding* observations is lower than in winter, perhaps the period spent at individual gumlicks (giving-up time) increases (as Harcourt found, comparing winter to summer), or gum is being taken in droplet form from branches of small diameter higher up, which would be less observable. But gum remains the major dietary element even when in short supply.

Discussion

The results show clearly that support use and locomotion in the Transvaal galagos are strongly influenced by the structure of the arboreal zone they occupy. Moreover, the results indicate that seasonal changes in locomotor behavior could be predicted to some extent from inferred changes in resource distribution and consequent changes in the frequency of use of the different arboreal zones. In this section I discuss some hypotheses concerning relationships of body size, diet, support use, and locomotion and then show how a consideration of the relationships of body size, energetics, and diet allows us to predict the broad patterns of locomotor behavior in the two Transvaal species.

Support Diameter and Body Size

Cartmill and Milton (1977) proposed that while large-bodied animals can be at an advantage in use of large-diameter supports because of their proportionately large cheiridia, small-bodied animals that developed claws could do as well or better. In the case of the tamarins, Garber (this volume) has suggested that claw-like tegulae arose specifically in connection with the vertical clinging postures adopted on large trunks for sap-eating. While the larger bushbaby species did use larger supports in most cases, one large support category was more frequently used by the smaller species. This is a consequence both of the generally lower stratum occupied by the small species, where tree trunks are a common support, and of the high frequency of gum-feeding on the tree trunks in *G. senegalensis*. The two are of course linked. *Small* cheiridia, in the absence of claws, are the key in this case to successful use of large supports by a small animal. Surface roughness of the bark, considerable in *Acacia,* favors small digits that can make use of cracks for purchase. The other exception to the general rule is that *G. crassicaudatus* uses *foliage* more often than the small species. As Fleagle and Mittermeier (1980) point out, one of the main reasons for a large species' using a large support would be stability, which is answered as well by using multiple small supports. But the result underlines the fact that a simple prediction of support diameter from body size cannot be made and that stratum occupied is as good a predictor, at least in this case.

Diet and Locomotion

Fleagle and Mittermeier (1980) point out correctly that we cannot expect to find simple relationships between food choice and locomotion. Frugivores do not have to be suspensory. Some specialized behaviors—such as *cantilevering,* which is rare in *G. crassicaudatus* but quite frequent in the small species—are specializations toward food-getting. *G. senegalensis* uses this mode for insect-catching in more than 80% of occurrences. But it is not because it does not eat as many insects that we do not find *G. crassicaudatus* using this mode as often. In *G. senegalensis,* a stylized two-handed predatory snatch is used that is not observed often in the large species. Cantilevering and the snatch are reflections of each other, and both are associated with the hindlimb domination of *G. senegalensis.* There are species-specific behavior patterns that do not reflect immediate behavioral relations. Such a pattern is the high frequency of *leaping* in *G. senegalensis,* which remains constant from season to season and from site to site. Less direct links must be sought to explain such constant behavioral characteristics, although this does not mean that they are not linked to diet, as I show below. In other cases certain locomotor modes are linked fairly closely to diet, albeit through the medium of the characteristics of supports used to search for and harvest food. For example, both Garber (this volume) and I found that *climbing* is the mode most closely associated with arthropod foraging, because supports tend to be small and unstable in the areas where insects are hunted (which is not necessarily the same from forest to forest). Similarly, exudate eating and vertical clinging are associated simply because exudates often occur on tree trunks. The fact that support characteristics are the medium does not mean that animals do not develop adaptations for such postures and movements when access to an arboreal zone is necessary for their survival, however.

Body Size and Locomotion

Cartmill and Milton (1977) argue that *climbing* should increase against *leaping* in the behavioral repertoire the larger the body size, because larger animals must either use disproportionately larger supports or several supports at once to sustain their weight and because the impact from a fall increases rapidly with increasing size. I do not think that either factor accounts for the difference in body sizes of *G. senegalensis* and *G. crassicaudatus* to any great extent. Substrates available support *G. crassicaudatus* adequately, except in the extreme periphery of the canopy, where it often uses multiple supports from which it leaps. At its body size, it can and does walk away from falls of 7 m with only a slight ruffling of its pride. The problems, however, may become very serious at larger body size ranges. Fleagle and Mittermeier (1980) suggest that small animals "see" small gaps as discontinuities that can be leaped, while larger bodied forms "see" them as short gaps that can be bridged. To some extent this must be true for these two animals, but because the mean *leap* in *G. senegalensis* is longer than that in *G. crassicaudatus,* different perception cannot be the main reason for the difference in the frequency of saltation between the two Transvaal species.

Locomotion and Stratum

Fleagle and Mittermeier (1980) found significant correlation between the frequency of leaping and use of the understory in the Surinam monkeys. They point out, however, that walking and running are equally effective in the canopy, and so the relationship between the continuity of supports and locomotor modes used is not a complete answer. I do not think we can expect ties of support use to locomotion to be as tight as that. Neither is the frequency of *leaping* in *G. senegalensis* solely determined by its use of the low zone. *Leaping* is an invariable characteristic that has probably evolved for more complex reasons but reasons very much tied to diet and to stratum, as I show below.

Energy Budgets and Body Size

The smaller animals are, the higher their metabolic requirements per gram weight, for metabolic energy requirements are scaled to the two-thirds power of body weight (Kleiber 1961). The costs of retaining a proper body temperature in a colder environment are also relatively higher, because of the low body mass/surface area ratio. Further, the energetic costs of locomotion are greater per unit of body mass in smaller animals, and in particular, the costs of increasing velocity of movement in quadrupedal gaits are relatively large for a small animal (Taylor et al. 1970). Despite these relations, the absolute costs of movement will be greater in a large-bodied animal because of its high mass.

Energy Budgets and Diet

Pough showed neatly (1973) that limits on the size of arthropods available to a predator limit insectivory in large-bodied animals. While a large animal has an absolutely larger energy budget, the costs of capture per prey item are not lower for a larger animal, and the energy to be derived from a capture is smaller in relation to energy needs. Although a diet of arthropods may provide twice as much caloric energy per gram as a plant diet, the time costs of failure in a hunt, the relatively low reward for a successful hunt, and the high total daily energy needs make insectivory an inappropriate strategy for a large-bodied animal. The reverse is true for small species, and additionally their high metabolic requirements make the acquisition of protein more important. An individual insect forms a high proportion of total daily energy requirements.

Therefore large-bodied forms may not be able to fill their daily energy budgets from arthropods alone, unless they take them in especially large individual packages such as termites or perhaps milipedes. Empirically, it appears that the largest primate insectivores weigh around 200 g while the smallest folivores weigh some 800–1000 g (*Loris tardigradus* and *Arctocebus calabarensis* versus *Lepilemur mustelinus* and *Avahi laniger;* Kay and Cartmill 1977). The breakpoint between the two strategies falls between 200 and 800 g.

Diet and the Volume of the Gut

Parra (1973) showed that gut volume varies directly with body size. Smaller animals do not have proportionately larger guts. Therefore, the volume available for fermentation of structural carbohydrates becomes smaller in smaller animals. Since, however, metabolic requirements increase relative to body mass with decreasing body size, the smaller animal must achieve a faster turnover of gut contents. Janis (1976) observed that it is not the digestive process that changes but the choice of food item. The smaller an animal, the higher the quality of food it ingests. Small animals apparently minimize fibrous matter and structural carbohydrate such as cellulose in the diet.

Implications of Energetics, Body Size, and Diet Relationships for the Transvaal Galagos

Dietary Composition. As the smaller of a pair of closely related species with a qualitatively similar diet, *G. senegalensis* will be expected to take a greater proportion of high-quality food and minimize the intake of structural carbohydrate. *Acacia* gums consist largely of structural carbohydrate, and clearly, from what we know of the diets of the two species, their roles in diets of galagos follow expectations from the relationships discussed. *G. crassicaudatus* takes more gum, *G. senegalensis,* more high-quality arthropod foods.

Foraging Strategy and Locomotion. For *G. senegalensis,* the need to catch and consume large quantities of arthropods puts a premium on high speed of locomotion. The energetic costs of acceleration are relatively high in small-bodied animals when they are moving quadrupedally. Further, stride length puts restrictions on maximum quadrupedal speed. Saltation allows greater speeds, and moreover energy costs remain the same or actually decrease at high speed in large or small hopping animals (Dawson and Taylor 1973; Dawson 1976) because elastic savings increase in relation to kinetic energy costs. It is likely that these savings accrue to *G. senegalensis* in ricochetal leaping, as well as in hopping locomotion. Saltatory locomotion is, of course, more common in *G. senegalensis.* Further, saltation enables it to move in areas with large discontinuities between supports if necessary. *G. senegalensis* must move where arthropods are, despite changing resource distribution.

For *G. crassicaudatus,* which specialize on gum that changes little in distribution and has low harvesting costs, the problem is to reduce its large overall costs of locomotion and metabolism. It would be expected to move slowly, quadrupedally, avoiding large discontinuities. The reward accruing from capturing an arthropod, which is low in relation to total energy needs, is not such as to make it worthwhile to follow the changing distribution of arthropods if that means moving away from the shortest, most continuous, and level paths between gumlicks. Only when large aggregations of arthropod food are available should *G. crassicaudatus* invest much in arthropod foraging by leaving the canopy for more discontinuous zones.

Habitat, Foraging, and Locomotion

G. senegalensis' emphasis is very much on the hindlimb (in clinging and saltation) while *G. crassicaudatus* uses the forelimb more (in standing and quadrupedalism). *G. crassicaudatus* feeds largely on gum on the major branches or branchpoints of larger trees, *standing* on a branch rather than *clinging* to it. It moves slowly from gumlick to gumlick along relatively large, low-angled branches, crossing between the peripheral branches by short *leaps* or by *foliage crosses.* It usually takes arthropods opportunistically in the canopy of the tree. When avoiding the observer, it escapes upward into the canopy. *G. senegalensis* forages for arthropods in the periphery of trees, moving rapidly by *leaping* or *running* when not foraging for arthropods (*climbing* is the movement type most used by both species during *foraging*). When escaping, it moves down into the low zone. *G. senegalensis* may move down onto the tree trunks to feed on gum, to make a rapid cross by a *leap* into the next tree, or to move farther down and forage on the ground or cross an open space. During much of the year, however, *G. senegalensis* moves at the level of the broken canopy formed by smaller trees, using short *leaps* to cross between crowns.

In conditions of abundance, the animals should reduce effort and risk in foraging. With abundant arthropods, randomly located, foraging in the higher peripheral zone would maximize the space that could be explored while minimizing the size of discontinuities that have to be crossed. With no time pressure, the slow movement because of the many small discontinuities would not be a problem. The height would minimize the risks of predation. For *G. crassicaudatus,* as a gummivore, there would be less incentive to move into the periphery of the trees, where movement is hampered by discontinuities, but some change in this direction might be expected since both species prefer arthropods over gum in choice tests conducted in the laboratory. Slow quadrupedal progression between gumlicks in the high, central zone remains an appropriate strategy, since gum is an exhaustible but regularly renewed resource for which random searching is unnecessary under most conditions.

If the weather is cold and dry, the high peripheral zone should become unattractive since the number of flying arthropods would be reduced, and extensive lateral exploration would be slow. The reservoir of moisture and arthropods in leaf litter should make the ground attractive for arthropod foraging, and a large area can be rapidly explored in a short time in the low zone. Risks are, however, higher and gum less abundant than in the intermediate zone, and so if gum is reasonably abundant, the latter zone should be appropriate, especially at Mosdene, where there is no high, continuous canopy. Only when gum is not available should we expect the low zone to be used.

G. senegalensis uses a different arboreal zone in each of the three studies, since the abundance and distribution of food resources change. *G. crassicaudatus* does not, however, markedly change its diet or behavior. Under conditions of abundant food resources the behavior of the two species converges, and under conditions of food scarcity it differs most. In summer, both species use *trees under 12 m* most frequently, presumably because these trees provide the maximum access to space in which to forage for flying arthropods, while minimizing the number of large discontinuities that have to be crossed and minimizing the risk of predation. At this season both species use small supports of low angle more often than at any other time. In winter, while *G. crassicaudatus* merely uses the periphery of the trees less and the high, central zone more, *G. senegalensis* shifts its activity downward into a smaller tree class, not only to allow passage from gumlick to gumlick in the broken low canopy of smaller bushes and trees (in a scaled-down version of *G. crassicaudatus* activity), but also to search the ground for arthropod food. Rather than behavior's becoming more alike at a season when both species are dependent on gum, it differentiates more. Although both species increase the use of *vertical* and *angled* supports, the change in *G. senegalensis* greatly outweighs that in *G. crassicaudatus.* When gum becomes scarce, the species again move in opposite directions: *G. crassicaudatus* upward and into a small group of reliable gum trees; *G. senegalensis* farther down toward the ground and out of the canopy of even small trees. At this dry season, the saltation and affinity for supports of high angle that typify *G. senegalensis* and differentiate it from *G. crassicaudatus* become most relevant. Here saltation allows fast lateral movement between ground and trunk, a survey of open ground for prey and predators from the safe vantage point of tree trunks, rapid escape from threatened attack on the ground or on the trunk, and rapid crossing of open country.

Although the species differ greatly in the strata they occupy, in their diets, and in their locomotion, they are rarely found sympatrically. It would be unlikely, given the constraints of body size/energetic relationships, to find *G. crassicaudatus* in open *bushveldt.* Why, however, does *G. senegalensis* not "move in underneath" *G. crassicaudatus?* While it is possible that the saltatory species requires high illumination to locate targets while moving and to spot prey and predators that are not found in the denser riverine woodland that *G. crassicaudatus* inhabits (and indeed Bearder [personal communication] found that *G. senegalensis* moves farther on moonlit nights), the answer is more likely suggested by the collapse of prey-size distinctions that Harcourt (1980) found. Under conditions of scarcity, competition for scarce arthropod prey may be too intense, especially since *G. senegalensis* could not take the dry, crystalline gum that *G. crassicaudatus* is often found eating under such conditions. In Gabon under the different conditions of tropical forest with its relative abundance and diversity of food species, several lorisids do coexist. There are lorisid species together in Gabon, and in coastal Kenya, *G. zanzibaricus* and the small-bodied *G. crassicaudatus* variant share the forest.

Conclusions

For surviving in a seasonally unstable subtropical environment, the two Transvaal species exhibit strategies that contrast notably with respect to body size and speed of movement. Differences in locomotion between the species seem to reflect the difference in body weight, which determines the foraging strategy appropriate for each species.

Large body size in *G. crassicaudatus* is a defense against predators, helps maintain body heat, and allows use of relatively bulky, low-quality foods, and if necessary, the taking of large quantities at any available gum site, for slow digestion. Slow quadrupedal locomotion is quite adequate to the demands of gum collecting, and large body size reduces the need for rapid escape from predators. The same body size relationships keep *G. crassicaudatus* in the canopy. The rewards offered by ground foraging for arthropods are small in relation to daily energy needs and to the costs of the saltatory locomotion necessary for safe, rapid movement in the discontinuous understory. Unless large packets of arthropod food, such as millipedes or termites' nests (or perhaps fallen fruit in other areas) are available, *G. crassicaudatus* should and does avoid the ground.

For *G. senegalensis,* small body size brings both the necessity and the possibility of obtaining arthropod food all year round. Movement in the low zone does pay off because saltatory locomotion makes it safe and minimizes search time. Kinetic heat from the rapid locomotion seems a requisite for use of the low arboreal zone by an arboreal primate of small body size. Indeed, all the small-bodied prosimian specialists in saltation—*G. senegalensis, G. alleni,* and *Tarsius* spp. (and probably *G. demidovii*)—use the low zone at least part of the time. It seems likely that the Eocene primates with similar limb morphology had a similar adaptation.

Thus viewing locomotion as a solution to problems of survival—particularly foraging and predator avoidance—rather than as an autonomous entity may help us considerably in an understanding of its evolution.

Acknowledgments

The project and this paper would not have been possible without the kindness and help of many people: the Galpins and the Thompson-Vise family gave study sites and hospitality; Simon Bearder and Anne B. Clark showed me where and how to look at bushbabies; Prof. G. A. Doyle gave me laboratory facilities, and continual assistance and advice there came from Craig Bielert and Theo. Bakker. Plant identification was provided by the University of Witwatersrand Botany Department. Financial support was provided by the L. S. B. Leakey Fund, the Department of Anthropology, Harvard University, and by my parents. Advice, encouragement, and necessary discouragement came from, among others: Farish A. Jenkins, Jr.; Alan Walker; Erik Trinkaus; Christine Janis; Maryellen Ruvolo; John Fleagle; Caroline Harcourt; and of course the editors and contributors to this volume. Data were analyzed by use of the Statistical Package for the Social Sciences.

References

Bearder, S. K. 1972. Aspects of the ecology and behaviour of the thick-tailed bushbaby, *Galago crassicaudatus.* Ph.D. thesis. University of the Witwatersrand, South Africa.

Bearder, S. K., and R. D. Martin. 1980. *Acacia* gum and its use by bushbabies, *Galago senegalensis* (Primates: Lorisidae). *Int. J. Primatol.* 1:103–28.

Cartmill, M., and K. Milton. 1977. The lorisiform wrist joint and the evolution of "brachiating" adaptations in the Hominoidea. *Amer. J. Phys. Anthrop.* 47:249–72.

Charles-Dominique, P. 1977. *Ecology and Behavior of Nocturnal Primates.* New York: Columbia University Press.

Charles-Dominique, P., and R. D. Martin. 1970. Evolution of lorises and lemurs. *Nature* 227:257–60.

Crompton, R. H. 1980. Galago locomotion. Ph.D. thesis. Cambridge: Harvard University Press.

Crompton, R. H. 1983. Age differences in locomotion of two subtropical *Galaginae. Primates* 24:241–59.

Dawson, T. J. 1976. Energetic cost of locomotion in kangaroos. *Nature* 246:313–14.

Fleagle, J. G. 1977. Locomotor behavior and muscular anatomy of sympatric Malaysian leaf monkeys (*Presbytis obscura* and *P. melalophos*). *Amer. J. Phys. Anthrop.* 46:297–308.

Fleagle, J. G. 1979. Primate positional behavior and anatomy: Naturalistic and experimental approaches. In Morbeck, et al. 1979:313–26.

Fleagle, J. G. and R. A. Mittermeier. 1980. Locomotor behavior, body size, and comparative ecology of seven Surinam monkeys. *Amer. J. Phys. Anthro.* 52:301–14.

Harcourt, C. S. 1980. Behavioral adaptations in South African galagos. M. Sc. thesis, University of the Witwatersrand, South Africa.

Hill, W. C. O. 1953. *Primates.* Vol. 1, *Strepsirhini.* Edinburgh: Edinburgh University Press.

Janis, C. M. 1976. The evolutionary strategy of the Equidae and the origins of rumen and cecal digestion. *Evolution* 30:757–73.

Kay, R. F. and M. Cartmill. 1977. Cranial morphology and adaptations of *Palaecthon nacimienti* and other Paromyidae. *J. Human Evol.* 6:19–53.

Kingdon, J. 1971. *East African Mammals.* Vol. 1. New York: Academic Press.

Kleiber, M. 1961. *The Fire of Life.* New York: Wiley.

Lehninger, A. L. 1970. *Biochemistry.* New York: Worth.

McKenna, M. 1980. Eocene palaeolatitude, climate and mammals of Ellesmere Island. *Palaeogeogr., Palaeoclimatol., Palaeoecol.* 30:349–62.

Morbeck, M. E. 1979. Forelimb use and positional adaptation in *Colobus.* In Morbeck et al. 1979:75–117.

Morbeck, M. E., H. Preuschoft, and N. Gomberg, eds. 1979. *Environment, Behavior, and Morphology.* New York: Gustav Fischer.

Napier, J. R. 1966. Stratification and primate ecology. *J. Anim. Ecol.* 35:411.

Parra, P. 1973. Comparative aspects of the digestive physiology of ruminant and non-ruminant herbivores. In E. Stevens, ed. *Literature Reviews of Selected Topics in Comparative Gasteroenterology.* Ithaca, N. Y.: Dept. of Veterinary Science, Cornell University.

Pough, F. H. 1973. Lizard energetics and diet. *Ecology* 54:837–44.

Prost, J. H. 1967. A definitional system for the classification of primate locomotion. *Amer. Anthrop.* 26:149–70.

Ripley, S. 1967. The leaping of langurs, a problem in the study of locomotor adaptation. *Amer. J. Phys. Anthrop.* 26:149–70.

Rose, M. D. 1979. Positional behavior in natural populations: Some quantitative results of a field study of *Colobus guereza* and *Cercopithecus aethiops.* In Morbeck et al. 1979:75–93.

Taylor, C. R., K. Schmidt-Nielsen, and J. C. Raab. 1970. Scaling of the energetic cost of locomotion to body size in mammals *Amer. J. Physiol.* 219:1104–7.

4 Use of Habitat and Positional Behavior in a Neotropical Primate, *Saguinus oedipus*

PAUL A. GARBER

EACH SPECIES OF ANIMAL in a tropical forest faces a variety of potential food resources and a number of alternative ways these resources can be acquired. Fundamental differences in nutritional composition, toxicity, availability, abundance, and distribution of food items, coupled with physiological constraints associated with consumer body size, metabolic requirements, and locomotor ability, have resulted in the evolution of specialized adaptations for foraging that may enhance feeding efficiency and must provide a nutritionally complete diet. Foraging strategies include patterns of habitat use, resource monitoring, modes of positional behavior, and morphology and enable a species to maximize energy and nutrient uptake and/or minimize the time and energy devoted to food acquisition. As Pianka (1978:108) states, "there is considerable evidence that animals actually do maximize their foraging efficiencies."

PAUL GARBER has studied positional behavior of tamarins of Panama in considerable detail. Here he presents results of his observations and interpretations of the relationships among body structures and habitat characteristics. His description of specialized patterns of feeding on gums exuded from trees makes interesting comparison with the paper by Crompton, whose subjects (*Galago senegalensis* and *G. crassicaudatus*) make similar use of tree gums. Garber provides an enlightening analysis of the constraints and opportunities, and of locomotor solutions to the problems of canopy dwelling, for a small, nonvolant animal.

Tamarins of the genus *Saguinus* are small-bodied, highly insectivorous callitrichid primates that have radiated in the tropical forests of Central and South America. Adult body weight ranges from 350 to 650 g (Hershkovitz 1977). Selection in this taxon appears to have favored high reproductive output (twinning and the potential of producing two litters per year) and very substantial parental investment (adult males, as well as adult females, devote a great deal of time and energy to the transport and care of the young). Metabolic costs of maintenance and locomotor activity are relatively high in small primates (Pianka 1978; Hladik 1978; Martin 1979), and thus tamarins might be expected to forage for high-quality food. In *Saguinus* such a diet usually takes the form of insect protein.

Because storage and use of food as fat reserves in the body are more difficult for small animals than for large ones (Temerin et al., this volume), tamarins and other small, highly insectivorous primates must obtain sufficient numbers of arthropods each day. The energy and time costs of procuring insects are substantial, however.

At present, field studies of tamarin diet and food acquisition activities are incomplete. Important questions exist about relationships among diet, body size, locomotor morphology, and positional behavior, and about habitat structure as it affects small primates.

In this paper I demonstrate that different types of food used by tamarins are located in different parts of the habitat, posing different problems of access for tamarins. Food acquisition thus requires diverse postural and locomotor behavior patterns related to specific food types. Body size emerges as a crucial factor for understanding tamarin adaptations. Henceforth I use the term *foraging* in a restricted way to mean activities involving the search for and capture of potential prey. *Prey* is used in its broadest sense, referring to all sorts of food items. *Feeding* includes only behaviors related to actual ingestion and processing of food material.

The Nature of Positional Behavior

Locomotion is activity in which "body mass (as opposed to limb mass) is displaced relative to its physical surroundings" (Prost

1965:1200). Thus locomotion is associated with such activities as travel and movement. In contrast, *posture* refers to those positional activities during which the body remains relatively stationary. Postural activities tend to dominate the positional repertoire during feeding, resting, and various social behaviors (Rose 1974, 1977; Morbeck 1975; Ripley 1967, 1977). Posture does not imply either inactivity or immobility. It includes behaviors in which the limb segments proper, and not the entire skeleton, are being manipulated to some end.

Operationally one can distinguish between locomotor and postural activities. In mechanical terms, however posture and locomotion represent integrated points on a behavioral continuum (Prost 1965). During postural behavior many of the forces acting on the musculoskeletal system are identical to those involved in locomotion. Conversely, however, the skeleton may be exposed to mechanical loads that are fairly irregular or infrequent during locomotor but common during postural activities. For example, hindlimb or bipedal suspension in *Saguinus* imparts certain tensile loads and related hip and tarsal placements that occur rarely during travel but constitute 15% of the positional repertoire during fruit and insect feeding. Therefore, in an adaptive sense, distinctions between posture and locomotion are important. Different use of the limbs during locomotor and postural behavior can result in alternative musculoskeletal adaptations.

If resources in a tropical forest are restricted in their distribution to particular zones of the habitat, as well as to supports of specific sizes, orientations, and weight-bearing capacities, then positional behavior and morphology may have an important influence on feeding and foraging activities. Accordingly, in examining strategies of food procurement I am especially concerned with the relationships between substrate preference and positional behavior.

Morphology and Feeding of Callitrichids

The digits of all tamarins and marmosets have laterally compressed, pointed, elongated, claw-like nails. It has been suggested that claw-like nails prohibit prehensile monkey-like locomotion on thin, flexible arboreal supports (Hladik 1970; Hershkovitz 1977). In fact, in describing callitrichid digital morphology, Hershkovitz (1977:32) concludes not only that callitrichid hands and feet are "least specialized among living primates," but also that during platyrrhine evolution, "the principal elements of support shift from the grappling and clinging claws of small callitrichids to the grasping clutching digits of the larger cebids" (1977:49). This does not, however, appear to be an accurate assessment of the positional behavior of tamarins in light of the results presented below.

Study Site and Methods

Study Site

From January through August 1978, I spent 850 hours observing free-ranging Panamanian tamarins at the Rodman Naval Fuel Farm, located on the Pacific Slope of Panama approximately 8 km west of Balboa (8° 57′ N, 79° 57′ W). Rodman is a 600-acre lowland dry tropical forest (Holdridge and Budowski 1956) composed of secondary vegetation. In species composition and forest structure, the vegetation is similar to that "found all along the relatively dry west side of Central America from Panama northward into Mexico" (McCullough 1956:14).

The home range of the main study group (9.4 hectares) includes four distinct habitat types that differ significantly in species composition and forest physiognomy. The recognition of different habitat types is essential for understanding patterns of movement and behavior in the Panamanian tamarin. These small animals spend much of the day moving about the forest in search of food resources, and there appears to be an important relationship between resource availability and habitat use. Because tamarins restrict certain behaviors and activities to particular zones of the forest, I describe here the four primary habitat zones exploited by tamarins at Rodman.

Habitat Zones

Zone A. Adjacent to and surrounding stream beds and river tributaries are areas of gallery forest. The gallery forest consists of tall emergent trees reaching 20–30 m in height. Although few in number, these emergents (*Anacardium excelsum, Enterlobium cyclocarpum,* and *Spondias mombin*) serve as sleeping sites for *Saguinus oedipus.* In these gallery areas the undergrowth is sparse and dominated by various species of spiny palms. The paucity of emergents and the vertical discontinuity of the canopy appear to have a significant effect on the locomotor behavior of the tamarins.

Zone B. The most prevalent kind of habitat at Rodman is low secondary forest. The vast majority of fruit feeding occurs in this zone, which is dominated by young saplings and mature trees 3–15 m in height. Unlike gallery forest, Zone B exhibits a fairly regular and continuous canopy. Characteristic plant species are *Cecropia peltata, Brysonima crassifolia, Miconia argentea, Inga punctata, Antirrhoea tricantha,* and *Vismia guianensis.*

Zone C. Critical to the foraging and feeding success of the Panamanian tamarin is a well-developed understory composed of a dense layer of low shrubs and vine tangles. This vegetation zone ranges from ground level to approximately 8 m in height. Travel in this milieu is restricted to cautious movement on a network of thin, flexible supports. Although a dense undergrowth is characteristic of Rodman, the areas of vine tangles are distributed nonrandomly and tend to be found on the forest perimeter and in areas of tree falls. It has been suggested that the low shrub layer and margins of the forest understory support a rich arthropod fauna (Paulian 1946, cited by Cartmill 1974). It is not surprising, therefore, that most insect feeding and foraging by tamarins occurs in this zone of the forest.

Zone D. Rodman is divided by dirt roads and/or grassy areas into patches of forest where cover is sparse and reduced to grasses and shrubs ranging up to about two m in height. At least once, but often twice each day, a tamarin would bound on the ground while crossing from one forest patch to another.

Methods

Quantitative behavioral data were taken by instantaneous focal animal sampling (see Garber 1980a for a more detailed description of data-gathering methods). This method has been successfully employed by a number of field workers (Richard 1970, 1973; Chivers 1974; Pollock 1977; Hladik 1977). Focal animal sampling allows the observer to direct attention to and record the instantaneous behaviors of a single individual at predetermined and regularly timed intervals.

To evaluate patterns of habitat use, substrate preference, and positional activity, several behavioral and environmental variables were recorded on prepared data sheets at 2.5-minute intervals throughout the day. (These variables include positional mode, substrate circumference, and substrate inclination; they are described further in accompanying tables and in the Results.) Each hour, therefore, provides 24 individual activity records (IARs) of tamarin behavior. Observations on a single individual ranged from 1 to more than 100 consecutive IARs. When the subject animal was lost, the next adult animal in view became the focal animal.

The quantitative data presented in this paper contain information collected solely on adult animals (there were four adult, one juvenile, and two infant animals in the main study group). These data have been pooled for analysis. In conjunction with observations of positional behavior I collected information on general ecology, forest structure, and the tamarin diet.

Results

The Tamarin Diet

The diet of the Panamanian tamarin is composed of three primary, distinct types of food measured by time spent feeding: insects (39.4%), fruits (38.4%), and plant exudates (gums and resins, 14.4%; table 4.1); about 70% of the insects are orthopterans (Dawson 1976). These resources are located in different parts of the canopy, and access to each requires a different mode of locomotor and postural activity. For example, tamarins are visual predators on mobile insect prey. These small primates (mean adult body weight-508 gm, Dawson [1976]) spend from one to three hours each day foraging and

Table 4.1. Diet of Panamanian Tamarins

Food Type	Number of IAR's	Relative Frequency (%)
A. Time spent feeding and foraging (combined)		
Insects	597	39.4
Fruit	582	38.4
Exudates	218	14.4
Leaves	35	2.3
Buds	26	1.7
Seeds	22	1.4
Flowers	2	0.1
Unknown	32	2.1
B. Time spent foraging		
Insects	504	69.8
Plant parts excluding exudates	176	24.3
Exudates	42	5.8

feeding on insects in the low shrub layer of the forest understory. In exploiting this zone of the forest, tamarins move through a network of thin, flexible nonwoody supports and dense vine tangles. Cryptic movement in this part of the canopy would be difficult for primates of larger body size.

When feeding on fruits, however, tamarins range higher in the trees, exploiting both the continuous layer and perimeter of the canopy. Feeding there demands a wide range of sitting and grasping postures on small but fairly stable arboreal supports. In contrast to this small-branch feeding, acquiring plant exudates involves a very different set of woody supports. In more than 95% of large-branch feeding (supports greater than 50 cm) tamarins clung to vertical trunks while feeding on plant exudates. Claw-like tegulae (nails) enable these primates to secure a grip on vertical and sharply inclined supports that are otherwise too large to be spanned by their tiny hands and feet. Although information on substrate preference and positional behavior in most tamarin species is incomplete, observations on *Saguinus oedipus geoffroyi* (Garber 1980a, b) indicate that the use of large vertical trunks is associated principally with a gummivorous habit (see Crompton, this volume). Thus it appears that both dietary pattern and foraging activity are, in part, constrained by the tamarins' ability to travel efficiently through certain resource-bearing zones of the canopy. In this regard, distinctions between posture and locomotion are critical, for each contributes significantly to the total positional repertoire.

Positional Behavior During Feeding

Table 4.2 summarizes observations of positional behavior during feeding. Approximately 20% of all feeding occurred on supports of less than 2.5 cm in circumference. On these fragile supports tamarins frequently fed while suspended *solely by their hindlimbs.* About 40% of feeding on these tiniest supports involved hindlimb suspension. Feeding below branches in this manner appears to be an important component of the positional repertoire of tamarins, and my observations show that callitrichid primates are quite capable of supporting their full weight by hanging from hindlimbs only.

Grasping the support with three or four limbs was a posture frequently adopted on thin supports (29%). In these cases the body was oriented either horizontally or obliquely, and the forelimbs contributed to the support and stability of the animal. In this posture, the hindlimbs remain the prime elements of support, whereas the forelimbs are engaged principally in food manipulation. More than 65% of terminal branch feeding (supports less than 2.5 cm) involved these prehensile postures.

With increasing support size there were concomitant changes in positional behaviors. On supports of 2.5–5 cm there was a marked increase in the use of horizontal branches and a decrease in the use of vertical branches (table 4.3). This appears to have been associated with a preference for adopting a sitting posture during feeding. Approximately 37% of all feeding records on supports of this size involved sitting (table 4.2). These branches are small, however, and grasping continues to be the positional activity observed most frequently (46%).

The greatest number of feeding observations occurred on supports of 5–10 cm (27%, table 4.2). On branches of this size class, sitting ("sit-lay" in table 4.2) replaced grasping as the primary pos-

Table 4.2. Relative Frequencies (%) of Selected Positional Behaviors During Feeding and Foraging

Support Size [b]	Feed/ Forage	Quad.-Wk.	Sit-Lay	Climb	Grasp	HL-Grasp	Jump	Bi.-Stand	Q.S.	Vert.-Cl.	Totals
<2.5	Feed	1.3	20.8	0.7	28.9	39.6	—[c]	6.7	2.0	0.0	20.7
	Forage	3.7	3.7	35.3	27.0	11.6	17.2	—	1.4	0.0	36.9
⩾2.5, <5	Feed	1.5	37.8	0.7	20.0	25.9	—	9.6	4.4	0.0	18.2
	Forage	12.3	7.1	33.5	23.9	11.6	8.4	—	3.2	0.0	25.9
⩾5, <10	Feed	1.5	52.9	0.5	25.0	12.7	—	4.4	2.9	0.0	27.2
	Forage	26.5	15.4	17.6	16.9	8.8	8.8	—	5.9	0.0	22.0
⩾10, <25	Feed	0.0	63.2	0.0	12.3	0.0	—	0.0	10.5	14.0	7.7
	Forage	30.2	37.2	2.3	9.3	2.3	9.3	—	7.0	2.3	6.8
⩾25, <50	Feed	0.0	50.0	0.0	0.0	0.0	—	0.0	7.7	42.3	3.5
	Forage	20.0	0.0	0.0	10.0	0.0	10.0	—	10.0	35.7	1.7
⩾50, <100	Feed	0.0	42.9	0.0	0.0	0.0	—	7.1	0.0	50.0	1.8
	Forage	7.5	2.5	0.0	0.0	0.0	0.0	—	0.0	90.0	1.3
⩾100	Feed	0.0	0.0	0.0	0.0	0.0	—	0.0	0.0	100.0	20.7
	Forage	—	—	—	—	—	—	—	—	100.0	4.6
Totals	Feed	0.9	33.0	0.4	17.2	16.2	—	4.4	3.1	24.8	
	(N)	(7)	(245)	(3)	(128)	(120)		(33)	(23)	(184)	(743)
	Forage	13.2	9.5	25.5	20.5	9.3	11.2	—	3.3	7.0	
	(N)	(79)	(57)	(153)	(123)	(56)	(67)		(20)	(42)	(599)

[a] Cell entries in body of table are percentages within support size class. Marginal percentages are of totals for feeding and foraging separately. "Quad.-wk." = quadrupedal walking, "HL-grasp" = grasping with hindlimbs only, "Bi.-stand" = bipedal standing, "Q.S." = quadrupedal standing, "vert.-cl." = vertical clinging.

[b] Support sizes are circumferences in centimeters.

[c] Cells with — indicate nonapplicable categories.

Table 4.3. Support Preference According to Branch Size and Orientation[a]

Support Circumference (cm)	Relative Frequency (%) of Use					
	Feeding			Foraging		
	H	V	O	H	V	O
2.5	33.5	15.6	50.7	16.8	34.4	48.7
2.5, 5	45.0	5.3	49.6	22.0	22.0	55.9
5, 10	41.2	6.3	52.4	29.3	10.7	60.0
10, 100	35.3	9.0	55.5	21.7	17.4	60.8
100	0.0	92.4	7.6	0.0	100.0	0.0
Numbers of observations	225	198	318	100	123	245
Percentage	30.3	26.7	42.9	21.3	26.3	52.7

[a] H = horizontal (within 15° of precise horizontal),
V = Vertical (within 15° of precise vertical), and
O = oblique (between 15° and 75° from precise horizontal).

Entries are percentages of row totals within feeding and foraging separately; e.g., 33.5% of all feeding on supports under 2.5 cm was on horizontal supports.

tural mode (53% vs. 38%). Supports of between 5 and 10 cm are quite slender, but they offer small primates like *Saguinus* a fairly wide support base from which to feed. In addition, the reduction in grasping activities is directly attributable to a decrease in hindlimb suspension (13%, table 4.2). On supports of this size tamarins are forced to secure a grip using at least three limbs and the frictional contact of the body against the substrate. In fact, once arboreal supports exceed 10 cm in circumference, the Panamanian tamarin exhibits a dramatic reduction in prehension. Less than 3% of all grasping postures occurred on moderate- and large-sized branches.

On supports of between 10 and 100 cm the Panamanian tamarin exhibits a sharp decline in feeding activity. Only 13% of feeding occurred on supports of such moderate to large size. Thus feeding occurs in what is best described as a small-branch setting. What is most interesting, however, is support preference on these boughs. Combining the data for supports of 10 to 100 cm reveals that oblique (56%) and horizontal branches (35%) were used frequently, but that vertical boughs (9%) were used infrequently. On supports of the next size class (greater than 100 cm) a very different pattern of support preference and habitat use emerged. On these largest branches support preference was as follows: oblique, 8%; horizontal, 0%; and vertical, 92% (table 4.3). The nearly exclusive use of vertical supports during large-branch feeding is an outstanding feature of tamarin feeding behavior; 96% of all cases recorded as large-branch feeding involved clinging to tree trunks while feeding on plant exudates. During no other activity does the use of large vertical supports contribute significantly to the positional repertoire of *Saguinus oedipus*.

To summarize, my observations of support preference and positional behavior during feeding in the Panamanian tamarin indicate that (1) feeding is primarily postural (2) during feeding tamarins exhibit a bimodal pattern of support preference, taking fruits and insects on small, thin branches with the aid of well-developed grasping abilities, and exudates on a very different set of arboreal supports; and (3) tamarins cling to large vertical trunks when feeding on plant exudates, using their claw-like tegulae to exploit a set of supports and food resources that would otherwise be unavailable.

Positional Behavior During Foraging

Tamarins spend a great deal of time each day procuring food. As stated above, foraging consists of activities used to locate and capture desired prey. I recorded a total of 729 foraging IARs. Foraging accounted for 11% of the activity of this species.

The amount of time tamarins forage is not equally distributed among all classes of food items. For example, plant parts and exudates, which compose roughly 60% of the diet, account for only 30% of foraging activities. The remaining 70% of foraging is devoted to the pursuit and capture of insects (table 4.1B). This investment in time and energy assumes great significance in light of Milton's observation that, "In most models of foraging strategy, the efficiency of foraging is measured by the net energy yield/foraging time. The underlying assumption is that the amount of time spent foraging is indicative of the cost of foraging." (1980:125). I believe the search for insect prey is the most important single factor in determining patterns of group movement and habitat use in the Panamanian tamarin.

Most foraging, like feeding, occurs in a small-branch setting. Approximately 85% of all observed foraging occurred on branches less than 10 cm in circumference (table 4.2). On supports of the smallest size (less than 2.5 cm), grasping (39%), climbing (35%), and leaping (17%) are the primary positional behaviors. On slightly larger supports (2.5–5.0 and 5.0–10.0 cm) these same positional activities account for most of tamarin foraging behavior.

In contrast with preferences observed during feeding, tamarins do not use moderate and large supports very much. Whereas 34% of feeding occurs on supports larger than 10 cm, only 16% of foraging occurs on branches of similar size. Moreover, there is less foraging on horizontal and more on oblique substrates compared with feeding (table 4.3). These differences between foraging and feeding are significant (χ^2, $p < .001$ in each case) and appear to be related to preferred patterns of positional behavior and the nature of resource acquisition on these supports. From table 4.2, sitting, which occurs principally on horizontal supports, is a favored feeding posture (33%) but is adopted much less often during foraging (9%). On the other hand, climbing, which typically involves oblique and vertical supports, is an important activity during foraging (26%) but is less significant and less frequent during feeding (0.4%).

The Panamanian tamarin exploited vertical branches in equal frequency during feeding and foraging. This similarity may be misleading. The majority of feeding observations on vertical branches involve large trunks (more than 77% on supports larger than 25 cm in circumference). During foraging, however, the reverse is true. Only 30% of vertical supports employed in foraging are greater than 25 cm.

The difference in preference for large and small vertical supports can best be explained with reference to patterns of feeding on insects and exudates. To capture insect prey, tamarins are forced to search on thin branches in the forest understory. Exudates are more predictable in their distribution, and in contrast to insects, preferred gumlicks occur on large vertical trunks. Thus, in the case of this callitrichid primate, positional behaviors and substrate preferences are closely associated with particular dietary regimes.

To summarize, these observations of support preference and positional behavior during *foraging* in the Panamanian tamarin indicate that (1) significant differences in substrate preference and positional behavior occur during feeding and foraging; (2) during foraging, tamarins engage in a number of vigorous positional activities, most noticeably grasping (30%), climbing (26%), jumping (11%), and quadrupedal walking (13%); (3) unlike feeding, which is primarily postural, 48% of foraging involves locomotor activities; and (4) approximately 70% of all foraging is associated with the pursuit and capture of insect prey.

My observations of feeding and foraging show that postural and locomotor activities are strongly correlated with food types. When procuring insects, tamarins employ a variety of postures and locomotor modes, but what appear to be energetically expensive ones are most often seen in predation on insects: 75% of all climbing and 87% of all jumping occurs in pursuit of this food type (table 4.4). Most sitting (76%) occurs during the acquiring of plant parts other than exudates, whereas almost all vertical clinging occurs in procuring exudates.

Discussion

Significance of Insects in the Diet

As mentioned in the Introduction (and in most other papers in this volume), small primates have higher metabolic costs per unit body weight for both maintenance and movement than larger animals and are less able to store food as fat to live on at some later time. Thus we expect strong selection on tamarins to use high-quality food (especially with respect to energy content) and to ensure consumption of that food each day. Insects constitute such a resource for the Panamanian tamarins.

Hladik (1978) provides information on nutrient content of various food resources of primates. Most fruit is relatively high in readily available energy. Tamarins eat very little green, leafy material, probably reflecting the relationship between body size and potential use of foods high in fiber (Temerin *et al.,* this volume). Insects contain large amounts of protein and lipids. These are valuable because they

Table 4.4. Positional Behavior and Food Type[a]

Food Type	Quad. Walk and Stand, Biped. Stand	Sit-Lay	Climb	Grasp and HL-Grasp	Jump	Vert.-Cl.	Totals
Insects	(54) 11.0 28.7	(65) 13.3 21.6	(117) 23.9 75.0	(184) 37.6 43.1	(66) 13.5 86.8	(3) 0.6 1.3	489
Plant parts	(134) 20.0 71.3	(230) 34.5 76.4	(39) 5.8 25.0	(243) 36.4 56.9	(10) 1.5 13.2	(11) 1.6 4.9	667
Plant exudates	(0) 0.0 0.0	(6) 2.8 2.0	(0) 0.0 0.0	(0) 0.0 0.0	(0) 0.0 0.0	(212) 97.2 93.8	218
Totals	188	301	156	427	76	226	1374

[a] Plant parts exclude plant exudates. Each cell of table contains (absolute frequency of IAR's) % of row total, % of column total.

Table 4.5. Nature and Availability of Insects and Fruit Consumed by the Panamanian Tamarin

Insect	*Fruit*
1. Scattered or dispersed distribution.	1. Clumped distribution
2. Mobile prey	2. Stationary prey
3. Detection and capture of an insect prey may have a negative effect or at best a marginally positive effect on the likelihood of additional captures	3. Detection and capture of one fruit typically has a strong positive effect on the chances of acquiring additional fruit
4. High cost in both pursuit time and pursuit energy	4. Lower cost in pursuit time and pursuit energy
5. High cost in search time and search energy	5. A moderate or low cost in search time and energy
6. Resource is potentially available only during certain times of the day	6. Resource is potentially available during all hours of the day
7. High protein and lipid content	7. High carbohydrate content
8. Acquisition associated with both locomotor and postural behaviors	8. Acquisition associated primarily with postural behaviors

contain more energy than carbohydrates, the primary constituents of most plant parts (Kleiber 1961). In addition protein is necessary for growth and repair of tissues.

Insects, have, of course, been under strong selection to avoid being preyed on by tamarins and other animals. Thus they present consumers with problems that differ from those posed by other food types (table 4.5). In particular, certain qualities of insect prey necessitate substantial time and energy investment in their acquisition: insects are mobile and elusive, they appear to be limited to specific areas of the forest, and detection and capture of one insect are likely to lower the chances of the tamarin's capturing another insect. Obviously this is not the case for fruit or exudates, where acquisition is associated with increased likelihood of procuring more of the same food.

Measurement of energy costs associated with foraging is extremely difficult, but there is reason to believe that time expenditures can indicate the importance of activities. Searching for insects takes up about 40% of foraging time, and data presented above show that vigorous activities are required for their capture. Thus I propose that insect predation is the most important determinant of the tamarin's foraging strategy and that we should see special adaptations in positional behavior, searching behavior (e.g., a specific search image for orthopterans), and patch selection related to dispersion of insects in the habitat.

Tamarins are small enough to be very agile on small and flexible supports, especially vines, and although I have no quantitative information on dispersion of their insect prey in time and space, insects seem to be concentrated in shrubs and vines at tree falls and forest edges (Zone C).

There are other small primates that eat insects, *Saimiri* in the New World and *Miopithecus talapoin* in the Old World. These animals forage in large groups, and it has been suggested that joint foraging serves to disturb insects and thereby increase capture rates for the primate predators (Gautier-Hion 1973; Klein and Klein 1975). Group size in tamarins is less than in these other species (table 4.6), and

Table 4.6. Group Size in the Callitrichidae

Species	Group Size (range)	Group Size (mean)	Reference
Saguinus midas (Columbia)	2–6		Thorington (1967)
S. nigricollis	4–8[a]		Izawa (1976)
S. oedipus oedipus (Colombia)	3–13	8	Neyman (1977)
S. oedipus geoffroyi (Panama)	1–19	7	Dawson (1976)
S. oedipus geoffroyi (Panama)	5–10	7	Present study
S. mystax (Peru)	2–6		Castro and Soini (1977)
S. fuscicollis (Peru)	4–10		Izawa (1976)
	20–40[b]		
S. fuscicollis (Peru)	2–26	2–6	Castro and Soini (1977)
S. midas midas (Surinam)	6	6	Mittermeier and Van Roosmalen (1981)
Cebuella pygmaea (Peru)	2–11		Castro and Soini (1977); Soini (1982)
Leontopithecus rosalia (Brazil)	2–8	3–4	Coimbra-Filho and Mittermeier (1977)
Callithrix humeralifer (Brazil)	4–13		Rylands (1981)

[a] Groups may temporarily merge to form larger units. This may occur when a preferred tree is in fruit.

[b] Large groups were observed only during certain months of the year. At all other times group size was less than 10.

when hunting insects, tamarins move about the forest in a deliberate and relatively quiet manner. Group members appear to be foraging independently but often between 1 and 5 m from their nearest neighbors. When individuals are observed in close proximity, there is no indication of coordinated or cooperative foraging. The attention of each individual appears to be occupied solely with its own predatory efforts.

Significance of Exudates

Tamarins feed on exudates of one species of very large tree, *Anacardium excelsum,* and the evolution of tegulae appears related to clinging to supports that are large relative to the grasping capacity of tamarins (Garber, in press). Other small primates feed on exudates as well (Coimbra-Filho and Mittermeier 1977; Garber 1980a, b). It is possible that the importance of exudates as a primate food source lies in providing a rich supply of calcium to offset the high phosphorous content of a diet containing substantial insect material (Bearder and Martin 1981; Garber in press).

Figure 4.1 illustrates the locations of the seven *Anacardium excelsum* trees found in the home range of the study group. Of these seven, the tamarins fed on exudates of only three trees, and all these were located in the core area of the home range. The tamarins frequently use *Anacardium* for sleeping trees, and so exudate availability can be monitored with very little effort, and little travel is needed to exploit it. The predictability of exudates is enhanced by the fact that they are available for one season only (May through July). Obviously acquisition of exudates requires much less time and energy than acquisition of insects and involves very different positional behaviors and associated morphological traits.

Summary

The diet of the Panamanian tamarin comprises three sorts of food that differ in location in the habitat and in the obstacles they pose to

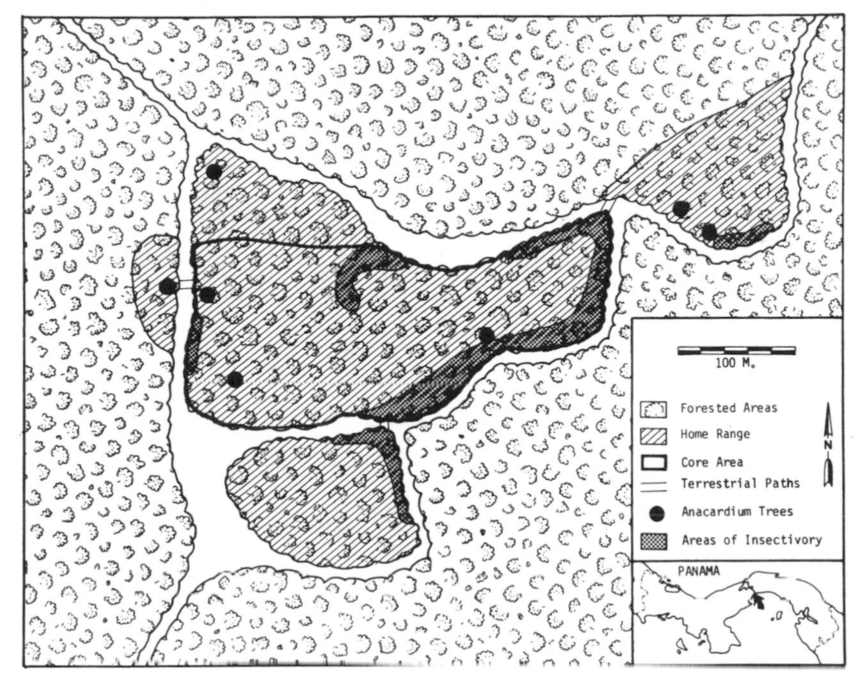

Figure 4.1. Map of the Rodman Field Site, Pacific Coast of Panama

exploitation by a small, nonvolant mammal. We thus expect, and find, that tamarins use markedly different positional behaviors to acquire insects, fruit, and exudates. Insects are found in shrubs and vines, and tamarins must move actively on thin, flexible supports to procure them. Much of the locomotion used in acquiring insects is energetically expensive. In contrast, fruit is located at the ends of terminal branches of trees and requires largely sedentary or suspensory postures involving little displacement of the animal's mass. Exudates are found on large vertical tree trunks, and tamarins cling vertically, employing tegulae, to exploit this resource.

Acknowledgments

I am grateful to Dr. Robert W. Sussman for his support, intellectual stimulation, and friendship during my graduate education. I am also indebted to Dr. Robert W. Sussman and Lynette Norr for their useful comments on earlier drafts of this manuscript. Finally, without the unselfish devotion of my father during his lifetime, and without the strength of his memory in death, all this would be meaningless.

This research was supported in part by a National Science Foundation Dissertation Improvement Award (Grant BNS77–24043); a Biomedical Research Support Grant, Division of Research Resources, N.I.H. (Grant 56219B); a Graduate Fellowship from Washington University, St. Louis, Mo.; and a Grant-in-Aid from the Scientific Research Society of North America—Sigma Xi.

References

Bearder, S. K., and R. D. Martin. 1981. *Acacia* and its use by bushbabies. *Int. J. Primatol.* 1:103–28.

Cartmill, M. 1974. Pads and claws in arboreal locomotion. *In* F. A. Jenkins, ed. *Primate Locomotion,* pp. 45–83. New York: Academic Press.

Castro, R., and P. Soini. 1977. Field studies on *Saguinus mystax* and other callitrichids in Amazonian Peru. In D. G. Kleiman, ed. *The Biology and Conservation of the Callitrichidae,* pp. 73–78. Washington, D.C.: Smithsonian Institution Press.

Chivers, D. J. 1974. The siamang in Malaya: A field study of a primate in a tropical rainforest. *Contrib. Primatol.* 4:1–335.

Coimbra-Filho, A. F., and R. A. Mittermeier. 1977. Tree-gouging, exudate-eating and the "short-tusked" condition in *Callithrix* and *Cebuella.* In D. G. Kleiman, ed. *The Biology and Conservation of the Callitrichidae,* Washington, D. C.: pp. 105–15. Smithsonian Institution Press.

Dawson, G. A. 1976. Behavioral ecology of the Panamanian tamarin, *Saginus oedipus* (Callitrichidae, Primates). Ph. D. thesis, East Lansing, Mich.: Michigan State University.

Garber, P. A. 1980a. Locomotor behavior and feeding ecology of the Panamanian tamarin (*Saguinus oedipus geoffroyi,* Callitrichidae, Primates). Ph. D. thesis. St. Louis, Mo.: Washington University.

—— 1980b. Locomotor behavior and feeding ecology of the Panamanian tamarin. (*Saguinus oedipus geoffroyi,* Callitrichidae, Primates). *Int. J. Primatol.* 1:185–201.

—— in press. Proposed nutritional importance of plant exudates in the diet of the Panamanian tamarin (*Saguinus oedipus geoffroyi*). *Int. J. Primatol.*

Gautier-Hion, A. 1973. Social and ecological features of the talapoin monkey—comparisons with sympatric cercopithecines. In R. P. Michael and J. H. Cook, eds. *Comparative Ecology and Behaviour of Primates,* pp. 147–70. London: Academic Press.

Hershkovitz, P. 1977. *Living New World Monkeys (Platyrrhini),* vol. 1. Chicago: University of Chicago Press.

Hladik, C. M. 1970. Les singes du nouveau monde. *Science et Nature* 102:1–9.

—— 1977. A comparative study of feeding strategies in two sympatric species of leaf monkeys: *Presbytis senex* and *Presbytis entellus.* In T. H. Clutton-Brock, ed. *Primate Ecology: Studies of Feeding and Ranging Behaviour in Lemurs, Monkeys and Apes,* pp. 324–54. London: Academic Press.

—— 1978. Adaptive strategies of primates in relation to leaf-eating. In G. G. Montgomery, ed. *The Ecology of Arboreal Folivores,* pp. 373–95. Washington, D.C.: Smithsonian Institution Press.

Holdridge, L. R. and G. Budowski. 1956. Report of an ecological survey of the Republic of Panama. *Carrib For.* 17:92–110.

Izawa, K. 1976. Group sizes and compositions of monkeys in the upper Amazon basin. *Primates* 17:367–99.

Kleiber, M. 1961. *The Fire of Life.* New York: Wiley.

Klein, L. L., and D. J. Klein. 1975. Social and ecological contrasts between four taxa of neotropical primates. In R. Tuttle, ed. *Socioecology and Psychology of Primates,* pp. 59–85. The Hague: Mouton.

Martin, R. D. 1979. Phylogenetic aspects of prosimian behavior. In G. A. Doyle and R. D. Martin, eds. *The Study of Prosimian Behavior,* pp. 307–57. New York: Academic Press.

McCullough, C. R. 1956. *Terrain Study of the Panama Canal Zone.* Raleigh, N.C.: Dept. of Engineering Research, North Carolina State College.

Milton, K. 1980. *The Foraging Strategy of Howler Monkeys: A Study in Primate Economics.* New York: Columbia University Press.

Mittermeier R. S., and R. G. M. van Roosmalen. 1981. Preliminary observation on habitat utilization and diet in eight Surinam monkeys. *Folia Primatol.* 36:1–39.

Morbeck, M. E. 1975. Positional behavior in *Colobus guereza:* A preliminary quantitative analysis. In S. Kondo, M. Kawai, A. Ehara and S. Kawamura, eds. *Symposium of the 5th Congress of the International Primatological Society,* pp. 331–43. Tokyo: Japan Science Press.

Neyman, P. F. 1977. Aspects of the ecology and social organization of free ranging cotton-top tamarins (*Saguinus oedipus*) and the conservation status of the species. In D. G. Kleiman, ed. *The Biology and Conservation of the Callitrichidae,* pp. 39–71. Washington, D. C.: Smithsonian Institution Press.

Pianka, E. R. 1978. *Evolutionary Ecology* (2nd ed.). New York: Harper and Row.

Pollock, J. I. 1977. The ecology and sociology of feeding in *Indri indri.* In T. H. Clutton-Brock, ed. *Primate Ecology: Studies of Feeding and Ranging Behaviour in Lemurs, Monkeys and Apes,* pp. 38–69. London: Academic Press.

Prost, J. H. 1965. A definitional system for the classification of primate locomotion. *Amer. Anthrop.* 67:1198–1214.

Richard, A. F. 1970. A comparative study of the activity patterns and behavior of *Alouatta villosa* and *Ateles geoffroyi. Folia primatol.* 12:241–63.

Richard, A. F. 1973. An ecological and behavioral study of *Propithecus verreauxi verreauxi* in different environments. Ph.D. thesis. London: University of London.

Ripley, S. 1967. The leaping of langurs: A problem in the study of locomotor adaptations. *Amer. J. Phys. Anthrop.* 26:149–70.

—— 1977. Gray zones and gray langurs: Is the "semi" concept seminal. *Yrbk. Phys. Anthrop.* 20:376–94.

Rose, M. D. 1974. Postural adaptations in New and Old World monkeys. In F. A. Jenkins, ed. *Primate Locomotion,* pp. 201–21. New York: Academic Press.

Rylands, A. R. 1981. Preliminary field observations on the marmoset, *Callithrix humeralifer intermedius* (Hershkovitz, 1977) at Dardanelos, Rio Aripuana. *Primates* 22:46–59.

Soini, P. 1982. Ecology and population dynamics of the pygmy marmoset, *Cebuella pygmaea. Folia Primatol.* 30:1–21.

Thorington, R. W. 1967. Feeding and activity of *Cebus* and *Saimiri* in a Colombian forest. In D. Stark, R. Schneider, and H. J. Kuhn, eds. *Progress in Primatology,* pp. 180–84. Stuttgart: Gustav Fischer.

5 Foraging and Social Systems of Orangutans and Chimpanzees

PETER S. RODMAN

FORAGING PATTERNS interact with morphology and social systems in a complex web of cause and effect, and their complexity may frustrate attempts to understand functional relations among the various elements of feeding ecology and sociality. One possible exit from the tangle may be via comparative analysis of social systems, morphology, and ecology of animals that differ in some, but not all, potentially interdependent variables. Crompton (this volume) presents an example of the utility of comparative analysis of morphology and foraging of congeneric species, and Waser (this volume) offers a convincing comparative analysis of foraging, feeding ecology, and grouping in two mangabey species.

Orangutans and chimpanzees provide another useful comparison. Here I examine their morphology (particularly sexual dimorphism in body size), social systems, and ecology in an attempt to "explain"

PETER RODMAN has studied orangutans and other primates of the Kutai Nature Reserve, East Kalimantan, intermittently since 1970. Here he presents data on diets, food dispersion, and ranging patterns of orangutans in comparison with similar data provided by Richard Wrangham on chimpanzees of the Gombe Stream Reserve, Tanzania. The paper develops a complex hypothesis about the ecological constraints and opportunities of the two forests that may have given rise to differences in social patterns, particularly differences in males, and in sexual dimorphism between chimpanzees and orangutans.

how these fit together for each and how similarities and differences in their social systems and sexual dimorphism relate to differences in their foraging regimes. Related discussions have been presented by Galdikas (1978) and Galdikas and Teleki (1981). I am quite aware that the following analysis is nothing more than an elaborate hypothesis that may or may not be tested (or testable) in the future. Given the unusual nature of these two apes, it seems worthwhile to formulate a scenario of their differentiation as a means of organizing current understanding of them in spite of limitations on testability and post-hoc reasoning. The data presented are results of Wrangham's (and others') study of chimpanzees of Gombe National Park, Tanzania, in 1972–73, and of my study of orangutans of the Kutai Nature Reserve of East Kalimantan, Indonesia, in 1970–71. Some results have been presented elsewhere (Wrangham 1975, 1977, 1979a; Rodman 1973a, b, 1977, 1979), but some new data are presented here that are results of collaboration with Wrangham aimed at extracting comparable measures of ranging and feeding from the separate studies.

Body Size and Sexual Dimorphism

Male and female body weights of the two species are presented in table 5.1. Note that weights given are only for fully mature adults

Table 5.1. Body Weights of Wild Orangutans and Chimpanzees

	Pongo pygmaeus [a]	*Pan troglodytes* [b]
Male weight (kg)		
Mean =	83.6	39.5
N =	6	9
Range =	72.6–90.9	36.2–43.0
Female weight (kg)		
Mean =	37.8	29.8
N =	9	6
Range =	32.7–45.5	26.9–32.9
Female/Male (%)	45.2	75.4

[a] Body weights are of specimens judged to be mature by eruption of the third molars and closure of cranial sutures, as reported by Lyon (1907, 1908, 1911) and revised by Eckhardt (1975).

[b] Body weights reported by Wrangham and Smuts (1980).

either measured on living animals or taken from wild, shot specimens whose age was determined by eruption of third molars and closure of cranial sutures. Based on these weights, which are of animals known to be or likely to be socially mature and which particularly do not include weights of subadult males, sex difference in body weight is marked for orangutans (female 45.2% of male body weight) but not so marked for chimpanzees (female 75.4% of male body weight). Some may note that these figures differ from other published figures, as for example, in Schultz (1940, 1941) and in Eckhardt (1975). This difference is due to reliance only on weights of wild, fully adult individuals for the calculations. It seems reasonably clear that although orangutans and chimpanzees are of roughly equivalent sizes relative to the range of body weights of animals, sex difference in body weight is extreme for orangutans and not extreme for chimpanzees.

Social Systems

Communal Chimpanzees

Early descriptions of the social system of chimpanzees concluded that populations were divided into subpopulations of 15 to 80 individuals that occupy a common range and assort more or less freely into smaller subgroups of varied age/sex composition (Goodall 1968, 1971; Sugiyama 1973; Nishida 1968; Nishida and Kawanaka 1972; Izawa 1970; Kano 1971; Reynolds and Reynolds 1965). Later work in the Gombe National Park revealed that adult males of one community defended their range against neighboring males, that they were aggressive toward other males to the point of invading neighboring ranges and killing neighboring males when possible, and that they did not leave their natal communities but grew to maturity where they were born and remained with the male community there (Bygott 1979; Goodall et al. 1979; Pusey 1979). On the other hand females were unlikely to participate in intercommunity aggression, and females often transfered out of the natal community at adolescence or later in life (Pusey 1979). Examination of grooming relations among chimpanzees showed a clear relationship of social rank to grooming among males of the community (Simpson 1973), but analysis of reproductive behavior revealed that rank relations were not so clear in access to females (McGinnis 1979). Matings were often promiscuous, with copulations between an estrous female and several adult males occurring in rapid succession (Goodall 1971). This is not to say that high rank among males was not related to mating success, but the observations reported suggest that differential success of males may not be as clear as in some other species.

Solitary Orangutans

This description of chimpanzee sociality contrasts sharply with my description of the social system of orangutans (Rodman 1973a). The adult population of orangutans of the Kutai Nature Reserve was divided into small units consisting primarily of adult females with offspring or of solitary adult males. Adult females occupied different home ranges that overlapped with each other, and neighboring females seldom interacted. Indirect evidence suggested that neighboring females were related, since movement of maturing females from the mother's range appeared to be conservative.

Adult males fell into at least two classes. *Resident* males occupied permanent ranges (at least for the duration of the 15-month study) that each overlapped ranges of several females but did not overlap with each other. Younger males (subadults) and some mature males appeared to be *wanderers* without attachment to a specific range. I proposed that wandering young males were in a stage of dispersal following which they would compete with other males for access to estrous females. If successful in competition for a female, a male would establish residence in the vicinity of that female and gradually expand his sphere of influence, if successful in each competition, as neighboring females came into estrus. Some males would, of course, never succeed and might remain wanderers throughout adult life; others might have only limited success and be displaced quickly after initial successful competition.

Long-term observations by Galdikas (1978, 1979) show some

agreement and some disagreement between this hypothetical scheme and behavior of adult males over a long time. Galdikas observed direct aggression between a resident male and another male in the presence of an estrous female and reported that only a resident male was known to have fathered an offspring after consortship with a female. Other observations show that, at least for some males, the pattern of residence over a long time may be more complex than I have proposed. Observations by Horr (1975) and Rijksen (1978) indicated that in some populations, ranges of neighboring resident males may overlap. Despite these differing observations, it is clear in all populations studied so far that there is strong male–male intolerance (Galdikas 1978, 1979; Horr 1975; MacKinnon 1974; Rijksen 1978; Rodman 1973a).

The critical differences in sociospatial systems of chimpanzees and orangutans emerging from the description so far are the following: (1) Female chimpanzees occupy a large communal range and emigrate from the natal community, whereas female orangutans occupy distinct but overlapping ranges and remain near the mother throughout life. (2) Male chimpanzees occupy and defend a communal range and are sedentary, whereas breeding male orangutans maintain individual ranges and emigrate from the place of birth. The comparison shows contrast in both male and female patterns, but recent analysis of female ranging patterns of chimpanzees by Wrangham (1979a; also Wrangham and Smuts 1980) dispels much of the contrast between qualitative characteristics of *female* patterns in the two species.

Solitary Female Chimpanzees

Using data from the general data pool of the Gombe National Park, Wrangham (1979a) found that anestrous females of a community had shorter daily path lengths and smaller four-day home ranges than adult males of the same community. The data indicated that on a short term, female chimpanzees were less mobile and occupied smaller parts of the community range than male chimpanzees. Data on year long ranges of anestrous females suggested that different females used *core areas* in different locations; in other words, although female ranges overlapped, they were different for each female and did not necessarily cover the entire community range. A more precise analysis of observations of the direction of entry of individual anestrous females into the banana-feeding area showed that each female entered from a characteristic direction that corresponded to the location of her core area. In addition, directions of entry by females were more clustered than directions of entry by males, suggesting that females used a smaller part of the communal range than males. Wrangham's analysis indicates that anestrous adult female chimpanzees occupy different but overlapping ranges within the communal range; in other words, dispersion of female chimpanzees is *qualitatively quite similar to dispersion of female orangutans.* Estrous female chimpanzees travel with males so that during estrus a female may use the entire range of the males. But estrus is an infrequent, transient condition for females, and the normal dispersion of females should be that of the anestrous females.

Early descriptions of chimpanzee sociality concentrated on characteristics of behavior when animals were together. The technique of provisioning the subjects to draw them into view facilitated gathering observations of social interactions. But provisioning artificially increased the proximity and therefore the rates of interactions of individuals, and the impression of highly gregarious animals, which contrasts strongly with descriptions of sociality in orangutans, may be incorrect. Halperin (1979) examined grouping patterns of adult male and anestrous adult female chimpanzees of the Gombe National Park by tabulating frequencies with which each of nine males and five females (with young) were found in various sorts of groups. The proportion of observations of a female *alone* with her young ranged from 55% to 77% with a mean of 65% for all observations for the five females. Taken in conjunction with Wrangham's analysis of ranging patterns, the picture of adult female life that emerges is of individual females who spend most of their lives alone (with young) in different, overlapping home ranges. This pattern clearly resembles the pattern of dispersion and sociality of females observed in orangutan populations.

Contrast between male patterns in the two species remains strong.

Adult male orangutans do not tolerate other males and spend most of their lives alone. Their loud vocalizations function at least in part to repel other males. Male chimpanzees, on the other hand, defend ranges communally, and within those ranges they are highly gregarious. Their loud calls may attract rather than repel other males (Ghiglieri, this volume). Comparison of the two species now leads to two central questions: Why do male chimpanzees defend ranges communally but male orangutans defend individual ranges? How is this difference related to the difference in sexual dimorphism between the two apes? Dispersion of females in space is qualitatively similar, and the female pattern makes sense. As Wrangham (1979b) has argued, in the absence of advantages to grouping, females should be solitary in order to minimize feeding competition and maximize benefits passed on to offspring through improved nutrition. Ultimately this pattern would lead to territoriality among females if resources were defensible (Brown 1964; Geist 1974; Wrangham 1979b), but characteristics of ranging and feeding of female chimpanzees and orangutans described below preclude defense of ranges (Mitani and Rodman 1979).

Since access to females limits reproductive success of males, dispersion of females is the critical factor affecting dispersion—and therefore social patterns—of males. As Wrangham (1979b) summarizes: "Females pursue strategies dependent on their nutritional needs and the distribution of food. Male strategies, in turn, depend on and may be constrained by the distribution of females." Given this basic principle, which has been stated in other forms by others (e.g., Geist 1974), why, given qualitatively similar dispersion of females, do orangutans and chimpanzees contrast so strongly in male social patterns and in sexual dimorphism? The answer is suggested by quantitative differences in foraging patterns of the two species.

Foraging Patterns

Methods of the two field studies are described elsewhere (Wrangham 1975, 1977; Rodman 1973b, 1977). Wrangham studied 14 adult males of the Gombe Stream Reserve and used observations of females taken by many observers at the Gombe Stream Reserve on standard check sheets and contributed to the general data pool. Additional observations of ranging by adult females were taken by B. Smuts in 1974–75. Rodman studied predominantly one habituated adult male orangutan, two habituated adult females, and one habituated adolescent female and made additional observations on an unhabituated male and two unhabituated adult females. Captions to figures 5.1–5.7 describe sampling procedures and specific methods.

Activity Profiles

Distribution of waking time to major activities was roughly equivalent for orangutans and male chimpanzees, although male chimpanzees spent more time feeding and traveling and less time resting than orangutans (fig. 5.1). Unfortunately no systematic data on activity profiles of female chimpanzees are yet available (Wrangham and Smuts 1980). I have analyzed sex differences in activity patterns of

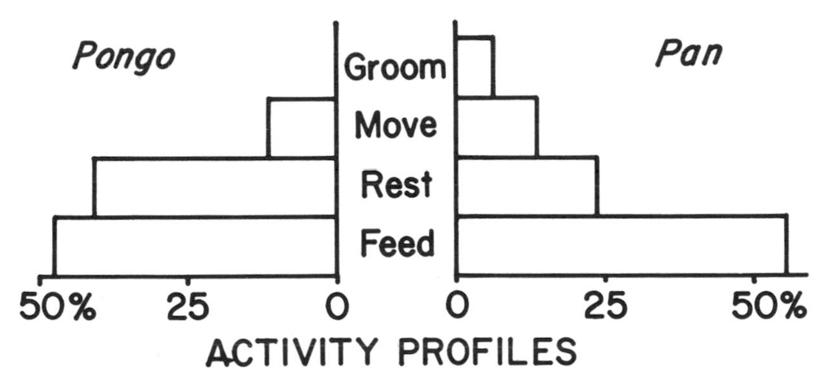

Figure 5.1. Proportions of observation time spent in four major activities during waking hours. *Methods:* 1. *Pan:* Observations were taken from May 1972 to January 1973, and from July to September 1973, on 16 adult males as targets. Activities were scored every 30 minutes during 54 all-day observations (N = 1,197 sample points). Proportions shown are of total observation points. 2. *Pongo:* Observations were taken from August 1970 through July 1971 on four habituated subjects, including two adult females with infants, one adolescent female, and one adult male. Records were made of transition times between activities on 42 whole days of observation (N = 28,728 min). Proportions shown are of total minutes of observation.

orangutans elsewhere (Rodman 1979); Galdikas (1978) has provided ample critique of that analysis.

Dietary Composition

Distribution of feeding time to different food types shows strong similarity between orangutans and male chimpanzees; data gathered without systematic sampling on female chimpanzees show a roughly similar dietary pattern (fig. 5.2). Both species were predominantly

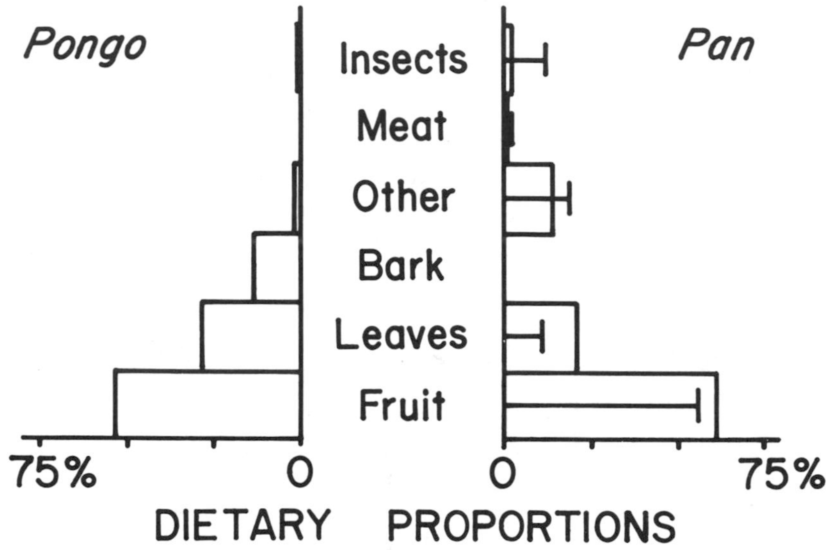

Figure 5.2. Proportions of feeding time spent on major food types. *Methods:* 1. *Pan:* Wide, open bars show proportions of foods for adult males; narrow, solid bars show proportions for adult females (data from J. S. Silk, personal communication; cf. Silk 1978). For males, data are on 16 targets observed for 780 hours from May 1972 through January 1973. Proportions shown are means for nine months of percentage feeding time on each food per month. Observations of females were taken on standard checksheets by a number of observers at the Gombe National Park for a total of 1,131 h of data from November 1969 through May 1975 (Silk 1978). Proportions shown are percentages of total recorded minutes of feeding on known foods (N = 29,204 min). Silk (personal communication) cautions that the data may be biased because observations are not from all-day follows, because not all foods were identified, and because there are unequal observations of each female, each time of day, and each month of the year. 2. *Pongo:* Data are mean percentages of feeding time per month from August 1970 through July 1971. Proportions are of a total of 40,022 min of feeding by subjects described in the caption to figure 5.2.

frugivorous and secondarily folivorous. Orangutans depended to a large extent on bark, which was not consumed by chimpanzees. Chimpanzees fed more heavily on other vegetation, including flowers and buds, than orangutans.

Time at Food Sources

Despite roughly equivalent diets, there is strong contrast between orangutans and chimpanzees in distribution of durations of feeding at food sources (fig. 5.3). Chimpanzees fed for less than 10 minutes at more than 50% of all food sources and for more than 90 minutes in only 4% of food sources; orangutans fed for more than 90 minutes in nearly 25% of food sources and for less than 10 minutes in only 11% of food sources. The comparison suggests that food sources of orangutans were effectively larger than food sources of male chimpanzees.

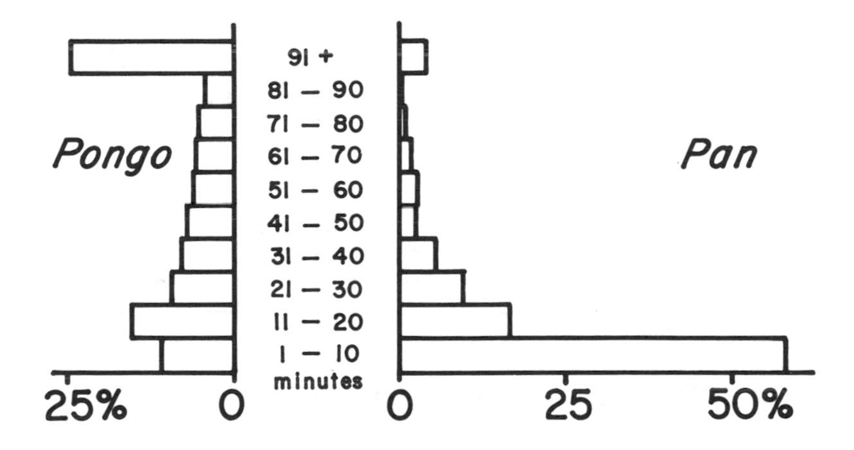

Figure 5.3. Porportions of food sources fed in for various durations. The figure indicates the percentage of all feeding durations at food sources that fall into the intervals indicated. *Methods:* 1. *Pan:* Data are from 18 whole days (two days from each of nine months) for nine male subjects (N = 396 food sources). 2. *Pongo:* Data are from 72 whole days of observation on three habituated adults from August 1970 through July 1971. Days were chosen in which observation began before 0630 h and ended after 1730 h with no more than one 30-min interruption in observation (N = 355 food sources).

In figure 5.3 I have neglected feeding party size, and so the times given do not reflect minutes of chimpanzee feeding per food source. This weakens my conclusion that the food sources are smaller for chimpanzees. Judged from mean male party size of the Kibale Forest (Ghiglieri, this volume), which is between 1.0 and 2.0, the effect of group feeding on the pattern shown would not, at its most extreme (doubling all times), completely vitiate the conclusion, since the sum of all observations of 40 or more minutes per tree is only 12.4%. Ghiglieri also shows that feeding group size is adjusted to the size of the food source. Recalculation of the data thus would have a complex effect on the distribution of time at food sources but probably would not change the large contribution of small food sources to the chimpanzees' foraging regime.

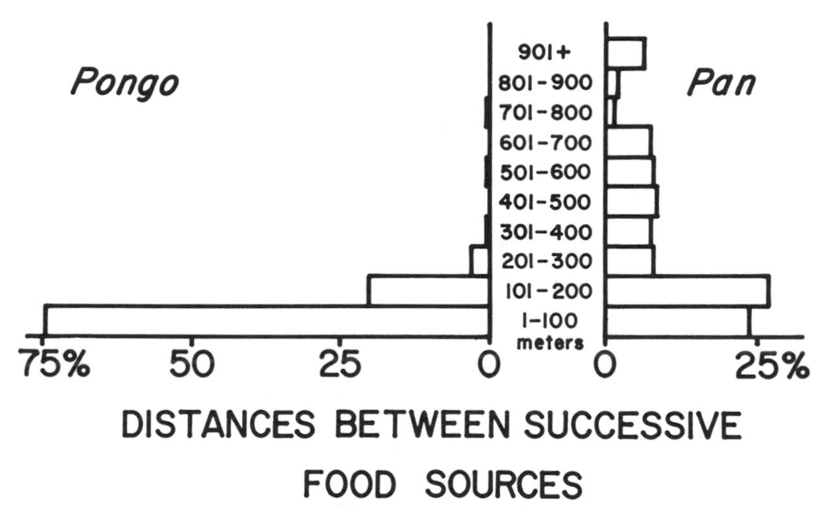

Figure 5.4. Frequency distribution of distances between successive food sources at which individuals fed for ten minutes or more. The figure shows the proportions of interfood source distances that fall into each of the intervals indicated. *Methods:* 1. *Pan:* Data are from 25 all-day follows of males from November 1973 through April 1975. 2. *Pongo:* Observations are of two habituated adult females and one habituated adult male on 55 days of observation beginning before 0630 h and ending after 1730 h with no interruption in observation, from August 1970 through July 1971.

Distances Between Food Sources

The path lengths between food sources at which subjects fed for 10 minutes or more are considered here (fig. 5.4). For male chimpanzees the median interpatch distance was approximately 200 m; the median interpatch distance for female chimpanzees was between 100 and 200 m; and the median interpatch distance for all orangutans was less than 100 ms. The comparison shows that distance between successive major food sources is larger for chimpanzees than for orangutans.

Day Ranges.

Median path length from start to end of the day was between 3 and 4 km for male chimpanzees, 2.7 km for female chimpanzees, and less than 0.5 km for orangutans (fig. 5.5). Galdikas (1978) re-

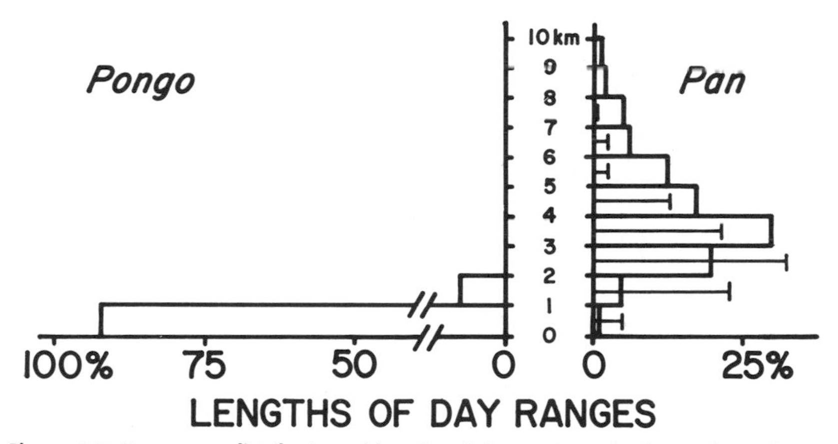

Figure 5.5. Frequency distribution of lengths of day ranges. The figure shows the proportions of path lengths from first movement to last movement that fall into each interval indicated. *Methods:* 1. *Pan:* Wide, open bars show male ranges; narrow, solid bars show female ranges. Data are for 196 days on 16 target males and 126 days on 14 target females from January 1972 through April 1975. 2. *Pongo:* Observations are of one habituated adult male and two habituated adult females on 76 days from August 1970 through July 1971. Days were selected on which observation began before 0630 h and ended after 1730 h and during which the location of the animal was known continuously although exact activity may have been obscured for any length of time.

ports a longer day range for orangutans at Tanjung Puting; others (MacKinnon 1974; Rijksen, 1978) report ranges similar to those of the Kutai Nature Reserve. It is clear that chimpanzees are more mobile each day than orangutans.

Home Range Size

The number of different 0.25 km² blocks entered by an animal in samples of whole days separated by at least one day is considered here (fig. 5.6). The median number of blocks used by adult, anestrous female chimpanzees was 16 with a range of 14 to 20. For male chimpanzees the median number of blocks entered was 29 with a

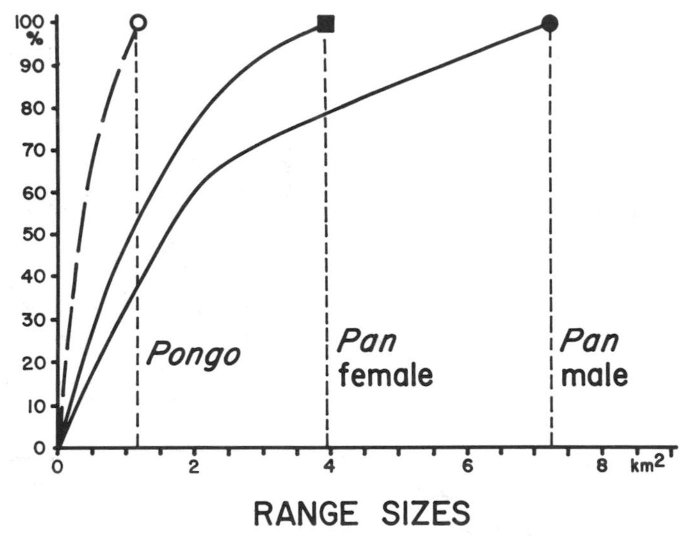

Figure 5.6. Cumulative use of 0.25 km² quadrats, arranged so that quadrats are added in decreasing order of frequency of use. Scores are numbers of entries into each quadrat expressed as percentage of total entries into all quadrats. Lines in the figure show the median values for the following samples. 1. *Pan:* Males—data are for five males, 10 whole-day follows per male, with sample days separated by at least one day. Females—data are for five females, 10 whole-day follows of four and eight of the fifth, again with successive sample days separated by at least one day. 2. *Pongo:* The line shows the median of six samples of whole-day follows separated by at least one day, including two samples on each of two adult females (N = 11, 10, 9, and 17 days) and two samples on an adult male (N = 9 and 10). Data were taken from August 1970 through July 1971.

range of 15 to 36. For two orangutan females and one orangutan male the median number of blocks used was 4 with a range of 3 to 6. The analysis indicates that chimpanzees use ranges that are several times as large as ranges used by orangutans.

The comparison above shows a consistent pattern. Although the two species are of similar body size and share a similar diet, chimpanzees use smaller, more dispersed food sources than orangutans. Given this difference, it is not surprising that individual chimpanzees travel farther each day—at about 10 times the speed—through larger ranges than orangutans. Compared to orangutans, chimpanzees are highly mobile, and adaptations for mobility—relative to orangutans—pervade morphology and behavior of chimpanzees. First, chimpanzees are terrestrial when traveling; they move on the ground in more than 70% of travel time, even in forest with closed canopy. Orangutans, on the other hand, are predominantly arboreal, and although adult males may descend to the ground to move over long distances, still more than 90% of travel time is spent in trees. On the ground orangutans seem to be clumsy quadrupeds. Their long, curved fingers and toes serve well in the canopy for suspension, but on the ground these are curled into what appear to be awkward fists for quadrupedal progression. Chimpanzees travel rapidly on the ground using relatively normal quadrupedal gaits and knuckle-walking with the hands. Both morphological and behavioral differences contribute to increased mobility of male and female chimpanzees compared with orangutans. Part of the significance of lower sexual dimorphism in chimpanzees lies in the relationship of body size to mobility, and the interrelationships of sexual dimorphism, mobility, and communal vs. individual defense of ranges by males are discussed below.

Mobility, Body Size, Male–Male Competition, and Sexual Dimorphism

Chimpanzees appear to face a foraging regime that demands more travel per day over larger ranges than orangutans (this argument depends on the premise that travel distances and range sizes tend to be minimized by selection favoring maximum investment in reproduction, at least for females). Female chimpanzees achieve the mobility demanded by this foraging regime by traveling terrestrially with appropriate morphological modification. Although their terrestrial travel is not efficient (Rodman and McHenry 1980; Taylor and Rowntree 1973), it is rapid and no doubt energetically efficient per distance traveled relative to ponderous arboreal progression by orangutans. Lighter bodies might also improve mobility and would reduce travel costs, as well as total metabolic costs, which would be advantageous as a means of reducing energetic expenses. But although female chimpanzees are somewhat smaller than female orangutans (table 5.1), the small difference in body size seems subjectively less significant than other morphological modification and terrestrial travel.

Male chimpanzees forage under the same relatively rigorous regime as female chimpanzees. They are faced with the additional problem of finding females and the problems of competition with other males for access to females. In comparison with male orangutans, male chimpanzees live with a population of females who move more rapidly over longer distances in larger ranges each day. The location of a female chimpanzee is less predictable than that of a female orangutan because of shorter persistence at any location, more possible locations in the larger ranges, and high rates of travel between locations. Thus, although both male and female chimpanzees must be more mobile than orangutans, male chimpanzees may face greater necessity of mobility than females. The males' need for mobility for both foraging *and* finding females may explain the considerably lower sex difference in body size of chimpanzees compared with orangutans.

Theory of sexual selection and mating strategies is constructed on the observation that normally one sex limits the reproductive potential of the other (Darwin 1871; Trivers 1972; Williams 1975). In mammals females normally limit males because of the small number of young a female can produce in a lifetime and the relatively small contribution that males can make to rearing of young. Although there are exceptions, male mammals commonly compete for the limited number of offspring of females, and social systems are polygynous.

Under these conditions, in order to achieve competitive fitness in the population, a male must normally father offspring of more than one female; selection consistently acts against those who do not. This mechanism produces male phenotypes suited to competition among males with two primary constraints: they must be acceptable to females as mates; and they must survive between conceptions. Large male body size often accompanies polygyny in mammals. Although there are few empirical observations to test the hypothesis, it seems logical that large body size of males results because larger body size will normally win in contests between individuals (there are, however, objections to this argument [Ralls 1976]).

The spatial system and sexual dimorphism of orangutans are a normal outcome of the system. Females are distributed evenly over the landscape, and association with males serves no function for them except for mating. Males are left to compete for access to females, and competition reduces to competition for space occupied by females. Critical contests occur in the presence of sexually active females. The contests are brutal (Gladikas 1978), and in such contests it is easy to see that larger body size is indeed advantageous.

The primary question to be asked is what *limits* body size of male orangutans? One critical limitation would be placed by the mobility necessary to maintain access to a sufficient number of females and "defend" them against access by other males. For the male orangutan, the females travel in relatively small ranges and move slowly through those ranges. Breeding populations of orangutans include approximately one male for two females, suggesting that the male is generally limited to defense of a range that includes about two female ranges. Given the slow movement of females through smaller ranges, I argue that the mobility necessary to monitor females is relatively low for the male orangutan.

Body size of adult male orangutans is "released" by the large size and relatively close spacing of food sources described above. If food sources were small, reduced yield per source would require greater travel to more feeding sites, which would increase costs of travel per day. As travel costs increased, large body size would become economically unfeasible. Similarly if food sources were much farther apart, increasing travel costs would lower the economic limits on body size. Body size of males is ultimately limited by requirements for mobility, but the limit is high.

Male chimpanzees face females dispersed similarly through the forest but females that move rapidly through large home ranges. At any time the location of the female is unpredictable, which makes defense against male competition difficult unless the male chimpanzee is highly mobile as well. Locomotor costs for high mobility in the male chimpanzee would increase rapidly with increasing body size, and it is argued here that the *required mobility of the male chimpanzee to defend an area occupied by females limits body size quickly.*

A second problem faced by male chimpanzees relative to male orangutans is the smaller food source. If a large proportion of food sources are small, larger body size may be impossible to support. It seems reasonable to argue from data presented above that in comparison with orangutans, male chimpanzees frequently face food sources small enough to limit body size.

Despite the limits to body size imposed upon male chimpanzees by required mobility and small food sources, there is still strong advantage to be gained by successful competition with other males for access to females. Larger body size would normally lead to advantage in competition between males, and it is not surprising that male chimpanzees are somewhat larger than female chimpanzees. In a sense, however, smaller body size limits individual power in conflicts, and there is an alternative means of increasing power in conflicts, which is cooperation to give *power in numbers.*

In order to examine this alternative, compare two hypothetical centers of power among male chimpanzees, one composed of a (hypothetical) large, single male and one composed of a pair of smaller males. First, what advantages and disadvantages accrue to each? The single male, if able to defend as many females as the pair, has the clear advantage of exclusive access to those females; the pair, on the other hand, must share the offspring. The pair have the advantage of greater mobility, which allows more rapid monitoring of positions of females and of potential competitors. The pair also have the advantage of being able to separate and forage independently when food sources are small or rare, with the possibility of rejoining strategically at focal points of competition if they are capable of signaling to each other.

A critical difficulty of communal defense is reduction of actual reproductive success (RS) of individuals as a consequence of sharing offspring of the females defended. In order to maintain equal RS per male—or to exceed it—the communal males should share at least as many females as could be defended alone by the same number of individuals. Part of this difficulty is overcome through mobility of the smaller communal males, who individually can spread over a larger area and then join together to defend when a competitor is detected. Long-distance signals such as the pant hoots of chimpanzees (Wrangham 1975; Goodall 1968; Ghiglieri this volume) would serve as the signal for strategic reunion, and there is some evidence that such behavior occurs (Goodall et al. 1979). Another part of the difficulty is offset by overlap between neighboring female ranges. Two males who individually might each defend the area occupied by two females need not defend twice as large an area in order to include four females in their shared communal range. So, for example, at the Gombe National Park, 12 females used ranges of approximately 4 km^2 (16 blocks of 0.25 km^2) each, but together occupied a range of only 12.5 km^2 defended by a community of 9 adult males. The implication is that a male does not need to range twice as far to cover the ranges of twice as many females. This implies that costs of defense may not increase as fast as numbers of females.

A final part of the males' solution to difficulties with communal ranging and defense is most intriguing. Relatedness may change the absolute number of females necessary to maintain a certain level of "effective reproductive success" (ERS). In order to explain this, take the initial condition to be that of successful, breeding male orangutans who occupy ranges that overlap approximately two females or who, on the average, control reproduction of two females for some variable time. RS and ERS are measured in numbers of females "controlled." Under these conditions, how many females must be included in the range to maintain an equivalent ERS if males join to share and defend a communal range? If males are unrelated, each male must have exactly two females in the range to maintain an ERS of 2, and communal range size must increase accordingly. But if males in a group of size N_m are related by some mean degree of relatedness, $\overline{r}$, then each male "shares" a proportion of the RS of other males equal to $\overline{r}$, and his total reproductive success is the sum of his own share and $(N_m - 1)\bar{r}$ of the reproductive success of his relatives in the male group (cf. Maynard Smith and Ridpath 1972). Figure 5.7 shows the effect of relatedness of as little as 0.1 on the number of females necessary to maintain an ERS of 2 as the number of males in the group increases. Note that with this level of relatedness, an ERS of 2 is maintained in a population that includes 11 males and 11 females—in which *the observed sex ratio is 1:1.* Given this argument and the possibility that males may recruit relatives, the difficulty of increasing range size to accommodate more adult female ranges as the male community increases in size is further offset. Males in the Gombe National Park communities are related, since none leave the natal group (Pusey 1979). If they have an average relatedness of 0.1 or more, the observed sex ratio of 9 males to 12

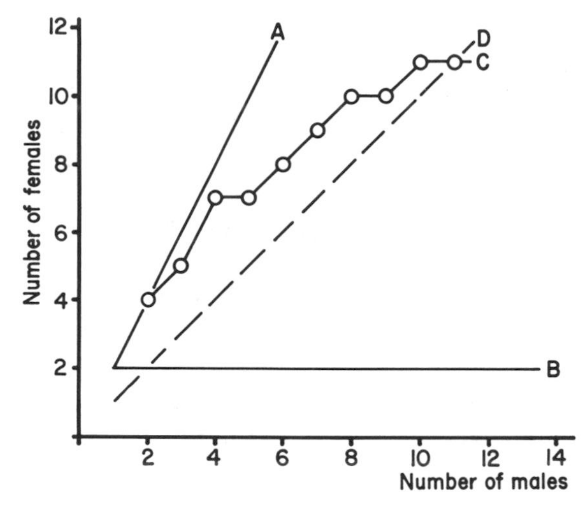

MALE RELATEDNESS and EFFECTIVE SEX RATIO

Figure 5.7. The figure illustrates the manner in which the number of females (N_f) must change in order to maintain an effective reproductive success (ERS) of 2.0 as the number of males (N_m) increases; ERS is measured by the ratio of breeding females to breeding males. A. Males are unrelated; thus $N_f = 2\ N_m$ for all N_m. B. Males are identical (relatedness between any two males, r_o, is 1.0) and $N_f = 2$ for all N_m. C. Males are related by $\bar{r}_o = 0.1$; thus $ERS = N_f/N_m + 0.1(N_m - 1)\ N_f/N_m$. C is treated as discontinuous since only whole numbers of males and females are possible, and at any N_m, N_f must provide *ERS* of at least 2. D. The dashed line indicates an observed sex ratio of 1:1; note that line A indicates a sex ratio of 1:2.

females provides an ERS of greater than 2 for each male. If average relatedness is higher than 0.1 among the males in the community, each male may have a considerably higher ERS than the arbitrary base of 2.0 suggested for orangutans.

The last point suggests a possible objection to the preceding argument: a male's ERS has been treated as composed of personal RS (PRS) and RS through relatives (RRS), but a male's RRS has been counted only if relatives are in the same group as he. Even a solitary male orangutan must have some RRS, although he cannot monitor or influence his relatives, since his brothers, cousins, etc., each have some likelihood of matings. To formalize the analysis, consider that the sex ratio of breeding males to breeding females is a measure of the probability that a relative will breed; in other words, consider that a relative has the same probability of breeding as any male in the population. If the population consists of highly competitive males who defend against all others, as in the case of the orangutan, and if the result is a decrease in the sex ratio, then another result is a decrease in the probability that any male will breed (even though the average RS of males remains constant). Now if a single male is successful as a breeder, what is his RRS if he is part of a highly competitive population of males compared with his RRS in a population in which most males accept male partners? On the assumption that the total number and mean relatedness of relatives are constant, RRS in the first case must be lower than in the second since the proportion of breeding relatives is lower in the first case than in the second. On the other hand, the male's PRS probably always decreases through cooperative breeding. The relationship of PRS to number of cooperating males (N_m) might be expressed as PRS = $(N_m)^c/(N_m)$. In this expression c is the rate at which total number of females increases as N_m increases. This total is divided by the number of males sharing the females to give the PRS of each individual. Clearly c must be greater than 1.0 for PRS to increase with cooperation, and PRS will always decrease if c is less than 1.0. It is difficult to imagine circumstances in which c would be equal to or greater than 1.0, which suggests that normally there must be cost to PRS of breeding communally. The exact cost $(1 - c)$ depends on ecological conditions in the population and on physical characteristics of individuals. In order for ERS to remain constant under conditions of communal breeding, let alone for ERS to increase, RRS must increase sufficiently to offset loss of PRS. Part of the cost is offset purely by the change in sex ratio, which leads to increased numbers of breeding relatives *regardless of whether they are in the same group or not.* Why, then, should male relatives remain *together* in *patrilineal* breeding groups? I argue subjectively that the patrilineal solution is likely because, although RRS is the same whether several relatives all breed together or all breed in separate groups, selection will always favor substitution of a relative for a nonrelative in the population of breeding males. Once the population is saturated (all breeding groups are full), related males can be added to the population only by substituting them for unrelated males, and this can be accomplished actively only in the *local* communal group. Thus, once communal breeding is allowed by sufficiently high c, clusters of relatives should develop and exclude nonrelatives. Such conditions occur when all male breeding locations are occupied by groups that together can repel newcomers. (An interesting alternative would be dispersal of groups of relatives who could carry power in numbers with them, as is the case in lions [Schaller 1972; Bertram 1976]. This system for chimpanzees appears to be precluded by irregular spacing of births and long lives so that conditions do not favor establishment of clusters of related males who might disperse together at an early age.)

This system has three additional articulations with the sociospatial system of orangutans and chimpanzees. First, relatedness among males will slow down evolution of sex differences by reducing the effects of female choice. Since the males are related and variance in male genotypes is thus reduced, choices between them have reduced selective effect. Second, although individual males always benefit by outcompeting others—even relatives—for offspring, the intensity of male/male competition within communities will be reduced because a successful competitor succeeds at some expense to a male relative. The cost to *inclusive* fitness reduces benefits to *individual* fitness in proportion to r. Finally, note that the male system of cooperation among relatives presents a problem to females. As they mature, they are faced locally with potential mates who are their relatives. As pointed out by Pusey (1979), the benefits of exogamy to females (through incest avoidance) may explain the usual pattern of female emigration from chimpanzee communities. Female emigration is therefore viewed as a consequence of adaptations by males to the foraging regime and to dispersion and ranging patterns of females.

Conclusion

A synopsis of the argument is simple. Foraging constraints interact with the social systems of orangutans and chimpanzees first by presenting females with an indefensible dispersion of food. Consequently single adult females of both species occupy overlapping but different ranges. Chimpanzees feed on relatively small, widely spaced food sources and adapt to this with rapid terrestrial travel. Low density of food leads to considerably larger range size of female chimpanzees. Male orangutans respond to female dispersion in normal mammalian fashion by competing for space occupied by females. The large size and high density of their food sources have allowed male body size to increase under selection for individual competitive ability. The relatively small ranges and slow movement of females have allowed some single large males to maintain control over more than one female's area and reproductive output. Male chimpanzees, constrained by small food sources, by the need for more travel between more dispersed food sources, and by mobile females, could not respond to sexual selection with increased body size but "opted" instead for power in numbers, with strategic deployment of genes among cooperating, related males who can disperse over sparse, small food sources when necessary and regroup for defense (or aggression) when opportune.

I have no pretense that this argument is unassailable. I suggest, however, that it is a reasonable interpretation of the limited data available and that it may raise sufficient questions—even objections—to generate more specific research on relations between foraging and social systems of these and other animals.

Acknowledgments

I am deeply indebted to Dr. Richard W. Wrangham for collaboration on analysis of data presented here, for comment on many of the arguments presented, and for permission to use his data in the paper. Although he has contributed substantially to the paper, I take full responsibility for weakness or errors that appear. I am indebted through Dr. Wrangham to all who facilitated and supported his research on chimpanzees of the Gombe National Park, including Dr. J. Goodall, Dr. D. A. Hamburg, Professor R. A. Hinde, the Grant Foundation of New York, and the authorities of the Tanzania National Parks. I thank Professor I. DeVore, the Department of Nature Conservation and Wildlife Management of Indonesia, the Indonesian Association of Zoological Gardens, the Indonesian Institute of Sciences, Mr. H. M. Kamil Oesman, Mr. C. L. Darsono, and personnel of the Pertamina Oil Company for sponsorship or assistance in my research in Indonesia. My research has been supported by NIMH Grant #13156 to I. DeVore, by grants from the National Geographic Society and the New York Zoological Society, by NSF grant #DEB 76-21413, and by USPHS Grant #RR00169.

References

Bertram, B. C. R. 1976. Kin selection in lions and evolution. In P. P. G. Bateson and R. A. Hinde, eds. *Growing Points in Ethology,* pp. 281–301. Cambridge, England: Cambridge University Press.

Brown, J. L. 1964. The evolution of diversity in avian territorial systems. *Wilson Bull.* 76:160–69.

Bygott, D. J. 1979. Agonistic behavior and dominance among wild chimpanzees. In Hamburg and McCown, eds. 1979:405–27.

Darwin, C. R. 1871. *Sexual Selection and the Descent of Man.* London: Murray.

Eckhardt, R. B. 1975. The relative weights of Bornean and Sumatran orang-utans. *Amer. J. Phys. Anthrop.* 42:349–50.

Galdikas, B. M. F. 1978. Orang-utan adaptation at Tanjung Puting Reserve, Central Borneo. Doctoral dissertation. Los Angeles: University of California.

—— 1979. Orang-utan adaptation at Tanjung Puting Reserve: Mating and ecology. In Hamburg and McCown, eds. 1979:195–233.

Galdikas, B. M. F., and G. Teleki. 1981. Variations in subsistence activities of female and male pongids: New perspectives on the origins of hominid labor division. *Current Anthrop.* 22:241–47.

Geist, F. 1974. On the relationship of social evolution and ecology in ungulates. *Amer. Zool.* 14:205–20.

Goodall, J. 1968. The behavior of free-living chimpanzees in the Gombe Stream Reserve. *Anim. Behav. Monogr.* 1 (3):161–311.

—— 1971. *In the Shadow of Man.* Boston: Houghton Mifflin.

Goodall, J. et al. 1979. Inter-community interactions in the chimpanzee population of Gombe National Park. In Hamburg and McCown, eds. 1979:13–53.

Halperin, S. D. 1979. Temporary association patterns in free ranging chimpanzees: An assessment of individual grouping preferences. In Hamburg and McCown, eds. 1979:491–99.

Hamburg, D. A., and E. McCown, eds. 1979. *The Great Apes.* Menlo Park, Calif.: Benjamin/Cummings.

Horr, D. A. 1975. The Borneo orang-utan: Population structure and dynamics in relation to ecology and behavior. In L. A. Rosenblum, ed. *Primate Behavior,* 4:307–23. New York and London: Academic Press.

Izawa, K. 1970. Unit groups of chimpanzees and their nomadism in savanna woodland. *Primates* 11:1–46.

Kano, T. 1971. The chimpanzee of Filabanga, western Tanzania. *Primates* 12:229–46.

Lyon, M. W. 1907. Mammals collected in western Borneo by Dr. W. L. Abbott. *Proc. U.S. Nat. Mus.* 33:547–72.

—— 1908. Mammals collected in eastern Sumatra by Dr. W. L. Abbott during 1903, 1906, and 1907, with descriptions of new species and subspecies. *Proc. U.S. Nat. Mus.* 34:619–79.

—— 1911. Mammals collected by Dr. W. L. Abbott on Borneo and some of the small adjacent islands. *Proc. U.S. Nat. Mus.* 40:53–146.

MacKinnon, J. R. 1974. The behaviour and ecology of wild orang-utans (*Pongo pygameus*). *Anim. Behav.* 22:3–74.

Maynard Smith, J. and M. G. Ridpath. 1972. Wife sharing in the Tasmanian native hen, *Tribonyx mortierii:* A case of kin selection? *Amer. Natur.* 106:447–52.

McGinnis, P. R. 1979. Sexual behavior in free-living chimpanzees: Consort relationships. In Hamburg and McCown, eds. 1979:429–39.

Michael, R. P., and J. H. Crook, eds. 1973, *Comparative Ecology and Behaviour of Primates,* London: Academic Press.

Mitani, J. C., and P. S. Rodman. 1979. Territoriality: The relation of ranging pattern and home range size to defendability, with an analysis of territoriality among primate species. *Behav. Ecol. Sociobiol.* 6:1–11.

Nishida, T. 1968. The social group of wild chimpanzees in the Mahali Mountains. *Primates* 9:167–224.

Nishida, T., and K. Kawanaka. 1972. Inter unit-group relationship among wild chimpanzees of the Mahali Mountains. *Kyoto Univ. African Stud.* 7:131–69.

Pusey, A. 1979. Inter-community transfer of chimpanzees in Gombe National Park. In Hamburg and McCown, eds. 1979:465–79.

Ralls, K. 1976. Mammals in which females are larger than males. *Quart. Rev. Biol.* 51:245–76.

Reynolds, V. and F. Reynolds. 1965. Chimpanzees in the Budongo Forest. In I. DeVore, ed. *Primate Behavior,* pp. 368–424. New York: Holt Rhinehart, Winston.

Rijksen, H. D. 1978. *A Field Study on Sumatran Orang-utans (Pongo pygamaeus abelii Lesson 1827*). Wageningen, The Netherlands: H. Veeman and Zonen B. V.

Rodman, P. S. 1973a. Population composition and adaptive organisation among orang-utans of the Kutai Nature Reserve. In Michael and Crook, eds. 1973:171–209.

—— 1973b. Synecology of Bornean Primates. Doctoral dissertation. Cambridge, Mass.: Harvard University.

—— 1977. Feeding behaviour of orang-utans of the Kutai Nature Reserve, East Kalimantan. In T. H. Clutton-Brock, ed. *Primate ecology: Studies of Feeding and Ranging Behaviour in Lemurs, Monkeys and Apes,* pp. 383–413. London: Academic Press.

—— 1979. Individual activity profiles and the solitary nature of orang-utans. In Hamburg and McCown, eds. 1979:234–55.

Rodman, P. S. and H. M. McHenry. 1980. Bioenergetics and the origin of hominid bipedalism. *Amer. J. Phys. Anthrop.* 52:103–6.

Schaller, G. B. 1972. *The Serengeti Lion: A Study of Predator-Prey Relations.* Chicago: University of Chicago Press.

Schultz, A. H. 1940. Growth and development of the orang-utan. *Contr. Embryology* 28:57–110.

Schultz, A. H. 1941. Growth and development of the chimpanzee. *Contr. Embryology* 29:1–63.

Silk, J. S. 1978. Patterns of food-sharing among mother and infant chimpanzees at Gombe National Park, Tanzania. *Folia primatol.* 29:129–41.

Simpson, M. J. A. 1973. The social grooming of male chimpanzees. In Michael and Crook, eds. 1973:411–505.

Sugiyama, Y. 1973. The social structure of wild chimpanzees: A review of field studies. In Michael and Crook, 1973:375–410.

Taylor, C. R. and V. J. Rowntree. 1973. Running on two or on four legs: Which consumes more energy? *Science* 179:186–87.

Trivers, R. L. 1972. Parental investment and sexual selection. In B. Campbell, ed. *Sexual Selection and the Descent of Man,* 1872–1972, pp. 136–79. Chicago: Aldine.

Williams, G. C. 1975. *Sex and Evolution.* Princeton, N.J.: Princeton Univ. Press.

Wrangham, R. W. 1975. Behavioural ecology of chimpanzees in Gombe National Park, Tanzania. Doctoral dissertation. Cambridge, England. Cambridge University.

—— 1977. Feeding behaviour of chimpanzees in Gombe National Park, Tanzania. In T. H. Clutton-Brock, ed. *Primate Ecology: Stud-*

ies of Feeding and Ranging in Lemurs, Monkeys and Apes, pp. 503–38. London: Academic Press.

—— 1979a. Sex differences in chimpanzee dispersion. In Hamburg and McCown, eds. 1979:481–89.

—— 1979b. On the evolution of ape social systems. *Soc. Sci. Infor.* 18:335–68.

Wrangham, R. W., and B. Smuts. 1980. Sex differences in the behavioural ecology of chimpanzees in the Gombe National Park, Tanzania. *J. Reprod. Fert.*, Suppl. 28:13–31.

6 Feeding Ecology and Sociality of Chimpanzees in Kibale Forest, Uganda

MICHAEL P. GHIGLIERI

CHIMPANZEES (*Pan troglodytes*) are intriguing creatures. Not only are they, along with gorillas, the closest living relatives of humans, as demonstrated by several lines of evidence (e.g., Huxley 1959; King and Wilson 1975; Sarich and Cronin 1976; Miller 1977; Fouts and Budd 1979), but also their social behavior approaches that of humans in its richness and complexity. Chimpanzees exhibit a rare and improbable constellation of ecological attributes: they are large-bodied, highly social, arboreal frugivores who travel terrestrially. Their primary habitat is the tropical rain forest of Africa, a relatively stable, natural "evolutionary laboratory." For the field ecologist and for those interested in the socioecology of humans and nonhuman primates, few other mammals evoke equal interest.

Despite this interest and several field projects of their socioecology, our understanding of the nature of wild, undisturbed chimpanzees is incomplete. Early, short-term investigations (3–12 months) of chimpanzees (e.g., Nissen 1963; Azuma and Toyoshima 1962; Kor-

MICHAEL GHIGLIERI is an ecologist who has studied chimpanzees of the Kibale Forest, Uganda, for two years. His work is rare because he did not provision his subjects to facilitate observation. Instead he capitalized on the natural provisions offered by large fruiting trees to habituate chimpanzees to his presence. Here he examines the relations between food dispersion and social patterns of the population. His analysis reports useful information on phenological patterns of Kibale Forest trees, as well as quantitative description of the grouping patterns of the chimpanzees.

tlandt 1962; Reynolds 1963; Sugiyama 1968, 1969; Albrecht and Dunnet 1971) yielded little or no data on the feeding behavior, sociality, or ranging of individually recognized chimpanzees, and their project designs ignored important ecological variables.

The two longitudinal studies of chimpanzees in Tanzania, at Gombe National Park (Goodall 1965; 1968; 1971; 1973a, b; 1979) and at the Mahale Mountains (Nishida 1968, 1970, 1979), have provided an excellent picture of individual behaviors. But because habituation of subjects was of superabundant "windfalls" that were replenished for years, the subjects' normal social patterns were modified greatly (Reynolds 1975). Wrangham (1974) noted that provisioning at Gombe promoted consistently large aggregations of individuals and warned, "Direct generalizations of the observational results gained at the feeding station to situations outside the feeding area should be avoided." Nishida (1979) admitted, "I do, of course, think that we should refrain from the provisioning method when doing purely ecological studies on a quantitative basis." Pusey (1979) suspected that provisioning may have had far-reaching effects on normal social development.

Salient features of chimpanzee sociality emerging from the Tanzanian studies were a fusion-fission society; female exogamy (Nishida 1979; Pusey 1979); and apparent territorial maintenance by parties of males who sometimes engaged in border clashes (Bygott 1979). After a reduction of the eight-year banana-provisioning program at Gombe, the study community of chimpanzees split unequally and became polarized in its ranging. By repeated attacks during two years, parties of males apparently annihilated the smaller community, killing three adult males and two females (Goodall et al. 1979). Trauma due to weaning the Gombe community from their permanent banana windfall was possibly exacerbated by increased population pressure caused by influx of chimpanzees from outside the park where agriculture was displacing them. The considerable influence of human activities at Gombe prevents confident conclusions regarding the normative pattern (if there is such a thing) of chimpanzee sociality.

The purpose of the present report is to examine social structure as a function of feeding ecology and sexual selection in an undisturbed population of unprovisioned chimpanzees in Kibale Forest, Uganda. Of the two influences, feeding ecology is the more straightforward and easily observed. The dispersion, density, size, and seasonality of patches of common food types may limit the number of chimpanzees who may forage efficiently in a single party and hence limit sociality. One aim here is to examine these constraints.

A second aim is to examine sex differences in membership of traveling parties and to interpret these differences in light of feeding ecology and sexual selection in the population. The results of sexual selection are expressed within the constraints of feeding ecology and are manifest in sex differences in behaviors. Observations from the two Tanzanian projects mentioned above lead to a picture of chimpanzee social structure dominated by male-male cohesiveness and territoriality. If males remain in their natal home range but females emigrate, the adult males within a community will be more closely related to one another than to the adult females (e.g., Glass 1953). Solidarity between males is predictable on the basis of increased inclusive fitness resulting from kin selection (Hamilton 1964, 1972), although reciprocal altruism (Trivers 1971) may also explain male-male cohesiveness in a territorial system. Whatever the balance of forces selecting for male-male cohesiveness, significant differences between the social relationships of males with males versus males with females should be apparent among the chimpanzees of Kibale.

Study Area

The Kibale Forest Reserve lies in the Toro District of western Uganda (0° 13′ to 0° 41′ N, 30° 19′ to 30° 32′ E; elevation 1,110–1,590 m) approximately 24 km east of the Ruwenzori Mountains. Its 56,000 ha of undulating hills and valleys support a complex mosaic of vegetational types. Grassland, woodland thicket, and colonizing forest types compose 40% of the cover, and the rest comprises tropical forest types in various seral stages and states of disturbance (Wing and Buss 1970). Elephant (*Loxodonta africana*), nine species of ungulate, and eight species of cercopithecoid primates are present. Large predators other than spotted hyena (*Crocuta crocuta*) are now quite rare, although illegal hunting of nonprimates by humans is common. Kibale Forest, lying at the eastern edge of the immense rain forest ecosystem of Zaïre, is a faunal ecotone for forest animals typical of Central and West Africa, and savanna species restricted to the east. Kingston (1967), Struhsaker (1975), and Ghiglieri (1984) give more detail.

Kanyawara

Compartment 30 of the Kibale Forest Reserve, extending southeasterly from the Kanyawara Forestry Station, is the site of Struhsaker's longitudinal study of red colobus monkeys (*Colobus badius tephrosceles;* Struhsaker 1975). Basically a peninsula of primary tropical forest surrounded by grasslands and disturbed or felled forest, compartment 30 is contiguous with an extensive block of primary forest to the southeast. Chimpanzees made seasonal use of Struhsaker's study area, and I made limited observations there (about 3% of my time).

Ngogo

Originally set aside as a 2-km^2 nature reserve but greatly expanded in 1975, Ngogo is the core of a region of pristine rain forest with no historical record of exploitation (10 km southeast of Kanywara; 0° 29′ to 0° 31′ N; 30° 24′ to 30° 26′ E). Ngogo was my primary study area; the majority of my observations of chimpanzees took place in a study area of 5.4 km^2. The site was prepared by cutting trails that ran roughly north-south or east-west at approximately 100- or 200-m intervals. These trails facilitated orientation, precise location of position, and my movements through the area.

Methods

I commenced field research on December 9, 1976, and spent most days of each month (except October 1977) engaged in research on chimpanzees until May 14, 1978. During this period I observed chimpanzees directly for 406.5 hours in Kibale Forest. Adult male-hours accounted for 37.4% of my observation of adults; immature individuals were focal animals infrequently. The rest of my time was spent in search for chimpanzees or in ecological observations related to them.

Observing Chimpanzees

Initially none of the Ngogo chimpanzees were habituated to a human observer. By stationing myself in an obscure but visible position about 50 m from large, fruiting trees of the genus *Ficus,* I was able to make repeated contacts with chimpanzees. Initial contacts almost always resulted in flight, but many individuals returned repeatedly to the same tree, or to other trees where I waited on vigil, resulting in multiple contacts. Repeated contacts with the same individuals led to a shift in their tendency to flee from me vs. their persistence at feeding in a "vigil" tree. Eventually several individuals ignored me and foraged, and some remained in my presence for hours of non-foraging behaviors. In short, they became habituated to me. Some individuals tolerated me in less than a month of first contact, others only after more than a year. Some never became tolerant. I openly observed habituated chimpanzees from as close as 10 m without disturbing them, but I used stealth to observe individuals whom I suspected would never tolerate me. Chimpanzees at Kanywara, unlike those at Ngogo, were tolerant even during their initial contacts with me.

Remaining on all-day vigils at large fruiting trees was my most effective means of contacting chimpanzees. I also located them by homing in on their vocalizations. Occasionally I initiated observations on chimpanzees in the early morning near their nests, where I had left them the night before. Quasi-random searches were the least successful way of finding chimpanzees, but I often resorted to that technique.

Upon contact with chimpanzees, I noted age-sex-reproductive classes represented in the party and the party's location. I usually chose the first adult that I saw as focal animal but switched to a new focal animal if the old one was not visible for 15–20 min. At 5-min intervals I recorded the location and the height above ground (if appropriate) of the focal animal, the tree species and/or food species if feeding, activity and posture, distance to nearest conspecifics and to monkeys (if present), and obvious responses to the presence of other primates. Weather was noted when it changed. Intraspecific or interspecific interactions, defecation or urination, and vocalizations were noted as they occurred. The activities recorded at each 5-min interval were those in which the animal was engaged at the precise ending of a 10-sec countdown interval, which occurred at the 5-min point. When the situation permitted, I recorded social interactions not including the focal animal.

Availability of Major Food Species

Phenology. Each month I monitored 107 specimens of 14 species of trees (79 of which were *Ficus* spp.) suspected or known to be important foods of chimpanzees. Struhsaker monitored 15 others in the same category; during five months of his absence I monitored these too. Characteristics observed for each tree included relative quantity and ripeness of fruit, quantities of young and mature leaves and blossoms, and extent of damage to foliage. Each characteristic was scored on a scale of 0 to 4, 4 being maximum.

Patch Size. I estimated the crown volumes of 25 vigil trees by measuring the east-west and north-south, horizontal diameters of each crown and estimating the vertical crown depth. Averaging the two, halved measures of horizontal crown diameter with the estimation of height yielded an estimate of crown radius, taking the crown as a hemisphere; computation of each volume estimate was simplified by using the formula 2/3 r^3. Lacunae within the crowns were not subtracted from the estimates of volume, because the lacunae seemed roughly equal among the trees. The resulting estimates of crown volume thus provided a valid relative index of patch size useful for comparisons. I did not attempt to estimate the actual amount of fruit present per unit volume, because I found no practical technique to sample huge forest trees with crowns 15 m above the forest floor. Sample counts on single trees suggested normal crops of several thousand fruits.

Patch Dispersion. During the initial months of study I walked the trail system (then nearly 70 km) to map it and the locations of all *Ficus* spp. visible from the trails. Species-specific patterns of dispersion, with evidence of fidelity to microhabitat, became evident. These data were of greater use in finding chimpanzees than in technical analysis of patch dispersion.

Patch Density. Data obtained during the mapping described above, restricted to *Ficus* specimens 10 m from the center of the trail, were used to estimate density of fig trees. Struhsaker (unpublished) enumerated trees (of $\geqslant$ 10 m in height) within 2.5 m of the center of the trails within the range of his study group of redtail monkeys (*Cercopithecus ascanius*). Although that home range was a small portion of the trail system at Ngogo, it seemed representative of the general area, and so I used these data in the analysis of patch density (Ghiglieri 1984).

Results and Discussion I: Feeding Ecology

The life history strategy of an animal results from natural selection favoring highest reproductive potential within the limits of physiology and ecology of the organism (Gadgil and Bossert 1970). Among chimpanzees, adult females give birth at intervals of about 5 years (Goodall 1977; Ghiglieri 1984). Such a low birth rate, coupled with a long life span and prolonged social and environmental learning, makes chimpanzees an extreme example of a K-selected species exhibiting a low r_m (Cole 1954; Emlen 1973). During any year, *most* adults fail to produce an offspring. In such a system, small increases in the efficiency of activities not directly concerned with reproductive biology may influence reproductive success.

It is especially true among K-selected species that reproductive success is only partly a function of sexual behavior because foraging efficiency in a competitive environment may be a deciding factor in individual fitness. As long as foraging is not efficient at the expense of reproductive activities, natural selection should favor higher efficiency (see Wrangham 1975). Among chimpanzees the foraging pattern appears to be a highly adapted behavioral complex that constrains sociality of the individual, though chimpanzees remain highly social. Because almost any reproductive success among chimpanzees is significant, subtle differences in the patterns of energy allocation and socialization may produce significant differences in reproductive success.

Activity Patterns

Because of differences in reproductive problems faced by each sex, we would expect sexual differences in activity patterns. A female does not compete for a mate but does need to forage and range so that her dependent offspring receive adequate nutrition. Males do compete for mates, may defend and monopolize estrous females (Tutin 1975) but do not need to range in accordance with the nutritional needs of their offspring.

Figure 6.1 shows the daily activity patterns of male and female chimpanzees beyond infancy in Kibale Forest. Data from all observations were tabulated for the categories foraging, resting, and traveling. Foraging includes picking, manipulating, eating, chewing, or carrying food items and scanning for or examining food items by vision, touch, or smell. Resting includes all activities not included in foraging and traveling (most social interactions, nonforaging self-maintenance activities, play, sleep, staring off into space, etc.). Traveling includes locomotor activities that took an animal from one general location in the forest to another. Displacements within a single tree during foraging or resting were not considered traveling.

Males as a class compared to females as a class spent more overall time foraging (62.1% compared to 52.4%), less time resting 25.8% compared to 37.6%), and more time traveling (12.1% compared to 10.0%). These sexual differences in activity budgets were significant (Wilcoxon rank sum test $Te = 120$, $Tr = 180$, $n = 12$, $p < 0.05$; data from 0700 to 1900 hours). Both sexes showed morning and late

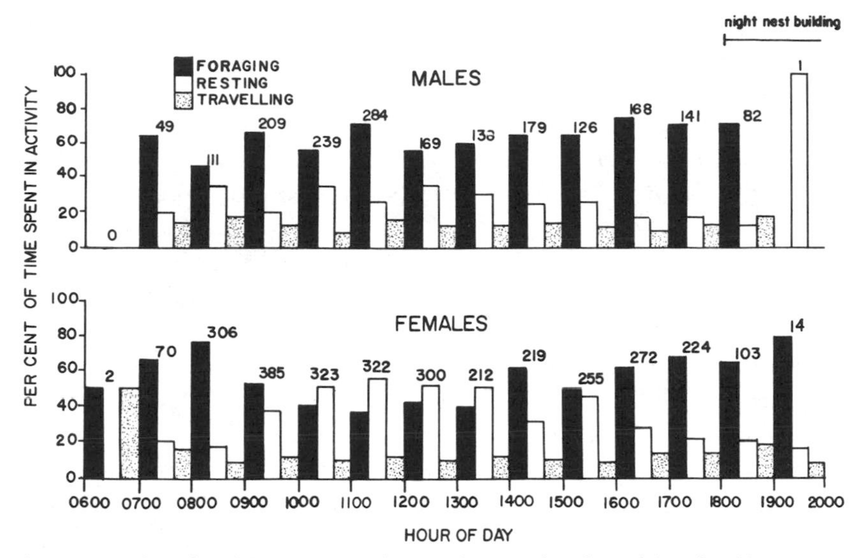

Figure 6.1. Diurnal activity patterns of post-infant-aged male and female chimpanzees in Kibale Forest. Entries are summations of instantaneous observations at 5-minute intervals. Number at top of column is sample size for each hour.

afternoon peaks in foraging activity separated by a midday peak in resting. Among males, foraging was the predominant activity during all hours of the day, whereas females rested more than half the time between 1000 and 1400 hours. Goodall (1968) reported only one daily feeding peak, between 1500 and 1800 hours, among Gombe chimpanzees.

My data are biased because approximately half of my observations were made during fruit tree vigils. This tended to undersample traveling time. The data on males differ little, however, from a similar activity budget reported by Wrangham (1975). He found that adult males spent 55.7% of their daylight hours foraging, 30.3% resting, and 13.8% traveling at Gombe. Although approximately midway between the activity budgets I recorded for each sex, Wrangham's data on adult males more closely resemble my data on males than my data on females. Predictably, the time spent traveling reported by Wrangham was 1.7% higher than my data.

The salient aspect of the activity budgets of Kibale chimpanzees is the significant difference between males and females. Among other great apes, Rodman (1979) found that an adult male orangutan (*Pongo pygmaeus*) spent more time feeding and less time traveling than each of three females in East Kalimantan. MacKinnon (1974), however, reported no differences in activity budgets of male and female orangutans in Sabah, perhaps because he included subadult males (who are of adult female size) in his analysis of male activity. Adult male orangutans are 2.2 times as heavy as adult females (Rodman, this volume). On the basis of the increased metabolic demand of increased body weight (Brody 1945), males require more food for self-maintenance. Adult male chimpanzees are only 1.3 times as heavy as females (Rodman, this volume), and so the difference in body weight is small. The metabolic demands of females, although smaller in body weight, may equal or exceed those of males because of the demands of pregnancy, lactation, and the transport of an infant (see Gunther 1971), although males' basal metabolic rates are higher (Brody 1945) and their metabolic demands are increased by increased travel (Taylor et al. 1970). I suspect that increased foraging time by males at Ngogo resulted from increased demands of travel, due possibly to nonforaging pressures such as the need to "patrol" their communal home range or to search for estrous females. This interpretation assumes that males at Ngogo behave similarly to those in Gombe (Wrangham 1975; Goodall et al. 1979).

Food Species and Types

Chimpanzees at Kibale used 50 plant foods while observed (table 6.1). They were highly frugivorous: 78% of feeding time was devoted to eating fruit pulp or fruit pulp and seeds (instances of seed eating only were not recorded as fruit consumption). It seems likely that seeds of most fruits (such as *Ficus* spp.) were ingested with the fruit either because to separate the seeds would greatly increase the handling time for each fruit or because seeds are nutritious. Digestion of seeds was not determined. Seeds of other species, such as *Monodora myristica* and *Mimusops bagshawei,* passed through the digestive tract of chimpanzees to emerge apparently undamaged in the feces. Young leaves eaten tended to be from tree species such as

Table 6.1. Summary of Opportunistic Observations of Chimpanzees Feeding
Note that data consist of numbers of on-the-five-minute-interval observations during which a chimpanzee was feeding on a food species and type. Data are random with respect to observer expectations; data from fruit tree vigils are not included (see text). Data were collected in Kibale Forest (and Budongo Forest where noted), Uganda, between December 1976–May 1978 and January–May 1981

	Food Type Eaten								
Food Species	Fruit[a]	Seed[a]	Blossom	Bark	Cambium	Wood	Leaf bud	Young leaf	Mature leaf
Aphania senegalensis								2	
Celtis durandii	2	2					9	19	
Celtis mildbraedii (Budongo Forest only)								2	
Chaetacme aristata								3	
Cordia millenii	28		13						
Cynometra alexandri (Budongo Forest only)		2							
Ficus brachylepis	13	13							
Ficus capensis	4	4							
Ficus cyathistipula	5	5							
Ficus dawei	25	25							
Ficus exasperata	21	21						11	
Ficus kitubalu	2	2							
Ficus mucuso	131	131							24
Ficus natalensis	49	49			1				
Markhamia platycalyx					2			1 (petioles)	
Mimusops bagshawei	19	19							
Monodora myristica	6	6							
Pseudospondias microcarpa	34	34	1	2				2	
Pterygota mildbraedii	95[b]	37[b]	11	7		1		2	
Treculia africana		4							
Uvaria sp.								1	
Uvariopsis congensis	34								
Tree, unidentified								2	
Vine, unidentified								10	
Animal foods:									
Red colobus monkey, apparently, 1 observation									
Termites, unidentified, 2 observations									
Total: 51 types & species	468[a]	354[a] 6[c]	25	9	3	1	9	55	24
Percent of all 600 observations:	78.0[a]	1.0[c]	4.2	1.5	0.5	0.2	1.5	9.2	4.0

[a] During 74.4% of fruit eating (such as *Ficus* sp.) fruit and seeds were ingested together. Where such overlap occurred, each feeding observation has been listed twice, but only the first figure was used to determine proportion of food type in diet. Digestion of seeds was undetermined.

[b] Food item eaten was wing of immature seed. Only one adult male was seen to ingest seeds.

[c] Seeds only eaten, not fruit.

Chatacme aristata and *Celtis durandii* whose leaves are unusually low in phenolics, a common chemical defense of plants against herbivory (McKey et al. 1978).

Reynolds and Reynolds (1965) reported Budongo chimpanzees as being 90% frugivorous. Hladik (1977) reported that provisioned chimpanzees at Ipassa, Gabon, ate 68% fruit by weight excluding provisioned foods. Jones and Sabater Pi (1971) wrote that chimpanzees were "mostly frugivorous" in Rio Muni. Of the 80 food types of Gombe chimpanzees reported by Goodall (1968), 60% were fruit. Wrangham (1975) reported that adult males at Gombe spent 59.5% of their foraging time eating fruit. Izawa and Itani (1966) reported that fruit composed 35.7% of food types of chimpanzees in the woodland savanna of Kasakati Basin. Although dietary composition reflects an interplay of preference and availability, these data suggest that a rain forest habitat may provide a richer, more stable source of fruit than more open habitats (see Gaulin 1979:16).

Other plant foods of Kibale chimpanzees included seeds, blossoms, bark, young and mature leaves, and leaf buds. My impression was that chimpanzees preferred fruit in most situations when they had a choice, but making an absolute statement concerning food preferences in the wild is not possible without knowledge of total food availability and the energetic costs associated with exploitation of each type. The few analyses of wild fruits that have been made (see Janzen 1979) indicate that *Ficus* fruits, the most common food of Kibale chimpanzees, are relatively high in protein and low in toxins when ripe. Such fruits are a superior food type that can be harvested and processed quickly once located.

Spacing of Foragers

Aggregations of up to 24 chimpanzees collected in large fruiting trees. Individuals (older than infants) consistently maintained appreciable interindividual distance during foraging ($\bar{X} = 9.45$ *m*, $N = 1{,}111$) that decreased markedly, *after* periods of intense foraging, when the animals formed small grooming clusters. As illustrated in figure 6.2, distances between nearest neighbors (excluding infants)

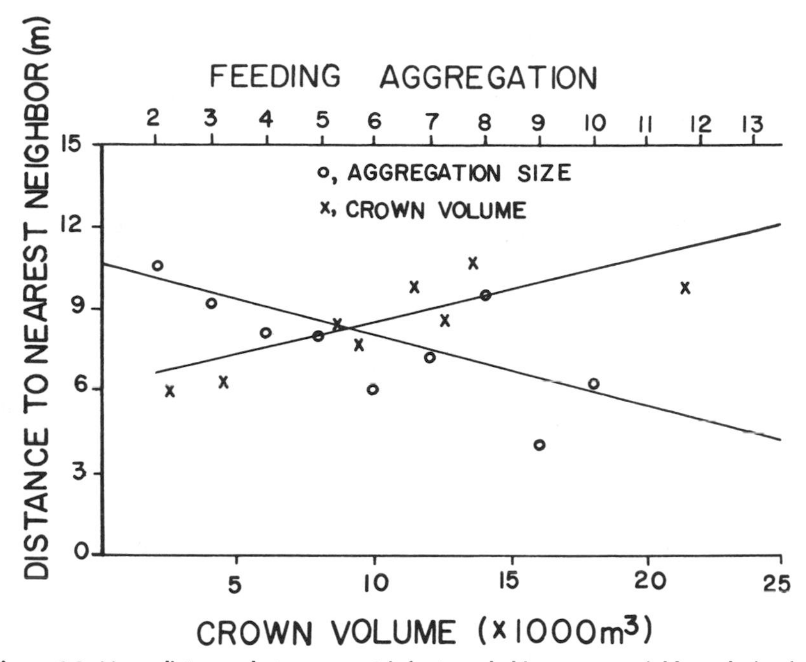

Figure 6.2. Mean distances between post-infant-aged chimpanzee neighbors during foraging in relation to size of feeding aggregation and to crown volume of food tree. See text for discussion.

during foraging decreased significantly with increasing size of feeding aggregations (r = .67, d.f. = 9, p ≤ .05). Conversely, distances between foraging neighbors increased with increasing crown volume of food tree (r = .82, d.f. = 8, p ≤ .01). These data suggest that chimpanzees forage farther from a foraging conspecific when conditions allow them to reduce proximity without hindering foraging. This tendency to disperse, rather than maintain or increase proximity during foraging, was not only statistically apparent, it was my strong impression during observation. Individuals appeared to watch each other in the crown and to move so that they did not increase their proximities. This pattern of reduced proximity may have been important in maintaining the low number of dominance interactions that I observed (Ghiglieri 1984), despite a potentially competitive situation.

Food Calls

Wrangham (1975) reported "food-calling" by adult male chimpanzees at Gombe. Independently, and unaware of Wrangham's report, I observed the same phenomenon. Food calls were defined as a specific, often distinguishable type of pant-hooting (see Marler and Hobbett 1975) given by an adult male as he approached a large food source. When food calls were given at Gombe, new individuals were significantly more likely to arrive (33%) during the caller's feeding bout than if he had not called (6%).

I defined a food patch as one of "vigil size" if it was a tree containing a superabundance of fruit and was visited repeatedly by several primate species and other frugivores (see discussion of patch size below). In this sense, superabundance implies nothing about presence or absence of competition among consumers, only that fruit was, in my qualitative estimation, highly abundant. When arriving at vigil patches, one or more adult males of a party of chimpanzees sometimes (27.3% of 93 visits) vocalized with long-range calls categorized as "pant-hoots" (Goodall 1968; see Ghiglieri 1984) characterized by wide variation in sound both between calls of the same male and among vocalizing males. Rhythmic hoots, grunts, shrieks, screams, wails, and an occasional roar were juxtaposed during vocalization sessions, and often the vocalizer pounded on the buttress of a tree, sometimes the food tree, just before climbing up to feed. I judged that these calls were audible within at least 1 to 2 km at Ngogo, and farther if the listener was in a high place. I never saw a female initiate such calls in a similar context.

Although I was never able to interpret distant pant-hooting as food-calling with certainty, H. Plooij (cited by Wrangham 1975) reported that the "food pant-hoot" at Gombe was distinguishable from other types of pant-hoots. Pant-hooting upon arrival at a superabundant food resource at Ngogo was a conspicuous signal that advertised the presence of the vocalizer over an area of 3–7 km^2.

Perhaps the initial question that arises in this context is whether alleged "food calls" attract other chimpanzees who subsequently share the food source of the caller. To test this question I pooled all data pertaining to arrival of parties containing one or more males at vigil

Table 6.2. Frequency of Arrival of Additional Parties of Chimpanzees at Superabundant Food Sources Following "Food Calls' and No "Food Calls" by Adult Males at Food Source, Ngogo, Kibale Forest, Uganda

		Arrival of Subsequent Chimpanzee Party During Feeding Bout of Caller, or Within 30 Min of Call		
		Yes	No	Total
Arriving party containing ≥1 adult male chimpanzees who:	Pant-hooted upon arrival	13 (59%)	9 (41%)	22 (100%)
	Did not pant-hoot	18 (34%)	35 (66%)	53 (100%)

trees in fruit, excluding cases when the party deserted the tree prematurely in response to my presence. Table 6.2 summarizes data on arrivals of parties of chimpanzees at vigil trees following the arrivals of parties containing one or more males who either gave food calls or did not. Arrival of later parties was significantly associated with food calls ($\mathbf{x}^2 = 4.049$, d.f. $= 1, p < 0.05$). Food calls evidently attracted other chimpanzees (both male and female) who subsequently shared the food resource of the caller, thus confirming Wrangham's earlier result (Wrangham 1975).

It is tempting to conclude from observations of interindividual distance in patches and from calling and responses to calling at food patches that chimpanzees are highly cooperative, perhaps even altruistic, in their foraging. Their calling behavior seems to increase the callers' competition for a known food source, and the changes in interindividual distance suggest that competition (use of a resource in short supply by two organisms so that consumption by one organism lowers the fitness of the other) occurs within food patches. The chimpanzee's behavior in food trees seems to minimize aggressive interactions, which lowers the interference between individuals while feeding. A chimpanzee acting in its own interest should not attract others to a food source in which it may need to reduce its own consumption as a result. If there is competition among individuals within patches, calling behavior takes on the aspect of altruism.

There may be, however, alternate explanations of these behaviors. Males may be pant-hooting to define territories. Marler and Hobbett

(1975) reported that pant-hoots were probably individually identifiable, and so it is likely that the chimpanzees of a community recognize one another's calls. Individuals who appear in proximity to a pant-hooting adult male shortly after his vocalization may be attracted to the individual himself rather than to his food resource. But even so, the result is that they consume resources that perhaps would have been consumed by the caller in their absence.

My observations suggest to me that food-calling produces little or no competition, and the benefits to the caller are potentially of three sorts: (1) the caller may benefit by social interaction with a subsequent arriver by mutual grooming or larger party size safer for territorial patrolling; (2) the subsequent arriver may be an estrous female with whom the caller may mate; or (3) the subsequent arriver may be related to the caller, which is likely if it is another male or an immature female (on the assumption of a female exogamous system), so that any increment added to the individual fitness of the arriver by improved nutrition is shared in the inclusive fitness of the caller (Hamilton 1964, 1972).

In summary, food-calling was a real phenomenon that likely increased individual and inclusive fitness in the caller.

Patchiness

Patch Size and Use. Although Kibale chimpanzees paused while traveling to harvest food en route, such as succulent new leaves of a vine or understory tree, my impression was their basic foraging pattern involved intensive use of isolated patches of fruit.

Patches, i.e., individual trees or groves of trees in ripe fruit, are irregularly distributed in the forest. Their size and density should impose bioenergetic limits on the number of chimpanzees that can efficiently exploit them; hence, patches limit sociality to an extent. A patch containing only enough to provide three adults with acceptable food for an hour of intensive foraging can provide only 6 min of foraging for the 30 adults of Ngogo community. In a homogeneous habitat, on the assumption of a model similar to that of MacArthur and Pianka (1966), the community traveling as a cohesive social group would need to travel 10 times farther than a party of three adults in order to visit a sufficient number of patches to feed everyone. A tenfold increase in traveling distance would greatly increase each chimpanzee's metabolic requirements and would virtually monopolize the daylight hours leaving no time for sedentary feeding or resting.

Adult chimpanzees seemed always to have the option to split away from a party to travel alone or with a smaller party. Such an open society allows individuals to reduce intracommunity competition for clumped but scarce resources simply by striking off on their own to forage solitarily on patches that would be energetically uneconomical if traveled to and exploited by a large party. Individuals who can split away from the community to forage on scarce resources and still maintain close social bonds have an advantage over those who persistently favor large parties when available patches are inadequate to support them. Thus, the fusion-fission social pattern results from individual adaptation for foraging efficiency.

Table 6.3 summarizes both my time spent on tree vigils and sizes of aggregations of chimpanzees seen in 35 trees of varying crown volumes. Although the largest aggregation in a single tree contained 24 individuals, the average size of all 687 feeding aggregations was only 3.6 individuals. Maximum aggregation sizes were significantly correlated with crown volumes of food trees (r = 0.5653, d.f. = 18, $p < 0.01$). Figure 6.3 shows the relationship between crown volume of vigil trees and the maximum aggregation of chimpanzees in each tree. Analyses of *average* size for all aggregations seen per tree (r = 0.2191, d.f. = 18, $p > 0.05$) and of average chimpanzee-hours per hour of vigil with patch size (correlation coeficient, r = 0.2516, d.f. = d.f. 18, $p > 0.05$), however, did not indicate significant relationships to crown volume. Larger crown volumes seemed to support more feeders over time than smaller crowns, but a bias in my sampling technique tended to diminish this difference. I usually spent a day or two more toward the end of the fruiting period of a tree of large crown volume than I spent at a small tree. The differential in cutoff times for durations of vigils emphasized the peak fruiting period of smaller trees and the peak plus the dwindling fruiting period of the larger trees, so that the use data per hour were not represen-

Table 6.3. Summary of Observer Time Spent During Vigils of Fruiting Trees at Ngogo and Presence of Chimpanzees per Tree Species

Species of Tree	No. of Trees Observed	Range in Crown Volumes (m^3)	No. of Days on Vigil	Hours of Vigil (hr:min)	Total Chimpanzee Hours in Tree per Vigil Hour	Aggregation sizes of chimpanzees		
						No. of Aggregations	Range in Aggregation Sizes	$\bar{X}$ no. per Aggregation
Cordia millenii	3	262–2,916	11	62:37	0.13	11	1–3	2.00
Ficus brachylepis	2	1,141–3,619	4	18:20	0	0	0	0
Ficus congensis	1	1,703	1	7:45	0	0	0	0
Ficus cyathistipula	1	884	9	42:49	0.06	5	1–2	1.20
Ficus dawei	4	8,183–13,261	49	478:14	0.81	279	1–14	3.23
Ficus exasperata	4	2,425–4,427	14	72:46	0.66	21	1–5	2.38
Ficus mucuso	14	4,781–21,303	51	393:10	1.06	352	1–24	4.12
Ficus natalensis	3	1,703–4,601	13	54:27	0.63	19	1–8	2.42
Ficus polita	2	3,470–4,965	3	17:05	0	0	0	0
Total 9 species	34	262–21,303	155	1047:13	0.78	687	1–24	3.60

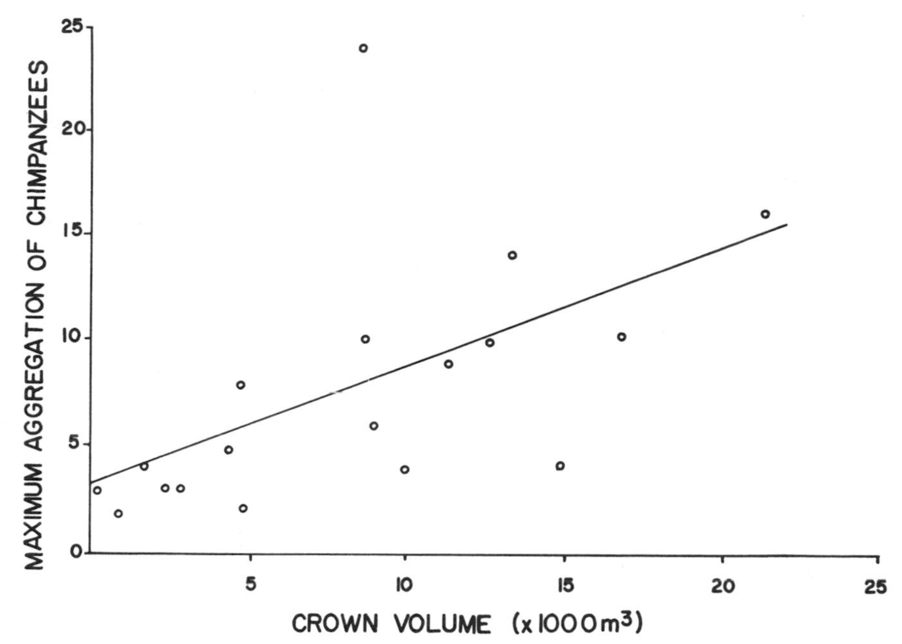

Figure 6.3. Patch size in relation to maximum aggregation sizes of chimpanzees. Patches are trees with very large crowns (larger than average) where I waited for and observed chimpanzees.

tative of the differences in crown volume *over time*. Another explanation (suggested by J. Cant) for the lack of significant correlations is that crown size remains constant while effective patch size (determined in part by fruit density) of the same tree decreases during the fruiting period.

To illustrate the differences in use of trees of large and small crowns I analyzed the patterns of visitation to a medium-sized *F. exasperata* (4,427 m^3) from July 3–10, 1977, and to a large-sized *F. dawei* (13,261 m^3) from April 16–26, 1978. These two vigils are used as examples because the influence of my presence was minimal and because the coverage of the fruiting periods of both was complete. The crown volume of the *dawei* was approximately three times greater than that of the *exasperata,* as were the respective maximum aggregations of chimpanzees who fed in each tree at one time (14 chimpanzees compared to 5). Use was also greater per unit time in the *dawei* (1.35 chimpanzee hours/vigil hour) than in the *exasperata* (0.84 chimpanzee hour/vigil hour). The number of different individuals who visited the *dawei* (36) was twice the number that visited the *exasperata* (17), and the number of repeat visits by individuals was almost three times higher in the *dawei* (1.11 repeat visits/vigil hour) than in the *exasperata* (0.39 visit/vigil hour).

Analysis of data from additional vigils indicates no significant differences in use of different species of tree by chimpanzees when crown sizes are not considered ($r = 0.4454$, d.f. $= 6$, $p > 0.05$), so that crown volume is probably the critical difference between the *dawei* and *exasperata* vigils discussed above. The unsurprising upshot of these vigil data is that larger crown volumes of fruiting trees attract more individuals, attract the same individuals for more repeat visits, sustain larger feeding aggregations, and are used more intensively over time than smaller crown volumes. Because of the tendency of chimpanzees to aggregate in them and to extend their visits for several hours of nonforaging activity, the density and distribution of large-crowned fruiting trees may be an important factor in limiting and/or facilitating interaction among community members.

Patch Density. The Ngogo study area contains nearly a hundred species of trees, yet relatively few were important food species of chimpanzees. Using Struhsaker's (unpublished) tree census and my census for Ngogo of *Ficus* spp. (Ghiglieri, 1984), I examined the relationship between relative density of food species and their importance as food. Table 6.4 compares the rank in relative density of the top 12 food species of Ngogo chimpanzees (as determined from opportunistic, nonvigil observations of feeding) with their rank of inclusion in diet. The difference between the two rank orders is significant (Sign test, $N+ = 10$, two-sided, $p < 0.05$).

This comparison illustrates an obvious but important aspect of foraging; i.e., chimpanzees are specialist feeders who perferentially exploit rare food *types* (see above) or rare *species,* rather than feed predominantly on common food types (e.g., leaves) of common food species as, for example, red colobus do in Kibale (Struhsaker 1975). A large-bodied specialist's diet of relatively rare food types dispersed in discrete patches implies the need for a large home range (McNab

Table 6.4. Top 12 Forage Species of Chimpanzees Compared With Relative Density Among Tree Species at Ngogo. Relative Density was Determined From Enumeration Data of Struhsaker (See Text)

Tree Species	Rank as Food Species	Rank in Relative Density (All Trees)
Ficus mucuso	1	87[a]
Pterygota mildbraedii	2	8
F. natalensis	3	35
Pseudospondias microcarpa	4	20
Uvariopsis congensis	5	2
F. dawei	6	63
Cordia millenii	7	40
F. exasperata	8	45
Mimusops bagshawei	9	18
F. brachylepis	10	34
Monodora myristica	11	14
Celtis durandii	12	3

[a] Note *F. mucuso* did not appear among the 86 species enumerated by Struhsaker (but did appear in my enumeration of *Ficus* spp.), and so I have arbitrarily assigned it the subsequent rank.

1963) and for increased travel between patches (MacArthur and Pianka 1966).

Patch Distribution. The existing data on distributions of tree species, arising from enumerations along transects, have not been analyzed statistically. Both my observations in the forest and the mapping of enumerated trees indicate a clear tendency toward nonrandom and irregular distributions among 9 of the 12 species in table 6.4. Most species exhibited fidelity to microhabitats characterized by specific topography and drainage; the effect of edaphic factors is unknown. Clumping into diffuse "grooves" was common, as was noted by Hubbell (1979) for 72% of 61 tree species in a dry tropical rain forest in Costa Rica. Hubbell reported that clumping increased with decreasing abundance. Species that apparently were eurytopic at Ngogo were *F. brachylepis, F. natalensis* and *M. myristica.* Subjectively, then, the forest at Ngogo was a complex mosaic of repeating floristic associations, which, from the perspective of a chimpanzee, might be viewed as coarse-grained and patchy.

Patch Phenology. A spatial assessment of patchiness with regard to the dispersion of fruit eaten by chimpanzees makes little sense without considering the phenology of the patches. While the spatial distribution of food species changes very slowly during a chimpanzee's life, edible fruit on the trees appears and vanishes on a weekly basis.

Figure 6.4 illustrates the fruiting periodicity of 12 species of trees used as food by Ngogo chimpanzees. A superficial glance reveals no marked season for all species combined but rather an unpredictable series of fruiting periods of mixed species. Pronounced peaks in periodicity were apparent in 10 species. Fruiting synchrony was most apparent in *F. dawei, M. muristica,* and *P. mildbraedii,* which, at approximately 12-month intervals, fruited during peaks centered around March-April, November-December, and March-April respectively. Other *P. mildbraedii* specimens at Ngogo fruited out of phase with those monitored for phenology ($N = 5$) and were an important food of chimpanzees. Individual *F. mucuso* fruited at odd, unpredictable intervals ($N = 20$ fruiting periods, $\mathring{X} = 6.55$ months, range = 4–13 months) with an interindividual fruiting synchrony possibly coinciding by chance. *U. congensis* showed a marked fruiting peak during June 1977 at Ngogo and Kanyawara, but the 15 phenological specimens monitored at Ngogo were relatively low producers. *F. brachylepis* and *F. natalensis* appeared to be asynchronous with intervals between fruiting periods varying from 5 to 17 months.

Phenological data from the above specimens indicate a general unpredictability for individual food trees on a fine level. But combining the data for all 101 sampled trees of species used by chimpanzees as food, for all 17 months, reveals a conspicuous pair of fruiting peaks from February to April in 1977 and in 1978. Figure 6.5 compares this composite furiting periodicity (and availability) for fruit at Ngogo with the month-by-month record of my sightings of chimpanzees. The average number of chimpanzees I encountered per hour at Ngogo during all of my field time was positively correlated with percent of maximum fruit available on a monthly basis ($r = 0.6486$, d.f. $= 17, p < 0.01$). My impression, unsubstantiated by data, is that many of the Ngogo chimpanzees migrated south, away from the study area, when fruit availability decreased within it.

Comparison of average size of traveling parties per month with percentage of fruit present per month reveals a trend that is not statistically significant ($r = 0.3374$, d.f. $= 17$, $p > 0.05$). I suspect

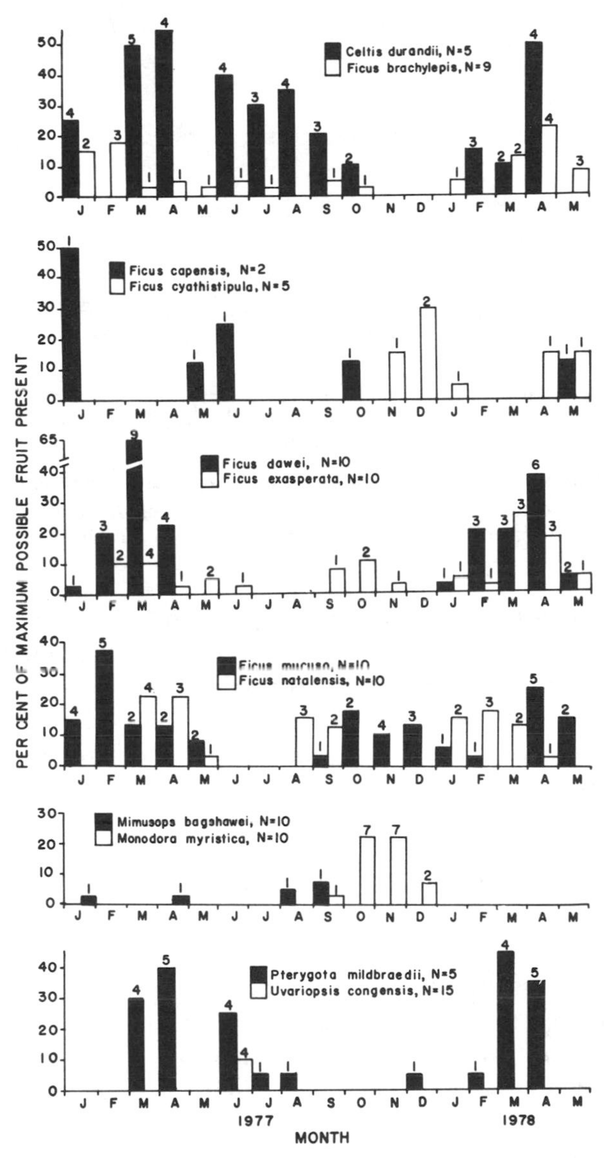

Figure 6.4. Fruiting phenology of some major forage species of chimpanzees at Ngogo. Number at head of column indicates number of trees with fruit for each month of fruiting. Note that sample sizes per tree species varied from 2 to 10.

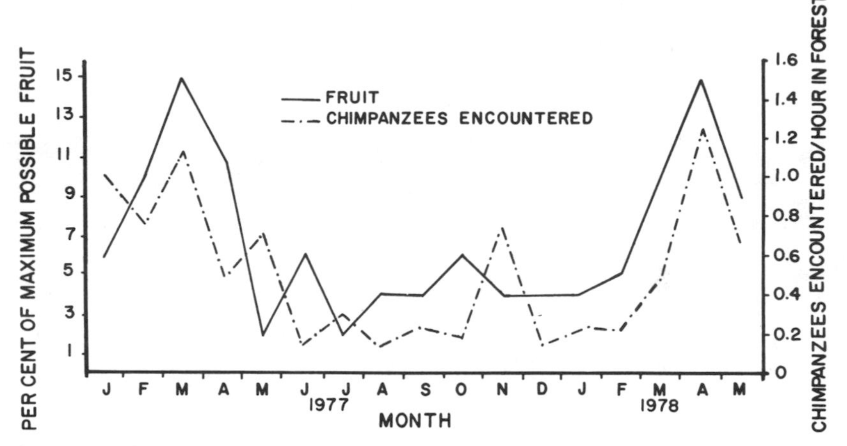

Figure 6.5. Chimpanzees seen per hour in the forest compared with the percentage of fruit present per month. Fruit presence is measured as a proportion of maximum possible fruit yield per total number (n = 101) of food trees in the phenology sample per month.

that if I had included a wider variety of food species (e.g., *Cordia millenii* and *Pseudospondias microcarpa*) in my monthly phenology samples, the greater representation of fruit availability would have yielded a significant correlation with size of traveling parties. My impression was that chimpanzees traveled in smaller parties during periods when fruit patches were less rich or more widely dispersed.

Results and Discussion II: Sociality

A fortuitous result of the banana-provisioning program at Gombe National Park was the experimental demonstration that chimpanzees travel in larger than normal parties when normal foraging constraints are lifted (Wrangham 1975). At Ngogo, affiliative behaviors (such as mutual grooming) were much more frequent than agonistic interactions (Ghiglieri 1984). Both observations indicate that companionship among chimpanzees may be favored over a solitary condition. This is not a trivial inference; it suggests that sociality is constrained by feeding ecology. If sociality is "expensive," we would expect individuals to direct their social activities to derive the maximum inclusive fitness from them. Traveling companions are probably the best indicators of a chimpanzee's choice of apportionment of social interactions because it must pay for companionship by increased feeding competition.

Traveling Parties

Size of Parties. A traveling party is defined here as one or more chimpanzees who arrived at or departed a particular place, from or to the same direction, within 3 minutes of one another. The average size of traveling parties was 2.6 individuals (N = 597), with sizes ranging from 1 to 24. Figure 6.6 summarizes frequencies of party sizes of mixed and unisexual composition. The modal party size for 1 for both unisexual parties. The largest parties were of mixed sexes. $\bar{X}$ = 5.4, N = 81), as would be expected by chance. The average size of parties containing males only was 1.6 individuals (N = 134), while

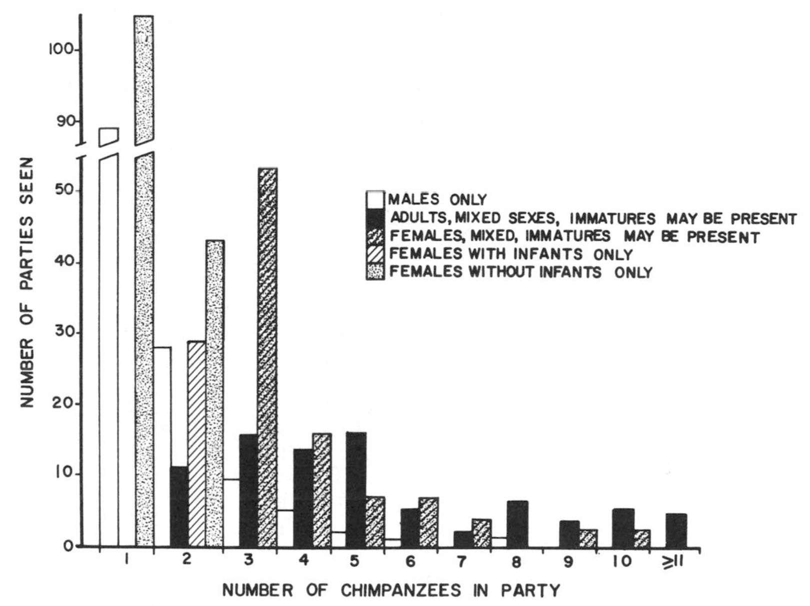

Figure 6.6. Frequencies of sizes and sexual compositions of traveling parties of Chimpanzees at Ngogo.

the size of those comprising one or more adult females, with or without attendant immature offspring but no adult males, was 2.4 (N = 336). The presence of immatures escalated the head count of parties of the latter type but added little biomass. If a mother and her immature offspring are considered as a *single unit,* 43.7% of all parties seen at Ngogo were solitary.

Solitary travel was commonly seen in other studies. Wrangham (1975) reported that 8.3%–41.8% of parties during two dry seasons and 36.2% during one wet season were solitary; party sizes varied seasonally and between habitats. Izawa and Itani (1966) reported 20% of parties in Kasakati Basin were solitary ($\bar{X}$ = 6.8 individuals per party). Nishida (1968) found that 30.7% of provisioned chimpanzee parties in the Mahale Mountains were solitary, noting that party size varied seasonally. Sugiyama (1968) reported that 28% of traveling parties at Budongo Forest were solitary ($\bar{X}$ = 3.9 individuals per party) and that mean sizes of parties varied seasonally.

Clearly sizes of traveling parties of chimpanzees vary greatly within a single habitat, vary between seasons, vary between habitats, and vary between observers. How much of this variance is an artifact of observers' investigational techniques is difficult to determine. Solitary chimpanzees are difficult to spot, for instance, and more reports of solitary individuals might be expected from observers whose bushcraft was superior. Chimpanzees in provisioned areas (Mahale and Gombe) seemed more likely to travel with companions.

Composition of Parties. Table 6.5 summarizes age-sex class associational tendencies for all sightings of sexed chimpanzees in traveling parties. Each individual of a specific age-sex class, in the company of another, counted as one incident. Two adult males traveling together received a total score of 2 because each of them was traveling with another adult male, and so on. Traveling parties were not random aggregations (see fig. 6.6).

Both adult and juvenile males were more likely to travel with other males than with females. Adult males preferred one another and differed significantly from adult females, not accompanied by infants, in the sex of their traveling companions (χ^2 = 56.793, d.f. = 7, $p < 0.005$), and differed in the same way from females accompanied by infants (χ^2 = 23.062, d.f. = 7, $p < 0.005$).

Table 6.5. Summary of Age-Sex Class Associational Tendencies of Traveling Parties of Chimpanzees at Ngogo, Kibale Forest, Uganda. Lower (Quantity) Represents Percent of Total Times Seen

		Seen in Party with One or More of the Age-Sex Classes Below						
Age-Sex Class	No. of Times seen Traveling	Adult Male	Subadult Male	Juvenile Male	Adult Female Without Infant	Adult Female With Infant	Juvenile Female	Juvenile, Undetermined Sex
Adult male	250	134 (53.6)	25 (10.0)	58 (23.2)	77 (30.8)	38 (15.2)	22 (8.8)	12 (4.8)
Subadult male	31	16 (51.6)	—	13 (41.9)	18 (58.1)	9 (29.0)	6 (19.4)	3 (9.7)
Juvenile male	104	46 (44.2)	17 (16.3)	16 (15.4)	30 (28.9)	38 (36.5)	13 (12.5)	6 (5.8)
Adult female without infant	298	75 (25.2)	24 (8.1)	41 (13.8)	139 (46.6)	57 (19.1)	63 (21.1)	24 (8.1)
Adult female with infant	167	33 (19.8)	10 (6.0)	38 (22.8)	54 (32.3)	44 (26.3)	69 (41.3)	25 (15.0)
Juvenile female	105	18 (17.1)	8 (7.6)	12 (11.4)	45 (42.9)	56 (53.3)	18 (17.1)	2 (1.9)

Surprisingly, even adult females with infants differed significantly from adult females without infants in their traveling companions. Females with infants were more likely to travel with other females with infants and with juvenile females than were females without infants, while the latter were more likely to travel with adult males and/or other adult females without infants than were those with infants ($\chi^2 = 34.519$, d.f. $= 7$, $p < 0.005$). All classes of females were significantly more likely to travel with other females than with males.

Azuma and Toyoshima (1962) reported that "familoid" group, containing one adult male plus one or more females with accompanying offspring, as the most common social grouping of chimpanzees at Kaboko Point, Tanzania. Goodall (1965) reported that 30% of groups ($N = 350$) were of mixed sexes, 28% were of males only, and 24% were of females with accompanying offspring. Because almost all chimpanzee observations at Gombe between 1961 and 1969 were at the provisioning station (Busse 1977), Goodall's (1965) mixed groups may have been overrepresented owing to the presence of large amounts of food. Halperin (1979) found that adult males and adult females with infants were more likely to be in unisexual parties (or solitary) than in mixed parties. Kortlandt (1962) distinguished between two major types of traveling parties: "sexual groups" containing adults of both sexes and "nursery groups" composed of females and immatures. Nishida (1968) reported that 10.6% of 218 subgroups were all male ($\bar{X} = 2.6$ individuals per party), 51.8% were mixed ($\bar{X} = 13.1$ individuals per party), and 13.3% were of mothers and offspring ($\bar{X} = 5.2$ individuals per party); here, again, provisioning may bias these data toward large mixed parties. Reynolds and Reynolds (1965) reported that 13.6% of 103 bands were of males only, 70.7% were mixed, and 15.5% were of mothers. These latter data are biased toward large parties because the chimpanzees usually were located by their vocalizations, and larger parties vocalize more than smaller ones.

Kortlandt (1962), Sugiyama (1968), Nishida (1979), Pusey (1979), and the present study all echo the conclusion that males exhibit more affiliative behavior toward one another than toward females. Parties made up of adult females and immatures are a second common pattern. The phenomenon of males' preferring other males as companions is expected within a kin-selected, male-retentive, territorial system, but why do females prefer one another?

There are several possible explanations. Females are of lower status than males and are likely to be cheated by them during mutual grooming or displaced by them during foraging. The general activity pattern of each sex differs so that a female who elected to travel with males would expend more energy and have less leisure time than otherwise, a situation that may be particularly disadvantageous for a female with dependent offspring (Rodman 1979). If males occasionally engage in territorial clashes, a female traveling with them may be exposing herself to danger unnecessarily. It may be in the interest of the social development of her growing offspring for a female to associate with other mothers so that the young can play together. This is especially important for young males, if indeed the system is male retentive, because they will be mutually dependent in adulthood. Some adult females may be related to one another as closely as the males are, either because a young female never emigrated and continues to socialize with her mother or because the females present at Ngogo immigrated to it but were born and raised together in the same neighboring community. I suspect that all of these factors may influence female choice of companions, but this choice may occur partly by default: if males do not favor nonestrous females as companions, these females are left only with one another as companions.

Conclusions

Chimpanzee ecology at Ngogo was dominated by one major aspect of their habits: the exploitation of irregularly distributed patches of rare, ephemeral resources by these specialized frugivores in an environment where I believe there to be considerable interspecific competition for resources (Ghiglieri 1984). Interactions among individuals occurred within the fluctuating structure of their fusion-fission society: an individual adaptation for foraging efficiency. The composition and duration of traveling parties followed a protean and ephemeral pattern. Party size was dictated to some extent by the density, size, and distribution of food patches. Party membership was nonrandom with regard to age, sex, and reproductive class of members and conformed to a model of sexual selection and territoriality in a male-retentive system.

Ecologically the chimpanzee is very similar to the orangutan (Rodman, this volume), a highly frugivorous great ape closely related to chimpanzees (Miller 1977). Patterns of diet seem similar for both species, and competition with sympatric frugivores may be equally severe. Orangutans are, however, solitary except during mating periods (Rodman 1973; MacKinnon 1974). Rodman (1973, 1979) discussed an ecological sexual dimorphism that may be responsible for males' and females' foraging independently. Yet in other ways orangutans and chimpanzees seem to face similar problems. The primary difference between these two species is mobility, particularly the ease and efficiency of terrestrial locomotion (see Rodman, this volume, for an independent discussion of this). As a behavioral indicator, day ranges, when only solitary individuals are considered, are much longer among chimpanzees (Rodman, this volume). The superior mobility of chimpanzees ameliorates foraging constraints enough for individuals to associate daily during favorable seasons. Sexual selection and kin selection are the probable forces that shaped the pattern of like sexes' ranging together among chimpanzees.

References

Albrecht, H., and S. C. Dunnet. 1971. *Chimpanzees in Western Africa.* Munich: R. Piper.

Azuma, S., and A. Toyoshima. 1962. Progress report of the survey of chimpanzees in their natural habitat. Kabogo Point area, Tanganyika. *Primates* 3:61–70.

Brody, S. 1945. *Bioenergetics and Growth.* New York: Hafner.

Busse, C. D. 1977. Chimpanzee predation as a possible factor in the evolution of red colobus monkey social organization. *Evolution* 31:907–11.

Bygott, J. D. 1979. Agonistic behavior, dominance, and social structure in wild chimpanzees of the Gombe National Park. In Hamburg and McCown, eds. 1979:402–27.

Cole, L. C. 1954. The population consequences of life history phenomena. *Quart. Rev. Biol.* 29:103–37.

Emlen, J. M. 1973. *Ecology: An Evolutionary Approach.* Menlo Park, Calif.: Addison-Wesley.

Fouts, R. S., and R. L. Budd. 1979. Artificial and human language acquisition in the chimpanzee. In Hamburg and McCown, eds. 1979:374–92.

Gadgil, M., and W. H. Bossert. 1970. Life historical consequences of natural selection. *Amer. Natur.* 104:1–24.

Gaulin, S. J. C. 1979. A Jarman/Bell model of primate feeding niches. *J. Human Ecol.* 7:1–20.

Ghiglieri, M. P. 1984. *The Chimpanzees of Kibale Forest.* New York: Columbia University Press.

Glass, H. B. 1953. The genetics of the Dunkers. *Sci. Amer.* 189(2):76–81.

Goodall, J. 1965. Chimpanzees of the Gombe Stream Reserve. In I. DeVore, ed. *Primate Behavior: Field Studies of Monkeys and Apes,* pp. 425–73. San Francisco: Holt, Rinehart and Winston.

—— 1968. The behavior of free-living chimpanzees in the Gombe Stream Reserve. *Anim. Behav. Monogr.* 1:161–311.

—— 1971. *In the Shadow of Man.* Boston: Houghton Mifflin.

—— 1973a. The behavior of chimpanzees in their natural habitat. *Amer. J. Psychiat.* 130:1–12.

—— 1973b. Cultural elements in a chimpanzee community. *Symp. IVth Int. Congr. Primat.* 1:144–84.

—— 1977. Infant killing and cannibalism in free-living chimpanzees. *Folia primatol.* 23:259–82.

—— 1979. Life and death at Gombe. *Nat. Geogr.* 155:592–621.

Goodall, J., A. Bandoro, E. Bergmann, C. Busse, H. Matama, E. Mpongo, A. Pierce, and D. Riss. 1979. Intercommunity interactions in the chimpanzee population of the Gombe National Park. In Hamburg and McCown, eds. 1979:13–54.

Gunther, M. 1971. *Infant Feeding.* Harmondsworth, England: Penguin.

Halperin, S. D. 1979. Temporary association patterns in free ranging chimpanzees. In Hamburg and McCown, eds. 1979:490–99.

Hamburg, D. A., and E. R. McCown, eds. 1979. *The Great Apes.* Menlo Park, Calif.: Benjamin/Cummings

Hamilton, W. D. 1964. The genetical theory of social behavior, I, II. *J. Theoret. Biol.* 12:1–52.

—— 1972. Altruism and related phenomena, mainly in social insects. *Ann. Rev. Ecol. and Syst.* 3:193–232.

Hladik, C. M. 1977. Chimpanzees of Gabon and chimpanzees of Gombe: Some comparative data on the diet. In T. H. Clutton-Brock, ed., *Primate Ecology: Studies of Feeding and Ranging Behavior in Lemurs, Monkeys, and Apes,* pp. 481–501. London: Academic Press.

Hubbell, S. 1979. Tree dispersion, abundance, and diversity in a dry tropical forest. *Science* 203:1299–1309.

Huxley, T. H. 1959. *Man's Place in Nature.* Ann Arbor: University of Michigan Press.

Izawa, K., and J. Itani. 1966. Chimpanzees in Kasakati Basin, Tanganyika. (I) Ecological study in the rainy season 1963–1964. *Kyoto Univ. Afr. Stud.* 1:73–156.

Janzen, D. H. 1979. How to be a fig. *Ann. Rev. Ecol. Syst.* 10:13–52.

Jones, C., and J. Sabater Pi. 1971. Comparative ecology of *Gorilla gorilla* (Savage and Wyman) and *Pan troglodytes* (Blumenback) in Rio Muni, West Africa. *Bibl. primatol.* No. 13. Basel: S. Karger.

King, M. C., and A. C. Wilson. 1975. Evolution at two levels in humans and champanzees. *Science* 188:107–16.

Kingston, B. 1967. Working plan for Kibale and Itwara Central Forest Reserves. Forest Department. Entebbe, Uganda.

Kortlandt, A. 1962. Chimpanzees in the wild. *Sci. Amer.* 206(5):128–38.

MacArthur, R. H., and E. R. Pianka. 1966. On optimal use of a patchy environment. *Amer. Natur.* 100:603–9.

MacKinnon, J. 1974. The behaviour and ecology of wild orang-utans *Pongo pygmaeus. Anim. Behav.* 22:3–74.

Marler, P., and L. Hobbett. 1975. Individuality in a long-range vocalization of wild chimpanzees. *Z. Tierpsychol.* 38:97–109.

McKay, D., P. G. Waterman, C. N. Mbi, J. S. Gartlan, and T. T. Struhsaker. 1978. Phenolic content of vegetation in two African rain forests: Ecological implications. *Science* 202:61–64.

McNab, B. 1963. Bioenergetics and the determination of home range size. *Amer. Natur.* 97:133–40.

Miller, D. A. 1977. Evolution of primate chromosomes. *Science* 198:1116–24.

Nishida, T. 1968. The social group of wild chimpanzees in the Mahali Mountains. *Primates* 9:167–224.

—— 1970. Social behavior and relationship among wild chimpanzees of the Mahali Mountains. *Primates* 11:47–87.

—— 1979. The social structure of chimpanzees of the Mahale Mountains. In Hamburg and McCowan, eds. 1979:73–122.

Nissen, N. W. 1931. A field study of the chimpanzee. *Comp. Psychol. Monogr.* 8:1–122.

Pusey, A. 1979. Intercommunity transfer of chimpanzees in Gombe National Park. In Hamburg and McCown, eds. 1979:465–80.

Reynolds, V. 1963. An outline of the behavior and social organization of forest-living chimpanzees. *Folia primatol.* 1:95–102.

—— 1975. How wild are the Gombe chimpanzees? *Man* 10:123–25.

Reynolds, V. and F. Reynolds. 1965. Chimpanzees in the Budongo Forest. In I. DeVore, ed., *Primate Behavior: Field Studies of Monkeys and Apes,* pp. 468–524. San Francisco: Holt, Rinehart and Winston.

Rodman, P. S. 1973. Population composition and adaptive organisation among orang-utans of the Kutai Reserve. In R. P. Michael and J. H. Crook, eds. *Ecology and Behaviour of Primates,* pp. 171–209. London: Academic Press.

—— 1979. Individual activity patterns and the solitary nature of orangutans. In Hamburg and McCown, eds. 1979:235–56.

Sarich, V. M., and J. E. Cronin. 1976. Molecular systematics of the primates. In M. Goodman and R. Tashian, eds. *Molecular Anthropology,* pp. 141–70. New York: Plenum.

Struhsaker, T. T. 1975. *The Red Colobus Monkey*. Chicago: University of Chicago Press.

Sugiyama, Y. 1968. Social organization of chimpanzees in the Budongo Forest, Uganda. *Primates* 9:109–48.

—— 1969. Social behavior of chimpanzees in the Budongo Forest, Uganda. *Primates* 10:197–48.

Taylor, C. R., K. Schmidt-Nielson, and J. L. Robb. 1970. Scaling of energetic cost of running to body size in mammals. *Amer. J. Physiol.* 219:1104–7.

Trivers, R. L. 1971. The evolution of reciprocal altruism. *Quart. Rev. Biol.* 46:35–57.

Tutin, C. E. G. 1975. Sexual behaviour and mating patterns in a community of wild chimpanzees (*Pan troglodytes*). Ph.D. dissertation. Edinburgh: University of Edinburgh.

Wing, L. D., and I. O. Buss. 1970. Elephants and forests. *Wildl. Monogr.* No. 19. The Wildlife Society.

Wrangham, R. W. 1974. Artificial feeding of chimpanzees and baboons in their natural habitat. *Anim. Behav.* 22:83–93.

—— 1975. The behavioural ecology of chimpanzees in Gombe National Park, Tanzania. Ph.D. dissertation. Cambridge, England: University of Cambridge.

7 Ecological Differences and Behavioral Contrasts Between Two Mangabey Species

PETER M. WASER

FORAGING BEHAVIOR and patterns of individual movement evolve, not in isolation, but in concert with such traits as body size, morphology, digestive physiology, and foraging group size and stability; selection acts on these traits through their effects on individuals' schedules of survivorship and fecundity. Yet, to make headway through the tangled web of relationships among these traits, we must somehow limit the number of variables, considering them only a few at a time.

One way of doing this is to consider some characteristics as fixed, to a first approximation at least. These can then be viewed as setting constraints within which other, more labile traits can evolve. For instance, we can take gut morphology as the independent variable and generate hypotheses concerning diet diversity, assuming that forag-

PETER WASER has worked in East Africa on various problems of behavioral ecology of mammals. In reporting his results, he has consistently applied creative analytical techniques through use of models drawn from diverse disciplines. Here he examines data on two species of mangabey, his own observations of *Cercocebus albigena* of the Kibale Forest, Uganda, and observations by Katherine Homewood on *C. galeritus* of the Tana River, Kenya. The results show some very clear differences in habitat and use of habitat between the two species, as well as some striking similarities in feeding, ranging, and spacing. Waser discusses these results and their implications for understanding the roles of phylogeny and environment in shaping behavior.

ing behavior is relatively labile and should be optimized within the constraints set by digestive physiology.

The problem then becomes one of deciding which aspects of foraging or ranging we view as the constraints. Is diet diversity fixed and pattern of movement a trait that must be modified to attain the right mix of food items—or vice versa? In this paper, I address this sort of question empirically, by examining the differences in foraging and ranging behavior of two closely related primates in radically different habitats. I compare aspects of feeding and ranging behavior in the two best known forest species of the African Papionini, *Cercocebus albigena* and *C. galeritus.* These two mangabey species share many details of social, as well as foraging, behaviors; biochemical studies suggest they have only recently diverged from each other and from *Papio* (Barnicot and Hewett-Emmett 1972; Cronin and Sarich 1976).

Study Areas, Subjects, and Methods

The data were collected during long-term observations of *C. albigena johnstoni* in the Kibale Forest, western Uganda (Waser 1974, 1975, 1976, 1977a, b, and unpublished data; Wallis 1979) and *C. galeritus galeritus* in the Tana River Forests of eastern Kenya (Homewood 1975, 1976, 1978). Methods in the two studies, including the categories of behavior recorded, were designed to allow direct comparison between them. The basic method of data acquisition was half-hourly scan sampling of the location, identity, and activity of all visible individuals in habituated groups followed continuously for 5–10 days monthly. Phenological data were recorded monthly from samples of marked trees, and spatial distribution of trees was determined from transects or quadrats throughout the study areas. Homewood (1978) and Waser (1977b) describe their methods in detail.

Both studies involved roughly one year of systematic observation, with short-term follow-up visits in subsequent years. Observations in each study are were concentrated on two groups: for *C. albigena* these were the 15-member "M" group (modal composition 3 adult males,

6 adult females, 2 subadult males, 4 juveniles and infants) and the 6-member "S" group (2 adult males, 2 adult females, 2 juveniles and infants). For *C. galeritus,* they were the 17-member "LD" (Mnazini N) group (2 adult males, 6 adult or subadult females, 2 subadult males, 7 juveniles and infants) and the 36-member "HD" (Mchelelo N) group (4 adult males, 16 adult or subadult females, 4 subadult males, 12 juveniles and infants). The two *C. albigena* groups inhabited overlapping home ranges; the two *C. galeritus* groups inhabited different forest patches, of which the Mchelelo (HD) area was considerably more productive and supported a much higher mangabey population density. Except where noted, comparisons are between the Kibale M and Tana LD groups, owing to their similar size and composition. Data from other groups or from the literature (particularly additional Kibale groups studied by Wallis 1979) are noted where available.

The two mangabey species are similar in weight and body length (Kingdon 1971; Napier and Napier 1967), but *C. albigena* is more slender in build and has a considerably longer tail, which it often uses as a balance or brace in arboreal locomotion. Field data confirm the generalization reflected by these anatomical differences, that members of the *albigena/aterrimus* species group are far more arboreal than those of the *torquatus/galeritus* group. In more than 18,000 individual scan samples, *C. albigena* was not once recorded on the ground; though these mangabeys did descend to drink, cross gaps in the canopy, or forage, they tended to avoid heights below 10 m (Waser 1977b). In contrast, *C. galeritus* was recorded on or within 2 m of the ground in 72.8% of all feeding observations (median of monthly values: Homewood 1978).

The areas used by the study groups differ strikingly in climate and vegetation. The Kibale is classified as moist evergreen forest, with an annual rainfall of 1475 mm (mean of 52 years' data, Kingston 1967). There are nominally two wet seasons annually, but rainfall is well distributed and peak rainfall months vary greatly from year to year. During the 12 months in which most of the *C. albigena* data were collected, rainfall at the study site totaled 1,658 mm, and the longest rainless period lasted only 10 days (Waser 1975). The relatively even distribution of rainfall throughout the year and the equatorial location of this area release many tree species from a synchronized annual pattern of phenology (Struhsaker 1975; Waser 1975). Of the five tree species most commonly used for food (primarily fruit) by *C. albigena,* the top three (*Diospyros abyssinica, Celtis durandii,* and *Ficus brachylepis*) have available fruit at least 10 months of the year; the fourth (*Pancovia turbinata*) fruits highly synchronously but at intervals of several years; and the fifth (*Ficus exasperata*) has a single annual fruiting peak. The study area as a whole shows a "predominance of phytophasic asynchrony" (Struhsaker 1975:134), a characterization that is particularly apt for those phytophases (especially fruit) of importance to mangabeys.

In the Kibale study site, forest is continuous over many square kilometers and vegetationally diverse; 67 tree species more than 10 m in height were identified in 5.78 ha of transect (Waser 1977b). In a sample of 1.43 ha, Struhsaker (1975) reports 51 species and an average density of 328 trees/ha.

In contrast, the Tana River Forests have been described as "patches of woodland, each of a few dozen hectares, linked by stretches of bush and grassland" (Homewood 1978:376). The Tana forests are lower in stature with a less continuous canopy than Kibale; individual forest patches are also less diverse, only 39 tree species of any size being enumerated in 0.78 ha in the Mnazini (LD) study area. The forest patches themselves are often zoned with large areas heavily dominated by single tree species (Homewood 1976). As in Kibale, there are two wet seasons annually, but rainfall is much lower and far more seasonally distributed. During the main *C. galeritus* study, rainfall was 396 mm, two-thirds falling in November and March–May. The forests are also inundated when the Tana River floods following the wet season in its upper catchment. As a result, fruiting is quite synchronous in many Tana trees, and fruit abundance peaks biannually during each wet season (Homewood 1978).

The dominant species in the LD group's diet, *Ficus sycamorus,* produces fruit year-round, but the second- and fourth-ranked species (*Phoenix reclinata* and *Saba florida*) fruit synchronously in both rainy seasons, whereas *Mimusops fruticosa* and *Acacia robusta* have well-defined annual fruiting peaks.

C. galeritus in the Tana forests regularly attain much higher population densities than Kibale *C. albigena* do; the Kibale supports less than 10 mangabeys/km^2, while parts of the Tana support several hundred. Even in the low-density Mnazini area there are roughly 33 *C. galeritus*/km^2. This seems unlikely to be due solely to differences in the intensity of interspecific competition, since the most important competitors, *Cercopithecus* spp. and other omnivorous primates, occur at densities exceeding 100/km^2 in both forests (Struhsaker 1975; Harcourt, unpublished ms.). The Tana forests are at much lower altitude (ca. 30 m vs. ca. 1,500 m) and have a correspondingly warmer climate (mean minimum temperature 22° C vs. 13° C); this, and the relatively fertile riverine soils, might contribute to their higher productivity. Another factor, possibly more important, derives from the stability of the two forest environments. Since the Tana River often changes its course, the vegetation "represents a dynamic mosaic in all stages of colonization . . . (and) the plant species are generally those typical of riverine and colonizing situations" (Homewood 1978:376). Homewood has argued that such plants, like colonizing species in general, allocate relatively more energy to reproductive parts (i.e., parts eaten by mangabeys) and invest less in their chemical defense than species do in more stable, saturated environments. Comparative data on trees' allocation of resources are not available, but it is known that the fruits of many Tana trees, such as palms and figs, are palatable to humans, whereas those of most Kibale trees are bitter, astringent, or otherwise distasteful (Homewood 1978; Waser 1977b). Thus the higher mangabey population density in the Tana River forests might reflect the higher productivity of reproductive parts of colonizing tree species.

Results

Feeding Behavior

Mangabey feeding behavior has been described in detail by Homewood (1978) and Waser (1975, 1977b). Both species spend similar proportions of time feeding. Eating and "foraging" (scanning substrates for food items) occupied *C. albigena* in 42.5% of activity scans;

C. galeritus were engaged in these activities 49% of the time. After the approximately two-thirds of insect feeding and foraging samples in which prey were not captured (Waser 1977b) are subtracted, *C. albigena* were observed to be eating 35% of the time, precisely the figure recorded for *C. galeritus.*

Both species use similar motor patterns in feeding and foraging; the similarities were particularly striking when mangabeys searched through rotten wood or dead leaves and sticks for invertebrate prey or in their handling of hard-walled or otherwise difficult-to-open fruits. Substrates for equivalent foraging methods, on the other hand, differed strongly, with *C. albigena* searching for invertebrates primarily on or inside branches, twigs, and epiphytes; *C. galeritus,* predominantly in leaf litter on the ground. *C. galeritus* also exploited fallen fruit and seeds, which *C. albigena* did not. Some rather specialized foraging techniques were used with quite different food types: *C. galeritus* harvested grass seeds "by running the cupped hand along dry, seed-laden heads" (Homewood 1976: 272), a technique that *C. albigena* applied to the midribs of young compound leaves to strip leaflets from *Millettia* and other trees.

Both mangabey species choose similar plant parts. Fruit and seeds compose 59% of *C. albigena* foraging observations, and 58% of those of *C. galeritus* (see Homewood 1978). In both species' diet fleshy fruits are more common than seeds. Fruits of the genera *Ficus* and *Diospyros* are conspicuous in both diets, and *Ficus* are the most highly preferred fruit for both species. Of the seven additional genera shared in common between the Kibale and Tana study areas, both Kibale M and Tana LD group mangabeys ate fruit of five (*Albizza, Blighia, Cordia, Mimusops,* and *Trichilea,* but not *Celtis* or *Markhamia*).

Young leaves and floral parts each accounted for 5% or less of each species' foraging observations. Neither species was observed to eat mature leaves. *C. albigena* feeds to a significant extent on the inner bark of several tree species, a mode of foraging less common in *C. galeritus.* On the other hand, roughly 10% of the *C. galeritus* diet consists of shoots, particularly of grass, a resource unavailable to an arboreal rain forest species.

Invertebrate foraging accounts for 25%–30% of foraging time in both species. Both ate eggs and small vertebrates (snakes, lizards, frogs, squirrels), although they rarely found them.

Feeding by the *C. galeritus* LD group was recorded on 61 plant species, excluding fungi, while the *C. albigena* M group fed on 63 (additional species were used as substrates for invertebrate foraging).

Over the year, the *C. albigena* diet is less dominated by particular species: the top-ranked five species contributed 25% of the *albigena* diet (M group) vs. 57% for *galeritus* (LD group). The list of five most heavily used species each month totals 21 different species over the year for *C. albigena;* only 16 species make the list for *C. galeritus.* These differences could reflect the higher tree species diversity and more complex phenological patterns in the Kibale forest, where the array of tree species in fruit changes more from month to month.

Over the short term, the picture is somewhat different. The median number of plant species utilized monthly was 22 (range 14–27) for the Tana LD group, and 17 (range 13–26) for the Kibale M group. The top five species each month accounted for a median of 78% (range 59%–89%) of the monthly plant diet for the LD group, 84% (range 74%–93%) for the M group. Thus, despite their simpler habitat, the Tana mangabeys manage to attain a monthly diet diversity equal or greater than that in Kibale. *C. galeritus* increase their short-term dietary diversity above what might be expected from the limited number of Tana tree species by exploiting a variety of plant life forms little used by *C. albigena,* including shrubs, climbers, herbs, and especially grasses.

To summarize, these two mangabey species maintain diets that are extremely similar in composition and short-term diversity. Striking differences in habitat do not prevent the expression of preferences for similar items and plant species, though they modify long-term diet diversity and force mangabeys to apply their similar foraging techniques to somewhat different substrates.

Ranging Behavior

C. albigena groups tend to move slightly farther per day than those of *C. galeritus* do. When measured by comparable means, daily distances traveled averaged 1270 m (range 580–2250 m) for *C. albigena,* 1098 m (range 615–1480 m) for *C. galeritus.* More striking were differences in the *pattern* of daily travel. Straight-line distances between nightly sleeping sites are influenced by the tortuosity of a group's path, as well as their rate of movement; by this measure, *C. albigena* move much farther (M group mean = 510 m, range 40–1360 m, *N* = 86) than *C. galeritus* do (LD group mean = 290 m, range 25–950 m, *N* = 62). Over the day, *C. galeritus* apparently backtrack more than *C. albigena,* a trend that must be even more prominent on a longer time scale, since *C. galeritus* remain within much smaller annual home ranges (53 vs. 410 ha, Tana LD vs. Kibale M group).

Over short intervals, however, the path characteristics of both species are quite similar. Ranging data throughout the year can be analyzed to determine the proportion of half-hour periods in which the group continued to move ahead (≤90° from the direction moved the half-hour before) or backtracked (>90°). From half-hour to half-hour, the *galeritus* LD group backtracked in 30.5% of all measurements, the *albigena* M group in 29.6%. A still finer scale comparison is possible between the M group and a sample of data from the Tana HD group during June-July 1976 (Homewood and Waser unpublished data). This group tends to backtrack more frequently than the LD group (see discussion), so that the comparison should emphasize the differences between species. Nevertheless, the distributions of angles of turn are rather similar (fig. 7.1). From scan to scan, groups of both species tend to continue straight ahead—they act as if they have inertia. The angles of turn of *C. albigena* are slightly more evenly distributed, but on this time scale, the differences are clearly minor.

In this sample, the mean distance moved per half-hour was slightly larger for *C. albigena* (56 m vs. 48 m). The main basis of the difference appears to be in the tail of the distribution of half-hourly distances (fig. 7.2); *C. albigena* groups could move as much as 50% farther. In addition *C. galeritus* rates of movement were somewhat less variable than those of *C. albigena;* when *C. galeritus* groups were moving, they were more likely to move at a steady, intermediate rate.

In summary, viewed from half-hour to half-hour, the search patterns of both species seem similar. However, viewed over a period of a day, differences emerge: *C. galeritus* returns to recently used areas more frequently. This trend becomes even clearer when viewed over a period of weeks. *C. albigena* tends to forage for up to a week

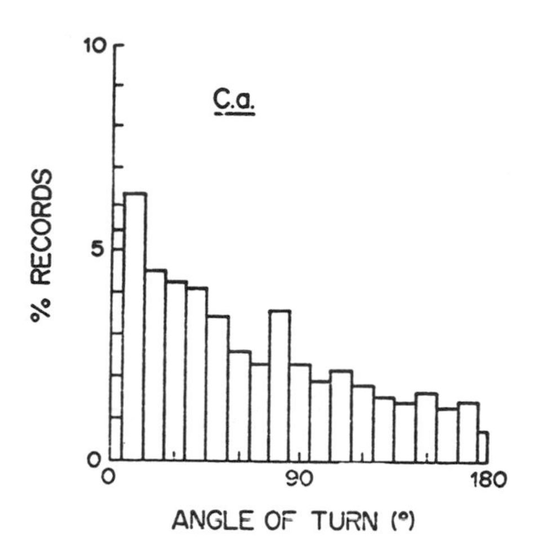

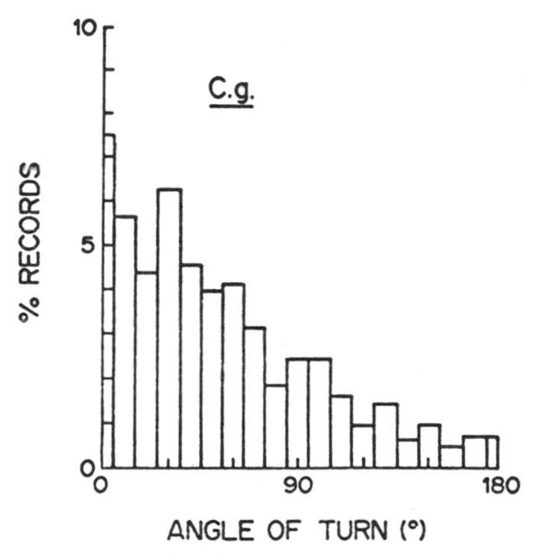

Figure 7.1. Angles of turn between half-hourly steps. *Cercocebus albigena* (top) and *C. galeritus* (bottom). Individuals' locations were mapped in half-hourly scans, and a center-of-mass location was calculated for each scan in which more than half of the group was visible. Measurement of an "angle of turn" required three consecutive scans with half or more of the group visible; if progression from scan to scan continued in a straight line, angle of turn was 0°. If the group was stationary over a half hour (i.e., center of mass moved less than 5 m), measurements were discarded. Data from dawn-1000 h and 1500 h-dark only were used, excluding the midday period of inactivity. Data are from the Kibale M group (120 days distributed across the year) and the Michelelo HD group (10 days in June–July). Homewood's (1976) more extensive observations indicate that June–July movements are not significantly different from those in other months.

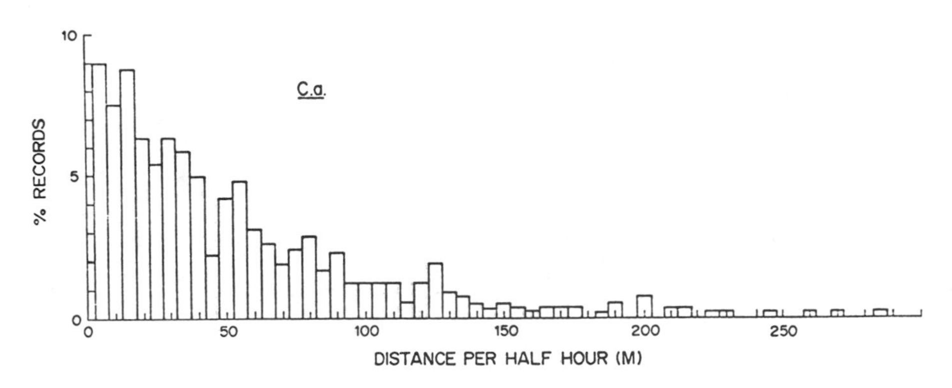

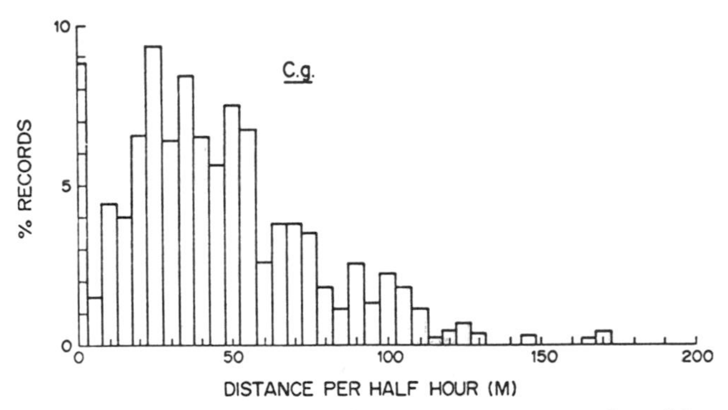

Figure 7.2. Rates of half-hourly movement in *Cercocebus albigena* (top) and *C. galeritus* (bottom). Data were collected as for figure 7.1.

or so within a few tens of hectares, usually centered on a large fruiting tree; often it then moves to an entirely new area. As a result, cumulative home range size measured for a *C. albigena* group grows in steps as local food patches are exploited and abandoned and can continue to grow even after a year of observation (Waser and Floody 1974; Waser 1976, 1977b). In contrast *C. galeritus* groups remain within their few tens of hectares year-round, using a significant fraction of the total area daily.

Group Size

Counts are available for 5 groups of *C. galeritus* and 15 of *C. albigena* (Homewood 1976; Wallis 1979; Waser, unpublished data). Some *C. galeritus* groups tend to fragment and may be joined only part-time by certain adult males (see below); their counts represent modal group size.

The ranges of observed group sizes overlap broadly, but the mean *C. galeritus* group (24.6 individuals, range 17–36) is nearly twice the size of the mean group of *C. albigena* (14.2 individuals, range 6–28: $p < .01$, $t = 3.16$, d.f. $= 18$).

Intragroup and Intergroup Dispersion

Within groups, individuals of both species tend to be widely scattered in space. A crude index of intragroup dispersion, the group's maximum dimension or "spread," confirms that members of Kibale *C. albigena* groups forage, move, and even sleep in a widely scattered array. Maximum spread of the M group was 270 m in 1972–73 ($\bar{X} = 90$ m). Group spread varies with group size and composition; during 1971 pilot studies, when the Kibale M group contained 2–3 additional males, the maximum dimension of the group ran as high as 480 m ($\bar{X} = 125$ m). Nevertheless, spread could be as high as 105 m ($\bar{X} = 40$ m) even in the 6-member Kibale S group (Waser and Floody 1974; Waser 1976).

C. galeritus group spread is comparable or greater. Homewood describes individuals in her LD group as spread over "50–500 m or more" (Homewood 1976:97).

The species appear to differ, however, in the *pattern* of dispersion of individuals within groups. Roughly 90% of the *C. galeritus* observed during scans had at least one neighbor within 5 m (Homewood, personal communication). In contrast, half of the *C. albigena* adults recorded in nearest neighbor scans had none within 5 m (table 7.1). The greater group spread and lower nearest neighbor distances of *C. galeritus* suggest that they are more often spatially subdivided than groups of *C. albigena*. Within groups, *C. galeritus*

Table 7.1. *C. albigena:* Proportion of Adults Recorded "Alone" (No Nearest Neighbor ≤ 5 m)

	Sightings Without Neighbors ≤5 m	Number of Sightings	Percent "Alone"
Adult females (6)	214	429	50
Adult females (3)	86	212	41
Subadult males (2)	90	149	60
Total	390	790	Mean: 49

individuals show a tendency to move in independent subgroups. In larger groups, such dissociation may be the rule (Homewood 1976). *C. albigena* groups may also split temporarily, but these subgroups generally form only to exploit medium-sized, separated fruiting trees and last only a few hours. Subgroups in *C. galeritus* are most frequent when fruiting trees are least common and may persist for days.

Not only subgroups but also individuals isolated by 100 m or more from the nearest group are sighted in both species. These are almost always males and are more common in *C. galeritus.* In 44 censuses 4 km long in the Kibale study area, Struhsaker (1975) sighted only two solitary mangabeys. Those isolated males that are observed are usually (though not always) group members temporarily foraging alone. Any of the three adult males in the M group might sometimes leave for a day or longer, during which time it could be sighted as much as 2 km from its group. The phenomenon is not restricted to large groups; in the S group, one of the two adult males was sighted more than 50 m from his nearest neighbor in nearly 10% of all scans. However, I sighted all 3 M group males in the group at least once on 88% of the days I followed it continuously, a proportion that showed no obvious seasonal variation.

In *C. galeritus,* isolated males are much more common, and the status of these animals is more complex. Data from a large Tana group, Mchelelo S, in June-July 1976 are illustrative (Homewood and Waser, unpublished data). One low-status male (L) was sighted in the body of this group on 22/23 of the days during which extensive observations were possible. The dominant male at the initiation of this study (FR) was present on at least 20/23 days. In early July, a new male (BT) immigrated and rapidly became dominant; he was present on every day following his initial detection. Five other adult males of intermediate status were observed in the group on 1–15 days only. Two of these five males were also sighted in other groups (one day each). The total number of adult males in the group during complete counts ranged from 4 to 9.

The HD group showed a similar pattern: on 22 days in 1976 with extensive observation, the dominant male (TR) was seen in the group on all 22. Two other males were sighted on 14 and 16 days; one of these visited at least two neighboring groups. The number of males seen in group counts varied from 2 to 4.

The smaller LD group, 2 males were tightly associated with the group both in the main study and in 1976. Nevertheless, additional males were sometimes seen on the group's periphery; in 1974, one of these eventually became the dominant after spending 6 months as an occasional visitor.

Thus it appears that, especially in high-density *C. galeritus* populations, many males are only loosely attached to social groups, sometimes moving independently or with neighboring groups. Isolated males include such individuals, as well as males in the course of permanent emigration; indeed, there may be no sharp distinction between these two extremes.

To summarize, groups of both species are spatially diffuse, but dispersion of individuals within groups is more variable in *C. galeritus,* among which both isolated individuals and discrete subgroups are more commonly seen.

Greater flexibility in *C. galeritus* grouping tendencies is also visible between groups. In both species, groups generally avoid each other (Waser and Homewood [1979] describe intergroup spacing more extensively), but in *C. galeritus,* this tendency is sometimes masked by the formation of temporary intergroup associations. Groups have been observed to forage and travel in close (<100m) proximity, sometimes with intermingling of members, for up to 2 days; most associations, however, last only a few hours. Such associations are most common in high-population-density areas. They may occur at any time of year but are particularly common during the wet season (Homewood 1976; Homewood and Waser, unpublished data).

Discussion: The Effects of Ecological Factors

The Tana and Kibale mangabeys differ ecologically in three clear-cut ways: (1) *C. albigena* is almost completely arboreal, *C. galeritus* primarily terrestrial; (2) *C. albigena* habitats are not strikingly seasonal, while those of *C. galeritus* show two strong fruiting peaks and two dry seasons annually; and (3) *C. galeritus* lives in forests more productive of exploitable food than *C. albigena* does and thus maintains a higher population density. The possible effects of each of these differences on diet, ranging, and spacing will be discussed in turn. Interspecific comparisons narrow the range of possible causal relationships (although the available data remain consistent with a number of alternatives) and suggest the sorts of ecological data required to test more explicit hypotheses. As with any attempt to infer the ultimate determinants of behavior from natural ecological variation, further effort ought best utilize additional mangabey populations in areas that share some but not all of the ecological characteristics facing the populations compared here.

Arboreality vs. Terrestriality

An arboreal existence, compared to a terrestrial one, clearly places different constraints on locomotion and, as a consequence, might influence abilities to manipulate food. Arboreality might be thought to restrict foraging to more discrete food sources (trees) that occur as a mosaic of different patch types (species). A mode of foraging that is likely to be energetically more demanding, it would also be expected to give access to categories of food items unavailable on the ground.

Available data do not, however, demonstrate clear differences attributable to most of these possible mechanisms. The more gracile build of *C. albigena* might indeed reflect different locomotory and manipulative capabilities, but (although these studies were not designed to examine motor patterns in any detail) it is the similarities that are subjectively striking. Both species tend to choose similar high-energy diets (Homewood 1978). With the exception of grass, neither species uses food types inaccessible to or rejected by the other.

Though degree of arboreality has surprsingly little influence on diet, the contrasts in mangabey ranging patterns could reflect differences in arboreal and terrestrial resource characteristics. For instance, a more uniform dispersion of food on the forest floor might favor more uniform rates of movement. Although both species exploit trees for fruits, *C. albigena* is also dependent on trees for invertebrate foraging, while *C. galeritus* uses the ground.

C. galeritus nearest neighbor distances increase during months of extensive invertebrate foraging (Homewood, 1976), a finding consistent with the possibility that terrestrial invertebrates are more evenly distributed than arboreal foods. One might expect that *C. galeritus* can forage at a relatively constant rate across large continuous areas, while *C. albigena* must move in more discrete steps from tree to tree. The low modal rate of movement by *C. albigena* groups (fig. 7.2) may then document within-tree foraging, and the long tail of rapid movements could reflect the occasional occurrence of large gaps between suitable trees.

Differences between arboreal and terrestrial substrates for foraging might also influence rates of backtracking. Homewood (1976) has suggested the pattern of backtracking by *C. galeritus* is linked to invertebrate foraging. Two lines of evidence support this contention: a seasonal correlation between backtracking frequency and percentage contribution of insects to the diet and a tendency for groups in areas of high fruit density to maintain travel patterns similar to those in low-fruit-density areas. I have previously argued that *C. albigena* rates of return to specific areas might also reflect insect renewal characteristics (Waser 1977b). Despite the fact that ecologists have often postulated backtracking rates to reflect resource renewal rates (Cody 1974; Clutton-Brock 1974; Gill and Wolf 1977), measurements of the necessary variables are virtually nonexistent. One might predict that resource renewal characteristics should affect not only foraging path characteristics but also home range size and overlap (Waser 1981; Waser and Wiley 1979).

Finally, larger and/or more flexible group sizes in *C. galeritus* may be allowed because a substantial proportion of foraging involves a uniformly distributed resource. Bradbury and Vehrencamp (1976), Kruuk (1978), and others have argued on theoretical grounds that group size should reflect the size of resource patches, a viewpoint early advocated for primates by Kummer (1971). Rates of daily movement in *C. albigena* increase with group size, a finding consistent with the hypothesis that larger groups must feed on more patches (trees). The similarity of food types exploited year-round by *C. albigena* suggests that such a constraint would act continuously. If mangabey group size is indeed limited by patch size, then the terrestrial habit might partially release *C. galeritus* from such constraints. The distribution of invertebrates on the ground seems unlikely to limit group size, and *C. galeritus* forage for these prey more intensively when food is least abundant—i.e., when selection on group size might otherwise be expected to be most intense. This being the case, the size of social groups should be free to increase beyond the numbers any single tree could contain although members could forage in temporary subgroups when the dispersion of arboreal food sources was such that this would increase foraging efficiency.

Seasonality

The relatively more seasonal climate and phenology of the Tana has pervasive effects on *C. galeritus* time budgets, diets, and ranging patterns. Time spent eating per se is roughly constant year-round, but during the dry season *C. galeritus* diversify their diet and spend more time foraging or searching for food items (Homewood 1978). These trends are those predicted by most variants of optimal foraging theory (e.g., Krebs 1979) and might be responsible for the somewhat higher overall proportion of foraging and eating in the time budget of the Tana LD compared with the Kibale M group.

Daily distance moved, nearest neighbor distances, and group spread, as well as diet diversity and foraging time, vary seasonally in the Tana LD group, increasing significantly in months with low fruit availability (Homewood 1976, 1978). No seasonal variation was detected in *C. albigena* diet characteristics, time budgets, or ranging patterns (Waser 1975).

Seasonality also influenced *C. galeritus* intergroup association tendencies. Intergroup mingling occurred primarily (though not exclusively) during periods of high fruit availability. Such a seasonal increase in intergroup tolerance might reflect food abundance either directly or indirectly (since births are seasonal and months with high fruit availability tend to be those with few opportunities for male-male reproductive competition).

Productivity

The higher population density (relative to *C. albigena*) that *C. galeritus* attains reflects both its smaller home ranges and its larger mean group size. Are both of these caused by a greater density of harvestable food in the Tana forests?

Homewood's (1976, 1978) intraspecific comparisons in the Tana forests elucidate this question. She determined relative productivity, as measured by cumulative percent cover of plant species used by mangabeys, in several groups' home ranges. Home range size for her study groups was smallest (17 ha) for the HD group, in the most productive forest patch; intermediate (53 ha) for the LD group; and largest (at least 100 ha) for the Mnazini S group, which used the least productive area. In both areas, mangabeys used larger areas during the dry than during the wet season. These trends are consistent with theoretical expectations (e.g., "the essential, slowly renewing resources whose distributions are sparsest relative to the needs of the animals set a lower limit on home range size," Altmann 1974:233). A comparable pattern has been reported for *C. albigena:* home range size is somewhat smaller in the southern Kibale forest, where population density is higher (Wallis 1979), and much smaller in eastern Ugandan forest patches, where mangabey density is extremely high, forests are secondary and probably very productive, and most interspecific competition is absent (Chalmers 1968).

The foraging path characteristics that underlie home range size differences also show intraspecific variation; straight-line distances between successive sleeping site locations are much shorter for the HD group ($\bar{X}$ = 140 m, range 20–465 m), in high-productivity habitat, than for the LD group ($\bar{X}$ = 290 m, range 25–950 m). Backtracking is also somewhat more frequent in the HD group (Homewood, 1976).

Patterns of intraspecific variation in group size are less clear. Larger *C. galeritus* groups inhabit both high- and low-productivity habitats, although the Mnazini S group (low-productivity forest) was typically dissociated into small, loosely coordinated subparties. On the other hand, group sizes are larger in the higher *C. albigena*-density areas studied by Wallis (1979: mean of 8 groups = 14.8 mangabeys) and Chalmers (1968: 1 group = 17 mangabeys) than in lower density areas (mean of 5 groups inhabiting disturbed areas of Kibale = 10.8 mangabeys).

While these findings could indicate that productivity per se influences home range size, travel routes, and group size, it is premature to conclude that differences between species reflect only this resource parameter. The Tana forests tend to be zoned in structure, with local areas dominated by small numbers of tree species. Combined with increased fruiting synchrony, zonation results in fruits being available in larger patches in the Tana forests. And, as has been indicated, invertebrates on the ground may be more evenly distributed than those confined to epiphyte-rich trees.

To disentangle the effects on behavior of food density, dispersion, and other resource characteristics, interspecific comparisons will require substantially more precise data on patch size, interpatch distance, and renewal rates of food within patches than are available in most contemporary field studies. To illustrate the problems of inference from "productivity" data alone, consider how an increase in productivity might be expected to influence group size. An increase in the rate with which fruit crops are produced, or an increase in the density of trees, would leave the number of individuals that could forage simultaneously in a patch the same. Only if fruit crops became larger per tree, and then only if monkeys reduced interindividual distances while foraging, would more animals be able to forage together without exploiting extra patches, thereby allowing group size to increase.

Conclusion

Foraging Behavior, Adaptation, and Phylogenetic Inertia

Despite the presence of what appear to be strikingly different selective regimes in the Kibale and Tana forests, many attributes of mangabey feeding, ranging, and spacing do not differ markedly. As discussed in the introduction, these attributes may be productively viewed as constraints, within which selection may shape other aspects of foraging.

Stable attributes of mangabey foraging behavior include foraging motor patterns, proportions of food types in the diet, and monthly diet diversity. The two species maintain similar diets, though they must alter ranging patterns to do so. Motor patterns in foraging remain the same between species, but foraging heights and foraging substrates differ. Over short periods of time, mangabeys manage to obtain diets of similar evenness and diversity despite contrasting habitats; over the year, differing spatiotemporal resource distributions impose differences in these dietary characteristics.

In ranging patterns, too, attributes on a fine scale are stable but are combined, in different environments, into different higher order patterns. Local patterns of food search—hour-to-hour rates of movement and patterns of turn—are similar in both species. Differences occur in larger scale concomitants of movement: rates of return to previously utilized areas and size of the home range. Finally, while the tendency to spread out and forage in a spatially extended array is common to both species, the details of individuals' dispersion patterns and the flexibility of those patterns are evidently influenced by the species' different surroundings.

Discussions of foraging and other behavioral adaptations often face the problem of phylogenetic inertia—is a particular trait there because it works best in the environment at hand or because it is somehow "in the nature" of the species that displays it? Comparisons like this one cannot determine why a particular *stable* trait is "inert"; we do not know, for instance, whether mangabeys retain similar diets because underlying genetic variation is somehow inadequate to allow evolution of this trait; because it once worked well enough to start mangabeys down a path toward an adaptive peak onto which they are now locked (though other patterns of diet, activity pattern, and digestive physiology might work equally well); or because, in some nonobvious but important way, the selective regimes the two species face with respect to this trait are the same. Such comparisons as I have made here can, however, identify those *labile* traits least subject to "phylogenetic inertia," which are presumably adapted to local conditions. These are then the behavioral characteristics with which we can formulate hypotheses about the workings of selection. Our ability to generate *general* hypotheses about the evolution of more labile behavioral traits in primates will depend on the extent to which characteristics that are stable among mangabeys are also stable and act as constraints within other primate radiations.

Acknowledgments

I thank Katherine Bulstrode, John Oates, and Simon Wallis for comments on the manuscript, and Greg Gardner for writing the foraging path analysis programs. Data analysis was supported in part by NSF grant SER 77-06731.

References

Altmann, S. A. 1974. Baboons, space, time, and energy. *Amer. Zool.* 14:221–48.

Barnicot, N. A., and D. Hewett-Emmett. 1972. Red cell and serum proteins of *Cercocebus, Presbytis, Colobus,* and certain other species. *Folia primatol.* 17:442–57.

Bradbury, J. L., and S. L. Vehrencamp. 1976. Social organization and foraging in Emballonurid bats. II. A model for the determination of group size. *Behav. Ecol. Sociobiol.* 1:383–404.

Chalmers, N. 1968. Group composition, ecology, and daily activities of free living mangabeys in Uganda. *Folia primatol.* 8:247–62.

Clutton-Brock, T. H. 1974. Ranging behaviour of red colobus (*Colobus badius tephrosceles*) in the Gombe National Park. *Anim. Behav.* 28:706–22.

Cody, M. L. 1974. Optimization in ecology. *Science* 183:1156 61.

Cronin, J. E., and V. M. Sarich. 1976. Molecular evidence for dual origin of mangabeys among Old World monkeys. *Nature* 260:700–2.

Gill, F. B., and L. L. Wolf. 1977. Nonrandom foraging by sunbirds in a patchy environment. *Ecology* 58:1284–96.

Harcourt, C. 1975. Ecology and social organization of the Sykes Monkey (*Cercopithecus mitis*) in the Tana River gallery forest. Unpublished ms.

Homewood, K. M. in press. Ranging patterns of the Tana mangabey *Cercocebus galeritus galeritus.*

—— 1975. Can the Tana mangabey survive? *Oryx* 13:53–59.

—— 1976. Ecology and behavior of the Tana mangabey. Ph. D. thesis. London: University College.

—— 1978. Feeding strategy of the Tana mangabey (*Cercocebus galeritus galeritus*). *J. Zool., Lond.* 186:375–92.

Kingdon, J. 1971. East African mammals: An atlas of evolution in Africa. Vol. I. New York: Academic Press.

Kingston, A. 1967. A working plan for Kibale and Itwara Central Forest Reserves. Forest Dept., Entebbe, Uganda.

Krebs, J. R. 1979. Feeding strategies and their social significance. In P. Marler and J. Vandenburgh, eds. *Handbook of Behavioral Neurobiology: Social Behavior and Communication,* pp. 225–70. New York: Plenum.

Kruuk, H. 1978. Foraging and spatial organization of the European badger. *Meles meles* L. *Behav. Ecol. Sociobiol.* 4:75–89.

Kummer, H. 1971. *Primate Societies.* Chicago: Aldine.

Napier, J. R., and P. H. Napier. 1967. *A Handbook of Living Primates.* New York: Academic Press.

Struhsaker, T. T. 1975. *The Red Colobus Monkey.* Chicago: Chicago University Press.

Wallis, S. 1979. The sociology of *Cercocebus albigena johnstoni* (Lyddeker): An arboreal rain forest monkey. Ph. D. thesis. London: University College.

Waser, P. M. 1974. Intergroup interaction in a forest monkey: The mangabey *Cercocebus albigena.* Ph. D. thesis. New York: Rockefeller University.

—— 1975. Monthly variations in feeding and activity patterns of the mangabey, *Cercocebus albigena. E. Afr. Wild. J.* 13:249–63.

—— 1976. *Cercocebus albigena:* site attachment, avoidance, and intergroup spacing. *Amer. Natur.* 110:911–35.

—— 1977a. Individual recognition, intergroup cohesion, and intergroup spacing: Evidence from sound playback to forest monkeys. *Behaviour* 60:28–74.

—— 1977b. Feeding, ranging, and group size in the mangabey *Cercocebus albigena.* In T. Clutton-Brock, ed. *Primate Ecology: Studies of Feeding and Ranging Behaviour in Lemurs, Monkeys and Apes,* pp. 183–222. New York: Academic Press.

—— 1981. Sociality or territorial defense? The influence of resource renewal. *Behav. Ecol. Sociobiol.* 8:231–237.

Waser, P. M., and O. Floody. 1974. Ranging patterns of the mangabey, *Cercocebus albigena,* in the Kibale forest, Uganda. *Z. Tierpsych.* 35:85–101.

Waser, P. M., and Homewood, K. M. 1979. Cost-benefit approaches to territoriality: A test with forest primates. *Behav. Ecol. Sociobiol.* 6:115–19.

Waser, P. M., and R. H. Wiley. 1979. Mechanisms and evolution of spacing in animals. In P. Marler and J. Vandenberg, eds. *Social Behavior and Communication,* pp. 159–224. New York: Plenum Press.

8 Body Size and Foraging in Primates

L. ALIS TEMERIN
BRUCE P. WHEATLEY AND
PETER S. RODMAN

AMONG THE MORPHOLOGICAL TRAITS that influence animal foraging capacities, body size has the most pervasive effect. Many parameters of morphology and physiology are subject to size-induced variation, and several—mass, linear dimensions, surface areas, volumetric capacities, standard metabolic rates, energy costs of locomotion, and rates of heat loss, for example—have an impact on three factors ultimately related to dietary content: an animal's energy (and other nutrient) requirements; the time and energy allocated to foraging activities; and the physical accessibility of food items.

Here we address the following question: How does size affect animal abilities to exploit food sources that differ in nutrient content, distribution, abundance, and accessibility? We consider interspecific contrasts in body size and foraging behavior and, as illustration, pre-

In the following paper, ALIS TEMERIN examines observations made by BRUCE WHEATLEY and PETER RODMAN on crab-eating macaques and orangutans of the Kutai Nature Reserve of East Kalimantan. These two species overlap considerably in dietary choice and differ by an order of magnitude in body weight. Temerin uses the comparative data to develop an analysis of the relationship of body size to foraging. She discusses interactions of body size with other important organismal characteristics as well.

sent behavioral observations drawn from long-term field studies of two sympatric Bornean primates, the crab-eating macaque, *Macaca fascicularis,* and the orangutan, *Pongo pygmaeus.* Initially, discussion focuses on how size contributes to differences in animal capacities to exploit diets of a particular quality and species composition. Because size has an impact on foraging in both ecological and evolutionary time, we then look at the relation between the two and in particular examine how size limits the possible directions of evolutionary change in diets. Finally, we consider whether it is appropriate to view body size as an adaptation for foraging.

The Comparative Sample

Study Species

Crab-eating macaques and orangutans differ greatly in size. *Macaca fascicularis* is a medium-sized primate, ca. 3–8 kg (Napier and Napier 1967), while *Pongo pygmaeus* is one of the largest, ca. 35–90 kg (Eckhardt 1975). In certain other respects the two species are similar, and this fact provides an element of control for the analysis of the influence of body size on foraging behavior. First, in regions of sympatry crab-eating macaques and orangutans overlap substantially in spatial distribution (Rodman 1973b, 1978; Rijksen 1978). Second, both species are highly arboreal: more than 95% of macaque activity is arboreal (Wheatley 1980; also see Rodman 1978; MacKinnon and MacKinnon 1978), and orangutans (sexes combined) spend about the same proportion of time in the trees (Rodman, this volume). (Marked size dimorphism in *Pongo* leads, however, to differences in the amount of terrestrial travel by adult males and other individuals [MacKinnon 1974; Rodman 1979; Galdikas 1979]. An adult male spent 20% of its travel time on the ground, for example [Rodman 1979].) Third, fruit comprises a major portion of the diet in both *M. fascicularis* (Wheatley 1980; MacKinnon and MacKinnon 1978; Rodman 1978) and *Pongo* (MacKinnon 1974; Rodman 1977; Rijksen 1978). In addition, both species exhibit dental (Kay and Hylander 1978; Kay 1981) and digestive tract (Chivers and Hladik 1980) traits consistent with a heavy reliance on fruit consumption.

Methods

The data that form our comparative sample were gathered during studies conducted at two sites located in the Kutai Nature Reserve, East Kalimantan, Indonesia: Hilmi Oesman Memorial Research Station (0°32′N, 117°20′E) and the Mentoko Research Station (0°24′N, 117°16′E). Both sites lie on the south side of the Sengata River and are about 17.5 and 25 km from the east coast of Borneo, respectively. The vegetation at each location is a combination of disturbed, streamside forest and primary (lowland dipterocarp) forest; secondary vegetation appears to be more abundant at the Hilmi Oesman site.

Wheatley conducted a 20-month study (1974–76) of a troop of *M. fascicularis* at the Hilmi Oesman station and also collected data on unhabituated orangutans. Data on macaque behavior were gathered through a combination of systematic sampling from 0600–1300 and 1600–1800 hours and a 3-week period of continuous, daylong observations. Information on orangutans was obtained by first-contact sampling and a small number of continuous observations. Rodman studied orangutan behavior during a 15-month period (1970–71) at the Mentoko site. Data on unhabituated *M. fascicularis* were also collected at this time and for 2 months in 1975. A continuous record of orangutan behavior, distributed throughout the day, was supplemented with first-contact samples. Data on macaques were collected through first-contact sampling. More detailed descriptions of study sites and observation methods are given in Wheatley (1978, 1980) and Rodman (1973a, 1977, 1978).

The long-term observations on habituated animals that form the core of our comparative sample are derived from research conducted at different sites and times and by use of sampling procedures that are not wholly comparable. To help control for the bias that may be introduced by these factors, data on both habituated and unhabituated animals were compared. Thus the behavior of habituated *M. fascicularis* was compared with that of unhabituated *Pongo* at the Hilmi Oesman site, and the behavior of habituated orangutans was compared with that of unhabituated macaques at the Mentoko location. We found that similar and often marked species contrasts occur at each site and parallel the intersite comparison of habituated animals. Accordingly, though our conclusions must be considered preliminary, we are confident that the data document real species differences. Studies of these primates at other locations (e.g., MacKinnon 1974; MacKinnon and MacKinnon 1978; Rijksen 1978; Aldrich-Blake 1980) have produced similar results, and this reinforces our confidence.

Size and Behavioral Capacities

The influence of size is mediated by its impact on the metabolic and physical traits of organisms. These traits help determine animal requirements for homeostasis, growth, and reproduction. They also help determine behavioral capacities (or what animals can and cannot do) and accordingly influence what in fact animals do.

Bounds on Dietary Quality

Several different classes of food items are potentially available to primates, including invertebrate prey, plant exudates, fruit, flowers, seeds, leaves, and bark. These classes vary in physical qualities, nutrient and toxin content, abundance, and spatiotemporal distribution (see fig. 8.1), and primate diets are characteristically mixed. One of the nutrient requirements that must be satisfied by dietary choice is an adequate energy supply. Food quality, defined as readily available energy and thus inversely proportional to plant fiber (or structural carbohydrate) content, is correspondingly a critical variable. Let us first consider how size may impose lower limits on dietary quality.

Lower Bounds. Two parameters subject to size-induced variation are especially important in the present context: standard metabolic rate,

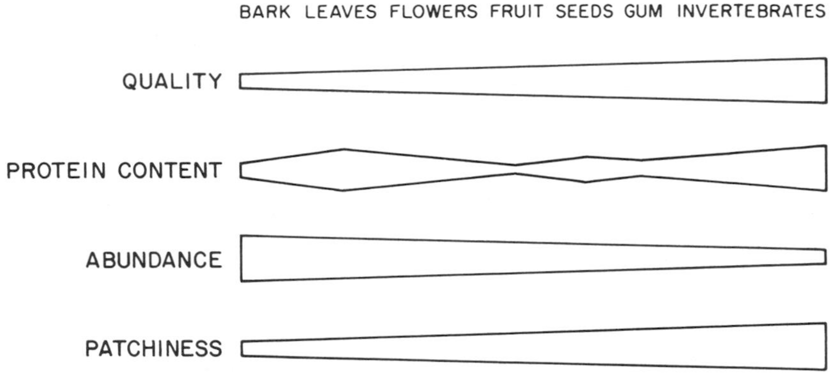

BARK LEAVES FLOWERS FRUIT SEEDS GUM INVERTEBRATES

QUALITY

PROTEIN CONTENT

ABUNDANCE

PATCHINESS

Figure 8.1. Variation in some attributes of primate foods. The width of a band corresponds with rank along a spectrum of values ranging from low to high.

which establishes baseline energy requirements and the rate at which these must be satisfied, and digestive capacities. Larger animals have greater energy needs, but since standard metabolic rates are proportional to body mass$^{.75}$ (or M$^{.75}$) (Kleiber 1975a), energy turnover per unit wt (Kleiber 1975b) decreases with size. As a consequence, larger animals can eat items that give a lower energy return per unit time, i.e., lower quality foods. When such items are eaten, a greater quantity must be processed for an energy gain comparable to that from higher quality items. Gut volume, a primary determinant of food processing capacity, scales roughly proportional to body mass (Chivers and Hladik 1980; Parra 1978). Correspondingly, as size increases, animals will be able to process greater quantities of food relative to their needs in a given time interval, and this permits the inclusion of—or increased dependence on—lower quality foods. (This assumes that smaller animals are unable to compensate fully for smaller gut capacities by increasing rates of energy assimilation. Few data are available to test this assumption; see Milton, this volume.)

On the basis of these considerations we would expect *M. fascicularis* to consume a higher quality diet than *Pongo*. This is precisely what we see (fig. 8.2): macaques devote more than 80% of their feeding time to fruit, and this is supplemented by short periods of insect, flower and bud, leaf, and grass consumption. Although fruit-eating is also important to orangutans, it is less so: slightly more than

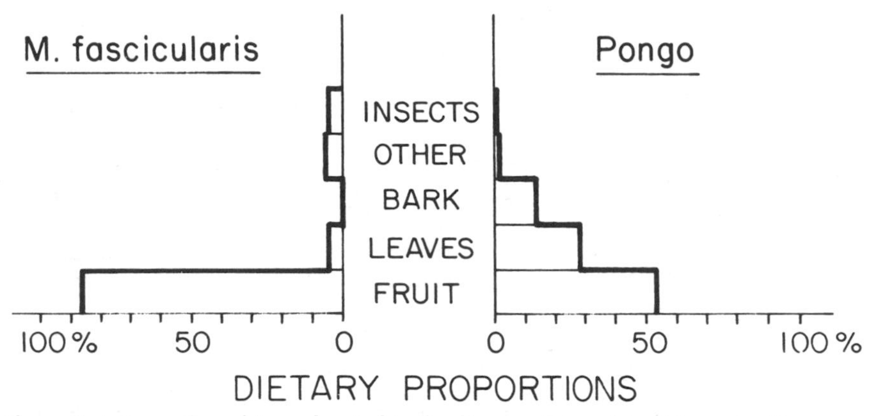

Figure 8.2. Proportion of time devoted to feeding on different food types by *Macaca fascicularis* and *Pongo pygmaeus*.

50% of feeding time is devoted to this dietary class. Leaves and bark, both low-quality items (Boyd and Goodyear 1971), are also prominent elements of the diet, comprising more than 40% of the time spent feeding.

Upper Bounds. While foraging, animals expend time and energy on activities that take place within food patches ("harvest") and on travel between food patches ("travel"). Dietary composition is ultimately the product of a compromise involving these costs, the need to allocate time and energy to other activities, and the benefits gained from foraging. Since size helps determine the magnitude of the time and energy costs associated with foraging, it plays a significant role in this context and helps set upper bounds on dietary quality.

Let us look at the time expenditures associated with harvest and travel. We know that energy requirements increase with size. If harvest rates do not increase at an equivalent rate, larger animals must spend more time feeding. Data on harvest rates in crab-eating macaques and orangutans are limited, but for the following reasons we think it safe to predict that, for high-quality items such as fruit, harvest rates do not increase with size at a rate commensurate with energy needs.

There are three components of harvest activity—acquisition, preparation, and mastication—and harvest rates of high-quality foods are generally limited by preparation or acquisition rates (Raemaekers 1979; Gaulin 1979). As body size increases, features such as greater mouth size, tooth areas, and masticatory force contribute to faster food preparation and mastication (see "p + m" in fig. 8.3). However, as the difference between animal and food size increases, this effect is probably progressively reduced. Food acquisition rates, on the other hand, probably reach an asymptote (see "a" in fig. 8.3). While such

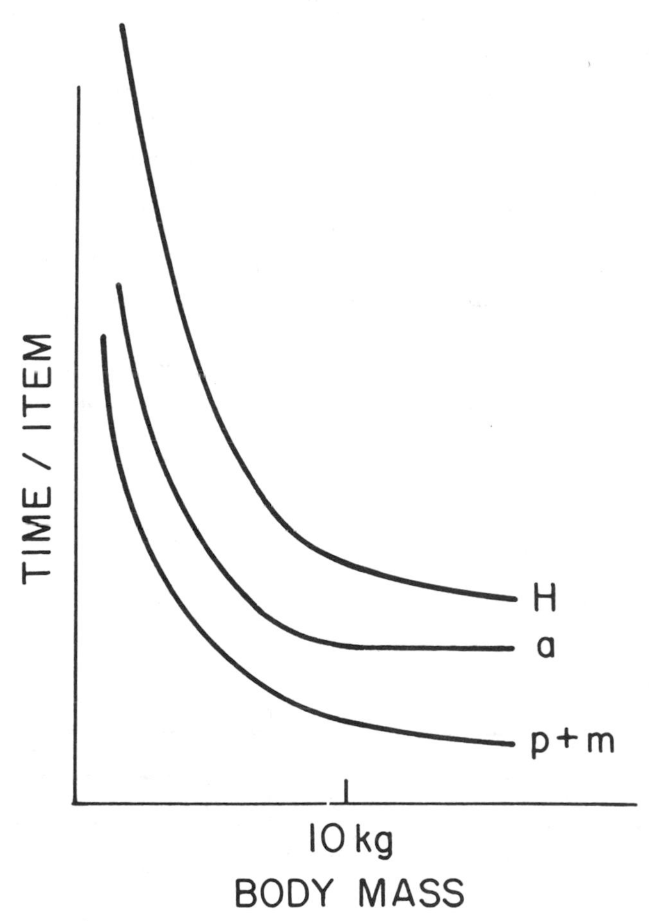

Figure 8.3. Postulated relation between harvest time per food item and body size in primates. H = total harvest time, a = time expended in the acquisition of food items, and $p + m$ = time to process and chew food.

features as longer limb length increase the size of feeding spheres (Grand 1972), increased weight can restrict immediate access to terminal branch areas, generally the regions of highest food densities, and thus effectively counter advantages gained in the first context. Similarly, potentials for greater speed of movement in larger animals (see below) are eventually counteracted by needs for stability. Raemaekers (1979) found that gibbons and siamang harvested fruit at similar rates on the average, suggesting that threshold values are approached at comparatively low body weights, ca. 10 kg or less.

Accordingly, to satisfy its energy needs by eating fruit (and/or other high-quality items), *Pongo* would have to spend more time feeding than *M. fascicularis.* Estimates based on activity profiles (fig. 8.4), mean day lengths (11.27 hr, orangutan; 11.37 h, crab-eating macaque), and time allocated to feeding on different food types (fig. 8.2) indirectly support this prediction: on the average, orangutans spend more time (2.8 h) feeding on fruit than macaques (1.38 h), even though fruit is a comparatively less important element of the diet.

Animal size has a direct influence on travel speeds, and these in turn have an impact on how much time is allocated to travel, and hence, on distances traveled. Biomechanical considerations suggest that habitual speeds of progression will increase as animals get larger (e.g., Alexander 1976; McMahon 1975), and we would thus expect orangutans to progress at faster rates on the average than crab-eat-

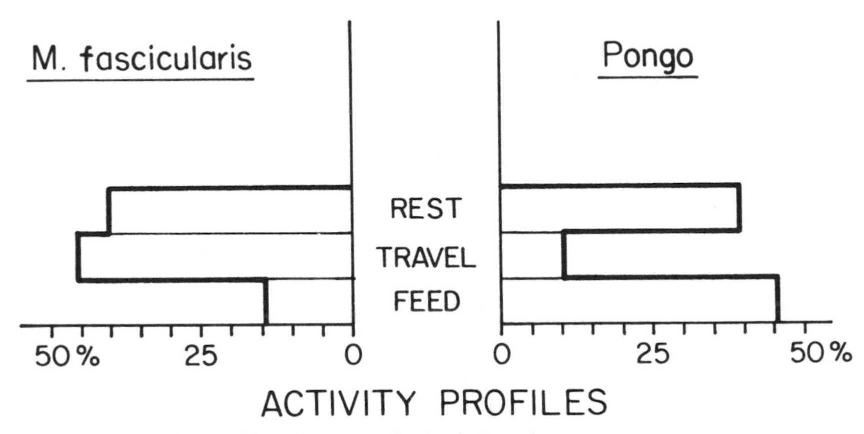

Figure 8.4. Activity profiles of *Macaca fascicularis* and *Pongo pygmaeus*.

Table 8.1. Average Travel Speeds [a]

M. fascicularis	*Pongo*	*Pongo* (predicted) [b]
0.40 km/hr	0.32 km/hr	0.57 km/hr

[a] Actual travel speeds are underestimated. Distances were measured primarily at ground level and thus do not take into account the often circuitous pathways necessitated by arboreal travel.

[b] Predicted travel speed is for a 35-kg orangutan. This has been calculated following the approach of Pennycuick (1979: 168): $v = k\, l^{.5}$, where v is speed, k is a constant, and l is limb length. Limb length is proportional to $M^{.34}$ in primates (Alexander et al. 1979), and k is determined by using the data on macaque travel speed (0.40 km/hr) and an estimate of body mass (5 kg). Thus, $v = 0.31\, M^{.17}$.

ing macaques. They do not, but in fact travel at somewhat lower speeds (table 8.1). The value expected if an orangutan were a scaled-up macaque is also included in table 8.1. It is not, of course, but this emphasizes the magnitude of the deviation from expected. One could conclude that in this instance size is not an important variable, but other considerations suggest that it is.

Arboreal travel is precarious at best, and increasing size makes compensatory demands for stability increasingly important. One response to this is to limit travel speeds. Correspondingly, we might expect that size increases will not result in increases in travel speeds that are comparable to those seen in terrestrial animals, and at some point, further size increases may necessitate their reduction (fig. 8.5). Impressions gained from watching orangutans locomote suggest that this is the case here.

When animals progress at lower rates they can either cover the same (or greater) distances and devote more time to progression or travel shorter distances with an equal or smaller time expenditure. The activity budget (fig. 8.4) and average day range length (table

Table 8.2. Average Values of Some Variables Related to Foraging

	M. fascicularis	*Pongo*
Day range (m)	1869	305
Feeding bouts/day	18.3	6.9 ♂
		7.2 ♀
Length feeding bouts (min)	10.5	50.9 ♂
		35.7 ♀
Feeding unit size (no. individuals)	30	1.83
Feeding unit biomass (kg)	105	70

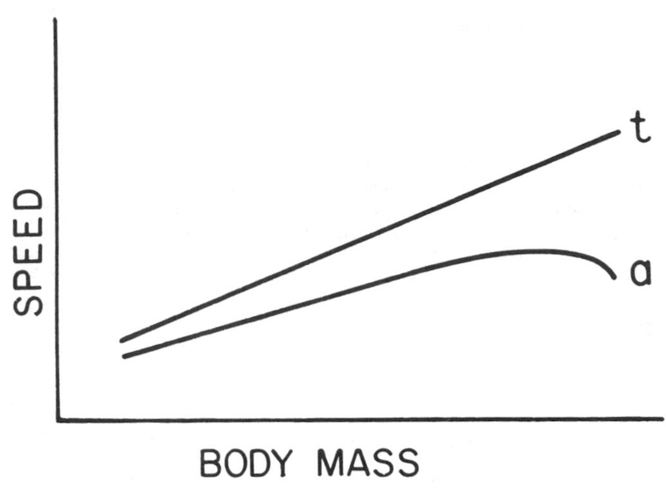

Figure 8.5. Proposed relation between animal size and travel speeds in terrestrial (t) and arboreal (a) progression.

8.2) of *Pongo* are consistent with the latter option: in comparison to *M. fascicularis,* orangutans spend less time traveling, and they travel much smaller distances per day on the average.

There are a couple of associated behavioral consequences. First, if a larger animal travels less distance than a smaller one, there will be less opportunity to exploit as many food patches. Data on the number of major feeding bouts per day engaged in by macaques and orangutans suggest that the larger species does visit fewer patches (table 8.2). Second, if larger animals visit fewer food patches, each patch must yield more food per individual. The average length of major feeding bouts is an index of the food available to a feeding animal, and bout lengths are four to six times longer in orangutans than crab-eating macaques. This contrast also exists when bouts of fruit-eating alone are compared: at the Hilmi Oesman site *M. fascicularis* averaged 13.8 min/bout while *Pongo* (unhabituated) averaged 60 min/bout.

Energetic considerations will of course also have a major impact on foraging capacities. We will focus on the energy costs of travel. (To a large extent the energy costs of harvest should parallel those of travel, albeit on a smaller scale.) Experimental studies have shown that the net energy cost per unit distance of level, quadrupedal locomotion tends to be roughly proportional to $M^{.7}$ (Taylor 1980). Thus while energy costs increase with size, they will increase at a progressively lower rate. This, when coupled with the general tendency for travel speeds to increase with size, suggests that larger animals should be able to travel greater distances on an equal proportion of total energy turnover. By extrapolating from data on several African ungulates, Pennycuick (1979) found that this distance scales with $M^{.4}$, for example. Accordingly, larger animals are in general less subject to energetic constraints on travel distance capacities, or mobility, than smaller ones.

Arboreal travel complicates matters, however. First, demands for progression along positive inclines are high during arboreal travel, and such locomotion involves an increment in energy expenditure that increases with size (Taylor et al. 1972). As a result, mobility will increase less rapidly with size in arboreal than terrestrial species, and at large sizes constraints on travel speeds will lead to a progressive reduction in the advantage size offers with regard to mobility (fig. 8.6). Second, locomotor repertoires vary in arboreal primates, and

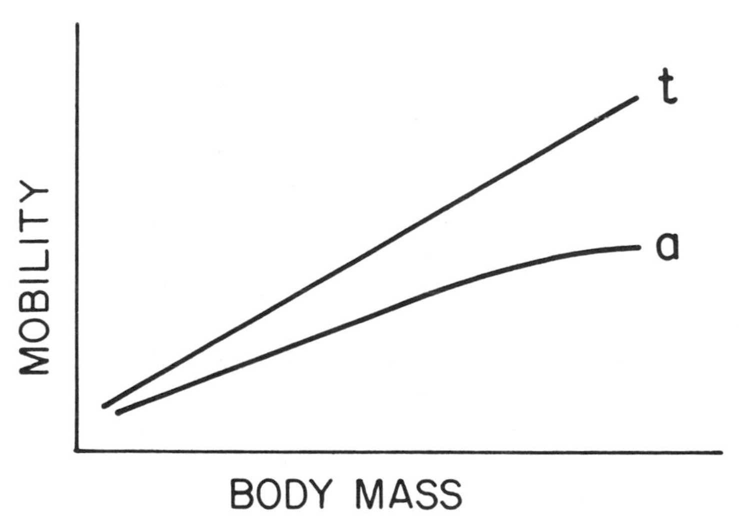

Figure 8.6. Hypothesized relation between travel distance capacities and body mass for terrestrial (t) and arboreal (a) progression.

these may contribute to different net costs of locomotion and thus confuse the influence of size. Species contrast in the frequencies with which they use different modes of progression, and the use of dissimilar locomotor modes may entail different expenditures of energy. In *Ateles,* for example, net energy costs of brachiation are higher than for quadrupedal progression (Parsons and Taylor 1977). Third, contrasts in size and locomotion will have an influence on the actual pathways traced during travel. There may often be more than one possible arboreal route between two points. Which is taken—and hence the actual distance traversed, as well as the time taken to do so—will depend on such factors as animal abilities to cross gaps of differing size and to move among unstable supports.

While there is no question that the energy costs of harvest and travel will be greater in *Pongo,* we lack the data to evaluate whether they are potentially more or less limiting vis-à-vis travel capacities than for *M. fascicularis.* That orangutans travel a much shorter distance per day than macaques strongly suggests that an equal percentage of energy turnover is not being allocated to travel. This could reflect the existence of stronger energetic constraints, but other interpretations also exist.

Thus far we have briefly considered how size may influence some of the behavioral variables that relate to foraging and have noted how differences between orangutans and crab-eating macaques tend to be consistent with size contrasts. It has probably become apparent that separating the putative influence of size from that of contrasting dietary compositions is difficult. For instance, primate day range lengths are negatively correlated with the proportion of foliage contained in diets (Clutton-Brock and Harvey 1977), and orangutans probably travel shorter distances per day because this is all they need to do to meet their dietary requirements. It is appropriate to ask then why *Pongo* must depend on a lower quality diet than *M. fascicularis.*

We know from macaque behavior and a comparison of orangutan and macaque feeding unit size and biomass (table 8.2) that sufficient fruit (and other high-quality food) is available to meet the energy requirements of *Pongo* if only it would move far enough on a daily basis. So why does it not? As noted, we do not have data to evaluate the importance of energy costs as a limiting factor, but data on time costs are good enough and suggest a simple answer: if orangutans were to travel the distances required, insufficient time would be left over for feeding and other needs. The slow speed of arboreal travel and long periods of feeding engendered by very large size thus appear to be critical factors imposing upper bounds on dietary quality.

Species Composition of Diets

Primate populations are found in areas where food species that are accessible to, and satisfy the nutrient requirements of, the animals are available in sufficient density. The same resources need not be present in different locations—intraspecific variation in primate diets has been documented frequently (e.g., Clutton-Brock 1977)—but their characteristics should be broadly similar, reflecting the existence of constraints on the set of species a primate can exploit profitably. Data on the food species of *M. fascicularis* and *Pongo* (as well as those of most other primates) are limited, but the species composition of primate diets is a topic worth considering, if only briefly. This is because comparative analyses of animals and their food species will further clarify the foundations of contrasts in resource and habitat use by primates and thus mechanisms leading to ecological segregation.

The plant and animal species used as food by a primate are typically a small subset of those available. Body size may influence a primate's choice of fruit (or leaf, or invertebrate, etc.) species through its effect on the physical accessibility of different items, the nutrient constraints on food choice, and capacities to rely on highly dispersed species.

Physical Access. Food harvest is contingent on animal abilities to acquire, prepare if necessary, and chew items. Plant and animal species frequently differ in characteristics that bear on their physical accessibility to consumers: arthropods occupy different microhabitats and vary in size; tree species differ in size, structure, and the disposition of their leaves and fruit; varieties of fruit contrast in dimensions, the nature of any protective covering, and hardness; and so on. Consumer size may have an influence on the time and energy expended to gain access to particular food items, and it may also determine whether access is possible at all. Three characteristics of primates—their large size relative to potential food size, good manipulatory abilities, and locomotor agility—suggest, however, that such effects will be critical only under rather circumscribed conditions.

In the context of food acquisition, primate size will probably have a more significant impact on capacities to enter, or move within, food trees than on abilities to remove (or capture) food items. For example, small size in a clawless primate can prevent its entrance into a food-bearing tree via a large vertical trunk (Cartmill 1979). If no other entry routes are open to the animal, it cannot exploit this food source. Dietary composition will be influenced if such limited access characterizes particular tree species, but the latter remains an unexplored phenomenon. Alternatively, large body size will restrict abilities to move among the small supports and dense tangles of branches and vines found at forest edges, tree falls, and shrub-layer height. Although foraging under such conditions need not be impossible for a larger primate, it may yield comparatively small benefits because harvest efficiencies are very low. Both *M. fascicularis* and *Pongo* forage in streamside secondary forest, but the former species uses it much more intensively (Rodman 1978; Wheatley 1980).

Once food items have been acquired, consumer size will be limiting when processing rates or access to edible (or digestible) portions depend on strength. Fruit species vary considerably in structural qualities, and not surprisingly, orangutans eat fruits characterized by hardness, tough husks, and so on, that crab-eating macaques and other smaller primates do not (personal observations; MacKinnon 1977).

Nutrient Content. The size-related traits that help set bounds on overall dietary quality, as well as other aspects of dietary content, may also be important in the narrower context of species choice. Relations between animal size and the nutrient content of acceptable foods are apt, however, to be subtle where they exist. This is because there is a narrow range of variation in nutrient composition, abundance, and spatiotemporal distribution among species of a particular food class. Then too, our principal concern in this instance must be with the role of size in setting *lower* bounds on nutrient content (e.g., quality, protein content). As illustrated in the earlier discussion of bounds on dietary quality, a general relation between the quality of food items and their distribution and abundance must hold if we are to generate predictions concerning the impact of size on upper bounds. There is presently no empirical support for the existence of such a relation for any aspects of nutrient content at the species level.

Metabolic turnover rates and digestive capacities are the principal size-dependent variables setting lower limits on food choice. Thus the low turnover rates and large gut capacities of large animals will permit them to consume fruits (or leaves, etc.) with more fiber (or less protein) than smaller consumers find nutritionally feasible to select. The systematic sampling of orangutan and macaque food species needed to evaluate whether these are significant factors in species selection has not yet been done. Nonetheless, some preliminary data suggest that in comparison with *M. fascicularis,* orangutans may eat fruits with a lower protein content (Wheatley 1978).

Spatiotemporal Distribution. Variation among food species in spatiotemporal distribution allows body size to influence selectivity, and this in turn may result in variation among primates in the details of habitat use. The following size-dependent variables are important in this context: capacity to survive periods when food intake does not meet requirements, mobility, and ability to eat low-quality foods.

Larger animals generally can withstand longer (or more frequent) periods of "suboptimal" food intake than smaller ones for two reasons. First, rate of fat metabolism scales roughly with $M^{.25}$ (Lindstedt and Calder 1981), and larger animals thus use fat at a lower rate per unit wt. Second, fat stores scale positively with body mass and probably at a rate more than commensurate with basal metabolic needs (e.g., proportional to $M^{.89}$—$M^{1.57}$ in birds [Calder 1974]). The impact of body size in this context is typically considered with respect to the foraging rigors engendered by marked, often cyclic variations in resource levels and/or climate (e.g., studies cited by Calder 1974; Downhower 1976). But there are also implications more directly related to the present discussion. Because rates of food intake can vary more as size increases (i.e., buffering against the vagaries of resource availability increases with size), larger size will permit the exploitation of food species characterized by comparatively low densities and/or temporal constancy.

A concomitant increase in both long-term mobility and the ability to consume lower quality foods with size also facilitates the use of widely dispersed food species by larger animals. They are generally able to range over larger areas of a habitat and thus can exploit species with low patch densities. Further, constraints imposed by food abundance (patch sizes) will be relaxed if animals supplement their intake of high-quality foods with more abundant low-quality items. That is, there need not be an inverse relation between patch density and patch size (or renewal rate) if other food sources are in plentiful supply.

Measures of the spatiotemporal distribution of, for example, the fruit species exploited by *Pongo* and crab-eating macaques are needed to test the prediction that large size in orangutans facilitates their use of more widely dispersed species. One set of existing data is suggestive. *M. fascicularis* and orangutans exploit both riverine and deep forest zones of the forest habitat, but the former zone is more intensively used by the macaques (Rodman 1978). Fruit productivity is higher in riverine zones (Rodman 1978; Wheatley 1980), a feature consistent with high patch densities and/or constancy of fruit species.

Size in Ecological and Evolutionary Time Perspective

When we examine the foraging behavior of primates to evaluate the impact of size we are viewing its role in ecological time. As noted earlier, animal size affects foraging in this time frame via a dual role: it is a major determinant of nutrient requirements, and it helps determine behavioral capacities to satisfy these requirements. Thus it contributes significantly to determining the set of behavioral responses—or tactics—an animal exhibits under circumstances comprising marked environmental heterogeneity and short-term (relative to generation length) variation in resource abundance and distribution. The influence of size extends beyond this to population and community levels as well. Events occurring at the level of the individual will eventually influence population sizes and the distribution of species among different habitats.

Through its impact on foraging in ecological time—whence the processes of natural selection occur—body size can also influence the evolution of foraging regimes. In evolutionary time the effect of size may be manifested as follows. A change in size can promote an evolutionary shift in dietary content, or it can increase the efficiency of a foraging regime (i.e., increase adaptation to a particular diet). Alternatively, size may be a source of strong constraints on evolutionary change of any sort. This is most likely in cases where species fall at the extremes of the size spectrum, particularly its upper end (Stanley 1973). These size-influenced phenomena are ultimately expressed through the production of phylogenetic trends and through the differential productivity of evolutionary lineages.

Evolutionary Change in Diets

The effect of a change in size on the parameters of time and energy integral to foraging is in many instances readily predicted. This permits us to consider on an abstract basis how size may limit the possible directions of an evolutionary change in dietary content. We will focus on the consequences of an increase in body size since this is the more common direction of change (e.g., Stanley 1973). In addition, it is a simple matter to adapt the model we present to explore other related topics. Two attributes of dietary content—overall quality and principal sources of protein—will be considered.

A model. It is convenient to focus on how size imposes constraints on the distance that a primate can travel on a short-term basis, because this contributes significantly to animal abilities to exploit food sources that vary in abundance, distribution, and coincidentally, quality. A series of schematic graphs (fig. 8.7) illustrates the pertinent variables.

E represents the energy contained in a diet of a certain average quality; as an animal travels farther it is able to eat more. Actual energy yield from a diet is a joint function of quantity eaten and assimilation efficiency, and it will be less than E by an amount dependent on dietary composition and animal identity. The slope of E increases when dietary quality declines (E^*, fig. 8.7) because more abundant and less patchy foods are available. But at the same time individuals must eat greater quantities (E must be higher) for an equivalent energy gain because assimilation efficiencies tend to decline as dietary quality does (Van Soest 1967). Eventually, limits on digestive capacity and/or time expenditures will result in a maximum level of intake. (Animals do not, at each instant in time, consume an "average" diet, nor are food sources [and energy] evenly distributed. E correspondingly departs considerably from reality, but it is nevertheless a useful approximation for this largely conceptual level of inquiry. Other variables [C, T] are similar approximations.)

C is an animal's energy expenditure on foraging. (Harvest and travel activities are combined in this instance.) It increases with the distance traveled and must be substantially less than E on the average.

T_h and T_t are the time expenditures for harvest and travel, respectively. Each is set against the same distance scale as the energetic parameters and increases with the distance traveled. Absolute limits on the amount of time that can be expended during a day's activity exist, and an upper horizontal line marks this limit in the graphs. Total time allocated to foraging must average less than this amount.

Both the energy and the time expended on harvest activities may vary with dietary quality. There is no firm basis on which to base predictions about C_h, and we assume that it remains constant. Likewise, it is difficult to predict the relation between dietary quality and harvest rates. Acquisition times may decrease as quality declines because lower quality items are more densely distributed within patches (e.g., leaves vs. fruit), and preparation times may also decrease. But mastication probably takes longer. In order to introduce another possible source of variation into our model, we will assume that the increased time spent on mastication does not completely counter decreased expenditures on the other components of harvest activity. Hence the slope of T_h should be less for lower quality diets (T_h^*, fig. 8.7c).

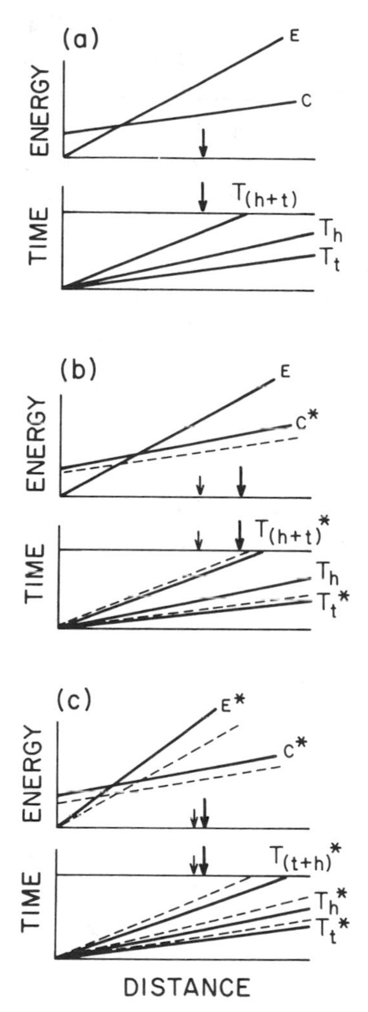

Figure 8.7. A schematic representation of the energy and time parameters related to foraging and how they can change with size. E = energy content of a particular diet, C = energy costs of foraging, T_h = time expended on harvest, T_t = time expended on travel, and arrows mark the distance traveled. Variables marked with an asterisk denote change from the ancestral condition, which in (b) and (c) is represented by faint lines and arrows. (a) is the ancestral condition; in (b) and (c) body size has increased and dietary content either remains the same (b) or changes to include more abundant, less dispersed items (c). See text for further discussion.

When an animal consumes a particular diet there will be a travel distance that results in an optimal energy gain. Arrows identify this in figure 8.7a, and their extension marks the energy and time expenditures involved, as well as the total energy ingested. Evolutionary change in animal traits related to foraging abilities or a change in environmental circumstances will lead to alterations in the parameters related to the time and energy costs of foraging. A shift in travel distances and/or diets may then be selectively advantageous.

Initially we proceed as if body size were the only trait subject to change. Also, we ignore factors that change with size and affect foraging indirectly, e.g., thermoregulatory requirements and risk of predation. These traits probably have greatest bearing on the scheduling of activity (viz., time allocated to foraging) and will be most important for smaller primates (see Crompton, this volume, for example). A size increase will tend to relax constraints on activity periods imposed by such factors and thus lead to greater flexibility in foraging options.

Increased size will result in the following changes in energy and time expenditures.

C increases. In arboreal primates the slope, as well as y-intercept, of C will become greater with size because the costs of vertical locomotion increase disproportionately in larger animals.

T_h either decreases or remains the same. If our earlier arguments about an expected leveling off of harvest rates with size are correct, harvest rates should increase with size only for the smaller primates. We will begin with a medium-sized primate and thus assume that T_h is unchanged.

T_t decreases, as long as locomotor speed increases with size. If locomotor speed does not increase (at larger body sizes; see fig. 8.5), T_t may remain the same or even increase.

Energy needs also increase, of course, for a larger animal, and the acceptable level of E must increase as a result.

Options. Under these circumstances there are two possible responses. First, animals may maintain the same diet and consequently, travel farther to meet their increased needs (fig. 8.7b). If the increment in the energy costs of foraging has not outpaced the added nutrient gain derived from more travel (i.e., if the slope of C^* is not greater than the slope of E), energy per se will not be limiting. But time will be limiting since more must be expended in harvest and possibly travel as well, leaving less for other needs. Hence, this is probably the less likely option. Second, animals may increase their dependence on more abundant and less dispersed items and thus consume a diet that is lower in quality (fig. 8.7c). In this context sources of protein will also shift proportionately in their contribution to the diet, from animal prey to more readily available vegetation (principally leaves). As a result, travel distances increase less than they would if dietary content remained unchanged, and total time expenditures do not change markedly. Constraints on the availability of time and energy that were intensified by the size increase are correspondingly relaxed somewhat, though as size becomes very large, these limits can become particularly stringent. Travel times may no longer decrease, for example, and this intensifies pressures to restrict increases in daily travel distances.

These considerations suggest a somewhat inexorable trend: the larger the animal, the lower the dietary quality and the greater the dependence on vegetation as a source of protein. A truly rigorous test of this thesis (recently termed the Jarman-Bell principle by Gaulin [1979], among others) is often impossible because few phylogenetic lineages are sufficiently well documented to provide information on both animal sizes and diets as they change through time. The trend has, however, been documented among closely related extant mammals, including primates (Gaulin 1979; Kay, this volume); rodents (Emmons 1980; Pizzimenti and DeSalle 1980); and artiodactyls (Jarman and Sinclair 1979; Gautier-Hion et al. 1980), implying that size does indeed exert a strong unidirectional pressure on dietary content.

The trend can be interrupted. A size increase need not be accompanied by a decrease in dietary quality (or use of animal prey as a protein source) if it follows or accompanies events that relax the constraints imposed by time and energy considerations. These include the following:

1. An increase in food density. A change in environmental conditions (e.g., the arrival of a more equable climate, loss of a competitor from the community, or a reduction in social group size) may lead to greater food availability and thus increase E.

2. The evolution of traits that (a) increase assimilation efficiencies (energy gain from E is correspondingly greater), (b) reduce the costs of travel and/or harvest (C is reduced), or (c) increase rates of travel and/or harvest (T_{t+h} is less).

3. A shift from arboreal to terrestrial travel (C and T_t reduced; see figures 8.5 and 8.6).

4. An increase in the length of the activity cycle, allowing more time for foraging.

5. A reduction in metabolic requirements, due to a reduction in basal metabolic rate and/or rates of growth and reproduction, such that energy can be allocated at a lower rate. In these circumstances ($E - C$) need not yield the same gain proportionate to needs as it did previously.

Such factors will help account for differences in scaling between animal size and diet. For example, among the primates it is clear that both orangutans and chimpanzees eat much more fruit than would be expected if we were to extend the trend displayed by Old World monkeys (Gaulin 1979). This apparent anomaly can probably be resolved if we attend to differences among monkeys and apes that could have displaced the scale relating size and dietary composition. Contrasts in locomotor traits (2b and c, above) are prime candidates. It has been hypothesized that the evolution of suspensory capacities in the common ancestor of modern apes led to an increase in harvest and travel speeds (T_{t+h} decreased) and, possibly, to a decrease in energy expenditures (a reduction in C) (Temerin and Cant 1981, 1983). For a given body size, hominoid abilities to exploit higher quality foods would have been correspondingly enhanced.

These factors, either singly or in combination, can also explain variance in the correlations observable among more closely related taxa. Among the apes, for instance, chimpanzees (*Pan troglodytes*) are roughly comparable to female orangutans in body mass, but they consume a higher quality diet on the average (e.g., proportionately more invertebrates and no bark; see Rodman, this volume). *Pan* also tends to travel on the ground (Wrangham 1977; Rodman, this volume), and it is likely that this substantially increases its ability to exploit both fruit sources and animal prey. Travel speeds in male chimpanzees average almost 3 km/h (Rodman and McHenry 1980), about 10 times the rate of orangutan travel (table 8.1). In addition, *Pan* commonly ranges 2–6 km/day (Wrangham 1979), a much greater distance than *Pongo* (table 8.2; Rijksen 1978; MacKinnon 1974). Other examples are discussed by Gaulin (1979).

Evolutionary Potential

At the beginning of this section we suggested that size, in addition to promoting evolutionary changes in foraging regimes, can restrict opportunities for such change. We shall briefly discuss how this may occur.

It is commonly recognized that frequently very large, and occasionally very small, members of higher taxa display specializations that may be viewed as representing accommodations to extreme size. That is, size itself can be an important source of the selection pressures resulting in the evolution of a specialized trait, or the evolution of such a trait may have served as a preadaptation for a change in size. (Strictly speaking, which situation pertains is of little or no import since it is likely that size and other traits often evolve in concert.) For example, two small prosimians, *Cheirogaleus medius* (ca. 120 g) and *Microcebus murinus* (ca. 60 g), can enter torpor (Petter 1978), a process that reduces energy requirements and thus enables survival through periods of low food availability and/or harsh climatic conditions. Animals possessing this ability tend to fall at the lower ends of size spectra (Whittow 1973). At the other size extreme, *Pongo,* the largest habitually arboreal primate, exhibits specializations of the locomotor system, e.g., extreme joint mobility and strong manual and pedal flexor capabilities (Tuttle 1969, 1970; Sigmon 1974; Lewis 1969, 1974), that allow animals to distribute their weight among several supports and, hence, sit and progress in relative safety.

In all cases where such specialized traits have an impact on variables related to foraging, the possible avenues whereby the constraints on diet imposed by size can be relaxed (alternatives 1–5) will be reduced in number. This is because each specialization represents one option already taken. (The basic premise here is that the number of avenues for evolutionary change decreases with increasing specialization. Eventually, intrinsic constraints on design or function prevent further change [Stanley 1973].) As a consequence, the frequency of evolutionary changes in foraging will generally be lower than in other instances. Since such changes often (if not always) accompany speciation, the evolutionary productivity of lineages should be less in such instances.

The existence of only a few large arboreal primates is of particular interest in this context. Only nine out of more than 150 habitually arboreal species exceed 10 kg, and all, with the exception of orangutans, are largely folivorous (Temerin 1983). In contrast, the smaller species span the full dietary spectrum. The example of *Pongo* is instructive because it is a clear case where large size has promoted locomotor specializations that restrict travel speeds and hence short-term mobility. This negative impact of large size on mobility probably exists in other large primates as well, though on a less dramatic scale. Siamang, for instance, travel at slower rates than sympatric gibbons (Raemaekers 1979). We emphasize, however, that limited mobility is only one of several factors that may have restricted options for dietary change and thus contributed to the low species diversity of large arboreal primates. Our repeated focus on locomotor capacities results from the obvious specializations (conveniently) displayed by *Pongo*. Other possibilities must await further empirical studies.

Is Body Size an Adaptation for Foraging?

Our answer to the question that heads this section is necessarily equivocal: "perhaps." This rather unappealing response has a simple explanation: animal size is important in several contexts besides foraging, including the scheduling of development and reproduction, reproductive effort, mating success, and predator avoidance. Thus a size shift may be a response to selection for alterations in the time and energy parameters related to foraging, or it may be a response to selection for changes in traits unrelated to foraging, such as life history variables. In the latter case any changes in foraging parameters that follow a size shift will reflect accommodation to the new size. Only in the first instance is size strictly an adaptation for foraging.

Whether or not we label size as an adaptation for foraging is, however, of little consequence. The value of inquiry into the selective bases of an evolutionary change in size lies partly in its contribution to a growing appreciation of the extensive, intertwined ramifications of such change. Assume, for example, that size changes in response to selection for increased foraging efficiency. Because life history parameters scale very clearly with size (e.g., Blueweiss et al. 1978; Western 1979; Tuomi 1980; Eisenberg 1981; Lindstedt and Calder 1981), this shift in size will be accompanied by changes in both foraging and life history traits. These are not separate domains of animal biology, however, and may exert a mutual influence: alterations in growth rates, reproductive effort, etc., entail changes in the allocation of energy and in time schedules, factors bearing significantly on foraging requirements. Changes in foraging patterns, on the other hand, may result in different rates of net energy gain and in different time expenditures and thus affect allocations to other requirements.

Clearly, the opportunities for complex feedback relations among size-related parameters are rife, and they should be no less numerous under other initial circumstances of an evolutionary change in body size. It is correspondingly appropriate to view the evolution of body size, whatever its selective basis, as inextricably involved with the evolution of foraging strategies.

Conclusions

The general correspondence between predictions of how size influences foraging parameters and naturalistic behavior patterns that is illustrated by our comparison of *M. fascicularis* and *Pongo* attests to the significant impact size can have on foraging. These results are not surprising, for the strong correlation between size and diet in primates generally, and the large size difference between our study species, lead us to expect them. Our ability to model the influence of evolutionary changes in size on dietary content, despite few empirical data, also indicates the powerful effect size exerts.

It is important, however, that we not fall prey to a false sense of complacency about how well we understand the role of size in the context of animal foraging. Until there are more nearly complete empirical data on the parameters of primate foraging behavior, it will be impossible to appreciate how variables operate in combination (e.g., size with food quality and protein content, or size with accessibility and food quality) to determine food choice. Furthermore, we are presently unable to evaluate rigorously whether, as size changes, energy, time, or both variables are the limiting factors. It has been very convenient to focus on time constraints in this context because most primates are restricted to about 12 hr of activity. But it is also quite possible that energy is a primary source of limits. Finally, size is only one of many traits that bear on foraging capacities (see other chapters in this volume), and its role as a causal factor in evolution will thus be variable. It can be fully explored only when size is studied in concert with other animal traits.

Acknowledgments

We are grateful to J. G. H. Cant for his critical and constructive appraisal of the ideas presented. Research by B. P. Wheatley was supported by NSF Grant #BMS 74-14190 and by grants from The Explorers Club and the Society of Sigma Xi. Rodman's research has been supported by NIMH Grant #13156, NSF Grant #DEB 76-21413, PHS Grant #RR00169, and by grants from the New York Zoological Society, the National Geographic Society, and the University of California. L. A. Temerin has been supported by NSF Grant #BMS 75-03130 and by grants from The Explorers Club and the Society of Sigma Xi. We have all benefited from the support of PHS Grant #RR00166 (Washington Regional Primate Research Center) through the enlightened interest of Dr. Orville A. Smith. We thank

C. L. Darsono, Kamil Oesman, the Indonesian Institute of Sciences, the Department of Nature Conservation and Wildlife Management, Indonesia, and the Pertamina Oil Company, for their sponsorship and invaluable assistance.

References

Aldrich-Blake, F. P. G. 1980. Long-tailed macaques. In D. J. Chivers, ed. *Malayan Forest Primates,* pp. 147–65. New York: Plenum.

Alexander, R. McN. 1976. Estimates of speeds of dinosaurs. *Nature* 261:129–30.

Alexander, R. McN., A. S. Jayes, G. M. O. Maloiy, and E. M. Wathuta. 1979. Allometry of the limb bones of mammals from shrews (*Sorex*) to elephant (*Loxodonta*). *J. Zool., Lond.* 189:305–14.

Blueweiss, L., H. Fox, V. Kudzma, D. Nakashima, R. Peters, and S. Sams. 1978. Relationships between body size and some life history parameters. *Oecologia* 37:257–72.

Boyd, C. E., and C. P. Goodyear. 1971. Nutritive quality of food in ecological systems. *Arch. Hydrobiol.* 69:256–70.

Calder, W. A. 1974. Consequences of body size for avian energetics. In R. A. Paynter, ed. *Avian Energetics,* pp. 86–144. Cambridge, England: Nuttall Ornithological Club.

Cartmill, M. 1979. The volar skin of primates: Its frictional characteristics and their functional significance. *Amer. J. Phys. Anthrop.* 50:497–509.

Chivers, D. J., and C. M. Hladik. 1980. Morphology of the gastrointestinal tract in primates: Comparisons with other mammals in relation to diet. *J. Morph.* 166:337–86.

Clutton-Brock, T. H., ed. 1977. *Primate Ecology: Studies of Feeding and Ranging Behaviour in Lemurs, Monkeys and Apes.* London: Academic Press.

—— 1977. Some aspects of intraspecific variation in feeding and ranging behaviour in primates. In Clutton-Brock, ed. 1977:539–56.

Clutton-Brock, T. H., and P. H. Harvey. 1977. Species differences in feeding and ranging behaviour in primates. In Clutton-Brock, ed. 1977:557–84.

Downhower, J. F. 1976. Darwin's finches and the evolution of sexual dimorphism in body size. *Nature* 263:558–63.

Eckhardt, R. B. 1975. The relative body weights of Bornean and Sumatran orangutans. *Amer. J. Phys. Anthrop.* 42:349–563.

Eisenberg, J. F. 1981. *The Mammalian Radiations.* Chicago: University of Chicago Press.

Emmons, L. H. 1980. Ecology and resource partitioning among nine species of African rain forest squirrels. *Ecol. Monogr.* 50:31–54.

Galdikas, B. M. F. 1979. Orangutan adaptation at Tanjung Puting Reserve: Mating and ecology. In Hamburg and McCown, eds. 1979:195–233.

Gaulin, S. J. C. 1979. A Jarman/Bell model of primate feeding niches. *Hum. Ecol.* 7:1–20.

Gautier-Hion, A., L. H. Emmons, and G. Dubost. 1980. A comparison of the diets of three major groups of primary consumers of Gabon (primates, squirrels, and ruminants). *Oecologia* 45:182–89.

Grand, T. I. 1972. A mechanical interpretation of terminal branch feeding. *J. Mammal.* 53:198–201.

Hamburg, D. A., and E. R. McCown, eds. 1979. *The Great Apes.* Menlo Park, Calif.: Benjamin/Cummings.

Jarman, P. J., and A. R. E. Sinclair. 1979. Feeding strategy and the pattern of resource partitioning in ungulates. In A. R. E. Sinclair and M. Norton-Griffiths, eds. *Serengeti: Dynamics of an Ecosystem,* pp. 130–63. Chicago: University of Chicago Press.

Kay, R. F. 1981. The nut-crackers—a new theory of the adaptations of the Ramapithecinae. *Amer. J. Phys. Anthrop.* 55:141–51.

Kay, R. F., and W. L. Hylander. 1978. The dental structure of mammalian folivores with special reference to primates and Phalangeroidea (Marsupialia). In Montgomergy, ed. 1978:173–91.

Kleiber, M. 1975a. *The Fire of Life.* Huntington, N.Y.: Robert E. Krieger.

—— 1975b. Metabolic turnover rate: A physiological meaning of the metabolic rate per unit body weight. *J. Theoret. Biol.* 53:199–204.

Lewis, O. J. 1969. The hominoid wrist joint. *Amer. J. Phys. Anthrop.* 30:251–68.

—— 1974. The wrist articulations of the Anthropoidea. In F. A. Jenkins, ed. *Primate Locomotion,* pp. 143–69, New York: Academic Press.

Lindstedt, S. L., and W. A. Calder. 1981. Body size, physiological time, and longevity of homeothermic animals. *Quart. Rev. Biol.* 56:1–16.

MacKinnon, J. R. 1974. The behaviour and ecology of wild orang-utans (*Pongo pygmaeus*). *Anim. Behav.* 22:3–74.

—— 1977. A comparative ecology of Asian apes. *Primates* 18:747–72.

MacKinnon, J. R., and K. S. MacKinnon. 1978. Comparative feeding ecology of six sympatric primates in West Malaysia. In D. J. Chivers and J. Herbert, eds. *Recent Advances in Primatology.* 1:305–21. *Behaviour.* London: Academic Press.

McMahon, T. A. 1975. Using body size to understand the structural design of animals: Quadrupedal locomotion. *J. Appl. Physiol.* 39:619–27.

Montgomery, G. G., ed. 1978. *The Ecology of Arboreal Folivores.* Washington, D.C.: Smithsonian Institution Press.

Napier, J. R., and P. H. Napier. 1967. *A Handbook of Living Primates.* London: Academic Press.

Parra, R. 1978. Comparison of foregut and hindgut fermentation in herbivores. In Montgomery, ed. 1978:205–29.

Parsons, P. E., and C. R. Taylor. 1977. Energetics of brachiation versus walking: A comparison of a suspended and an inverted pendulum mechanism. *Physiol. Zool.* 50:182–88.

Pennycuick, C. J. 1979. Energy costs of locomotion and the concept of "foraging radius." In A. R. E. Sinclair and M. Norton-Griffiths, eds. *Serengeti: Dynamics of an Ecosystem,* pp. 164–84 Chicago: University of Chicago Press.

Petter, J. J. 1978. Ecological and physiological adaptations of five sympatric nocturnal lemurs to seasonal variations in food production. In D. J. Chivers and J. Herbert, eds. *Recent Advances in Primatology.* 1:211–23. *Behaviour.* London: Academic Press.

Pizzimenti, J. J., and R. DeSalle. 1980. Dietary and morphometric variation in some Peruvian rodent communities: The effect of feeding strategy on evolution. *Biol. J. Linn. Soc.* 13:263–85.

Raemaekers, J. 1979. Ecology of sympatric gibbons. *Folio primatol.* 31:227–45.

Rijksen, H. D. 1978. A field study on Sumatran orangutans (*Pongo pygmaeus abelii* Lesson 1827): Ecology, behaviour and conservation. *Meded. Land-bouwhogeschool Wageningen* 78–2.

Rodman, P. S. 1973a. Population composition and adaptive organization among orang-utans of the Kutai Reserve. In R. P. Michael and J. H. Crook, eds. *Comparative Ecology and Behaviour of Primates,* pp. 171–209. London: Academic Press.

—— 1973b. Synecology of Bornean primates I. A test for interspecific interactions in spatial distribution of five species. *Amer. J. Phys. Anthrop.* 38:655–660.

—— 1977. Feeding behaviour of orang-utans of the Kutai Nature Reserve, East Kalimantan. In Clutton-Brock, ed. 1977:383–413.

—— 1978. Diets, densities, and distribution of Bornean primates. In Montgomery, ed. 1978:465–78.

—— 1979. Individual activity patterns and the solitary nature of orangutans. In Hamburg and McCown, eds. 1979:235–73.

Rodman, P. S., and H. M. McHenry. 1980. Bioenergetics and the origin of hominid bipedalism. *Amer. J. Phys. Anthrop.* 52:103–6.

Sigmon, B. A. 1974. A functional analysis of pongid hip and thigh musculature. *J. Hum. Evol.* 3:161–85.

Stanley, S. M. 1973. An explanation for Cope's rule. *Evolution* 27:1–26.

Taylor, C. R. 1980. Mechanical efficiency of terrestrial locomotion: A useful concept? In H. Y. Elder and E. R. Trueman, eds. *Aspects of Animal Movement,* pp. 235–44. Cambridge, England: Cambridge University Press.

Taylor, C. R., S. L. Caldwell, and V. J. Rowntree. 1972. Running up and down hills. Some consequences of size. *Science* 178:1096–97.

Temerin, L. A. 1983. Evolutionary sources of primate diversity. Ph.D. dissertation. Davis, Calif.: University of California.

Temerin, L. A., and J. G. H. Cant. 1981. An explanation of the evolutionary divergence of Old World monkeys and apes. *Amer. J. Phys. Anthrop.* 54:284.

—— 1983. The evolutionary divergence of Old World monkeys and apes. *Amer. Natur.* 122:335–51.

Tuomi, J. 1980. Mammalian reproductive strategies: A generalized relation of litter size to body size. *Oecologia* 45:39–44.

Tuttle, R. H. 1969. Quantitative and functional studies on the hands of the Anthropoidea I. The Hominoidea. *J. Morph.* 128:309–64.

—— 1970. Postural, propulsive, and prehensile capabilities in the cheiridea of chimpanzees and other great apes. In G. H. Bourne, ed. *The Chimpanzee.* 2:167–253. Basel: Karger.

Van Soest, P. J. 1967. Development of a comprehensive system of feed analyses and its application to forages. *J. Anim. Sci.* 26:119–28.

Western, D. 1979. Size, life history and ecology in mammals. *Afr. J. Ecol.* 17:185–204.

Wheatley, B. P. 1978. The behavior and ecology of the crab-eating macaque (*Macaca fascicularis*) in the Kutai Nature Reserve, East Kalimantan, Indonesia. Ph.D. dissertation. Davis, Calif.: University of California.

—— 1980. Feeding and ranging of East Bornean *Macaca fascicularis*. In D. G. Lindburg, ed. *The Macaques: Studies in Ecology, Behavior and Evolution*, pp. 215–46. New York: Van Nostrand Reinhold.

Whittow, G. C. 1973. Evolution of thermoregulation. In G. C. Whittow, ed. *Comparative Physiology of Thermoregulation*. 3:201–58. London: Academic Press.

Wrangham, R. W. 1977. Feeding behaviour of chimpanzees in Gombe National Park, Tanzania. In Clutton-Brock, ed. 1977:503–38.

—— 1979. Sex differences in chimpanzee dispersion. In Hamburg and McCown, eds. 1979:481–89.

9 The Role of Food-Processing Factors in Primate Food Choice

KATHARINE MILTON

MOST PRIMATES are strongly dependent on plant foods. Field studies show that primates do not feed on plant parts at random but rather have decided food preferences (Casimir 1975; Hladik 1977; Oates et al. 1977, 1980). The food choices of primates have generally been attributed to one of two principal factors: the nutritional and/or toxic content of the particular plant part or its relative availability in space or time (Hladik and Hladik 1969; Casimir 1975; Glander 1978; McKey et al. 1978; Milton 1979; Oates et al. 1980). It has also been suggested that body size may influence primate food choice (Hladik 1978; Gaulin 1979; Milton 1980). Though each of these factors would appear to play some role in determining primate food choice, very often sympatric plant-eating

KATHARINE MILTON has studied behavioral ecology of howler monkeys and spider monkeys on Barro Colorado Island, Republic of Panama, as well as at several sites in South America, and most recently has examined the ecology of tribal people in Amazonia and woolly spider monkeys in southeastern Brazil. She has successfully measured physiological variables of animals and plants that play important roles in theory but often receive short empirical shrift. She has carried out extensive analyses of nutrient and other components of howler monkeys' foods and has measured metabolic rates of wild howlers through imaginative solutions to this difficult problem. In the following paper she presents results of a new piece of research on gut passage rates in various primate taxa. Her results and discussion provide a unique perspective on the role of gut morphology in food choice by animals.

primate species of approximately the same adult body size show quite different patterns of food choice (Hladik and Hladik 1969; Clutton-Brock 1974; Hladik 1977; Sussman 1979; Milton 1981). At times a particular food appears hyperabundant, indicating that immediate competition for it may not be an issue. In such cases, none of the above factors, either singly or in combination, is sufficient to explain the feeding patterns observed. This strongly suggests that other factors are also important.

In examining factors related to food choice, the ecological literature has tended to focus primarily on features of external morphology. In birds, for example, food choice has been correlated with such features as body or bill size, and similar explanations have been suggested for food choices of fish and reptiles (Schoener 1965; Hespenheide 1971, 1975; Ricklefs 1972; Werner 1977). Most such studies, however, have examined food choices of secondary rather than primary consumers. Features of external morphology may well be of critical importance to secondary consumers since their items of diet, animal prey, are generally mobile and often protected by external defensive features such as spines or claws. Secondary consumers may therefore require corresponding features of external morphology to cope efficiently with the escapist or defensive tactics of particular prey types. As primary consumers eating sessile plant parts, primates face many problems quite different from those faced by secondary consumers. These problems would appear to call for somewhat different adaptive solutions.

Plant Defenses

In large part, the problems faced by plant-eating primates have to do with the types of defenses plants employ to protect their potentially edible parts from predation. Such defenses can be quite elaborate and range through several levels of organization; generally, plants employ several levels of defense simultaneously. The first level of defense may be mechanical. Mechanical defenses can include such things as thorns, spines, hairs, or hard seed coats; they can also include the height at which the potentially edible item, or the branching structure supporting the item, is found. Most such external defenses appear to pose little problem to primates (Milton, personal observation). The second line of defense can be the manner in which potentially edible parts are deployed in space and time. The low individual densities characteristic of many tropical tree species and their varied phenological patterns can function, at least in part, to lower predation on leaves, flowers, or unripe fruits (Janzen 1970; Milton 1980). The single most important defense employed by plants, however, would appear to be the chemical composition of the plant parts themselves. It is this third level of chemical protection that appears to be the most ubiquitous plant defensive strategy. Under the term *chemical composition* I include the nutritional content of the plant or plant part, its proportion of indigestible material, and its content of presumably defensive compounds such as phenolics or alkaloids. To appreciate better how these chemical features might function to deter primary consumers, it is useful to take a closer look at some of the problems associated with each.

Nutrient Content

Unlike many foods from the second trophic level, foods from the first trophic level show considerable variability in nutrient content. For example, leaves, particularly young leaves, may contain considerable protein but are usually low in ready energy (Milton 1979, 1981). Conversely, ripe fruit tends to be high in ready energy but notably low in protein (Hladik and Hladik 1969; Milton 1981). Flowers of particular species may be quite nutritious but, as a food category, show considerably more interspecific variability in nutritional content than either leaves or fruit (Milton 1980). Many primates must choose foods from more than one dietary category each day to get the balance of essential nutrients and energy they require. This in turn limits the amount of food that can be eaten from any one category per unit time. Potential nutrient imbalance is a real and persistent problem for many primary consumers and one that is generally foreign to secondary consumers (Maynard and Loosli 1969; Westoby 1974). This is not to imply that nutrient allocation patterns in plant parts reflect herbivore defenses. But the nutrient imbalance characteristic of most plant foods nonetheless poses a dietary problem for primary consumers that must somehow be overcome.

Indigestible Materials

A second major problem with plant foods, and one that deserves further study, is the fact that most plant parts are high in indigestible cell wall material. Plant cell walls are made up primarily of cellulose, hemicelluloses, and lignin. These three cell wall constituents are impervious to all known digestive enzymes of vertebrates. Thus the more cell wall material a primate eats, the more indigestible bulk is being passed through the digestive tract, and this material, in the absence of certain mitigating factors, provides no nutritional benefits to the feeder (Moir 1967; Parra 1978). Apparently in response to this problem, many plant-eating animals have enlarged sections in the gut that harbor vast colonies of bacterial flora with cellulolytic properties. This gut flora can degrade the cellulose and hemicelluloses (structural carbohydrates) of plant cell walls by fermentation. During fermentation various end products are produced, including energy-rich short-chain fatty acids (VFAs). These fatty acids can often be absorbed by the host and may make an important contribution to its energy budget (Bauchop and Martucci 1968; Parra 1978). In the absence of sufficient gut flora, however, animals eating diets high in cell wall material, particularly highly lignified material, apparently derive little nutritive or energetic benefit from passing this material through the gut.

The leaves eaten by primates are generally high in cell wall material. Some 30% to more than 50% of the dry weight of a given leaf may be made up of cell walls (Milton 1979). Therefore, we might predict that primates deriving an important part of the diet from leaves should show adaptations either to facilitate fermentation or to move indigestible matter rapidly through the gut.

Fruit, too, can be extremely high in indigestible material. This is because most edible fruit pulp is elaborated by trees to serve as a lure for seed-dispersal agents. Tree species have many techniques for ensuring that fruit-eating animals actually will disperse seeds away from the parent tree. Often nutrients adhere so tightly to the seed that it seems most efficient for the feeder to swallow the entire fruit—skin, seeds, and pulp—and digest off the pulp (e.g., *Spondias mombin; Doliocarpus* spp.). In other cases, seeds are so minute or so thoroughly mixed with the edible pulp that it seems most expedient to swallow the entire fruit rather than try to pick out the indigestible material (*Ficus* spp.; *Hyeronima laxiflora;* Milton, personal observation). In both cases, however, the feeder is filling up on seeds, as well as pulp, and, unlike leaves, seeds generally are not broken down by digestive enzymes or bacteria in the gut and are excreted intact (Hladik and Hladik 1969; Milton personal observation). Thus fruit-eating primates, whose foods typically contain seeds that are swallowed, should have some means of dealing with indigestible material such that it is passed from the gut as rapidly as possible once digestible nutrients have been removed from the seed surface. This poses an interesting paradox for primates eating both fruits and leaves since fruits, because of the seeds, would appear to call for relatively rapid food passage rates whereas the latter would be most efficiently digested with slow food passage rates (Milton 1981, and below).

Secondary Compounds

All plant parts contain the chemical constituents known as secondary compounds, some of which may function to deter plant-eating animals (Feeny 1971; Freeland and Janzen 1974; Ryan and Green 1974; McKey et al. 1978; Oates et al. 1980). Some secondary compounds are distasteful or malodorous; others can interfere with the digestion of nutrients in the gut or with the metabolic processes of the feeder, at times with fatal results (Freeland and Janzen 1974; Glander 1975). There is a vast and rapidly growing literature on the possible interrelationships between secondary compounds and primary consumers, but as yet there is no general consensus about the role of these substances in plant defense.

Few data are available on the role of secondary compounds in determining patterns of primate food choice. Available data suggest that in some cases such compounds do not exert any notable influence in food selection whereas in others they may well be of considerable importance (Glander 1975, 1978; Hladik 1977; Milton 1979; Oates et al. 1980). *In vitro* experiments show that leaf tannins present in species eaten by primates can bind with protein and lower digestive efficiency. If similar conditions exist *in vivo,* this too could influence food selection patterns (McKey et al. 1978; Milton et al. 1980; Waterman et al. 1980).

Solutions to Chemical Problems

Since many principal defenses of plants are chemical, they are internal to the plant or plant part whereas the principal defenses of foods from the second trophic level tend to be morphological and external. We might therefore expect that primary consumers, in contrast to many secondary consumers, have been under strong selective pressure with respect to the development of features of internal morphology particularly well suited to deal with the primarily internal defenses of their principal items of diet. In particular, the morphology of the gut is expected to show adaptations serving to counteract or resolve many of the nutritional and other chemical problems characteristic of plant foods. Such adaptations, in turn, are likely to play an important role in food choice by permitting particular primate species to specialize on some subset of the total range of plant resources. This could lower foraging costs associated with selective feeding on patchily distributed plant foods or facilitate use of more abundant plant foods high in fiber (e.g., mature leaves). It could also help to ameliorate possible competition between sympatic primate species eating plant-based diets.

In order to appreciate why the morphology of the gut might help to resolve some of the chemical problems inherent in most plant foods, it is useful to understand some aspects of the digestive process. The digestibility of food depends on two principal factors, namely, digestive rate and the amount of time food can remain in the digestive tract. The amount of time food remains in the digestive tract is determined by the passage rate of digesta and the capacity of the digestive rate (Van Soest 1977). This relationship can be expressed in a simplified fashion by the equation: $D = r_d\ (L/r_p)$ where D is digestibility, r_d is rate of absorption per unit area of gut, L is gut length, and r_p is passage rate for digesta in some unit such as millimeters per minute. Thus overall digestibility of a given meal is affected by the interaction between the amount of time food can remain in the digestive tract and the passage rate of food. In effect, these two factors represent opposite points on a continuum, and animal species can be ranked along this continuum, depending on whether they tend to maximize the efficiency of digestion or the volume of food processed per unit time (Bell 1971; Parra 1978; Milton 1981). Animals that pass food through the gut relatively slowly for their body mass presumably have adopted a solution emphasizing maximal nutrient extraction from each meal. In particular, we might expect such animals to show adaptations in the digestive tract for the efficient fermentation of plant cell wall material, which is generally a time-consuming process. Conversely, animals passing food through the gut relatively rapidly for their body mass might be feeding on a resource base that is generally of poor quality or low in one or more essential nutrients. Fruit would fall into this latter category since it tends to be high in nonstructural carbohydrates and low in protein. By passing large quantities of low-quality or imbalanced foods rapidly through the gut, such animals should be able to extract an adequate and balanced diet.

Therefore, a knowledge of the food passage rate characteristic of a given species can give insight into the type of foods the animal is best able to process and help to explain its pattern of food choice.

To investigate the role of gut morphology in primate food choice, I carried out a series of feeding experiments to determine food passage rates of two neotropical primate species, howler monkeys (*Alouatta palliata*) and spider monkeys (*Ateles geoffroyi*). Howler monkeys and spider monkeys occur sympatrically over much of their very wide geographical range, are approximately the same adult body size, and feed exclusively on foods from the first trophic level. Thus they are good subjects for an examination of this nature.

Feeding trials showed that howler monkeys, which eat considerable foliage, retained food in the digestive tract for a significantly longer time than spider monkeys (Milton 1981). As discussed above, leaves tend to be high in cell wall material, but if food passage rates are sufficiently slow and adequate numbers of cellulolytic gut flora are present, the structural carbohydrates in leaves can provide energy for the feeder indirectly. Data show that considerable cell wall material is degraded in the hindgut of howlers and indicate that some of the fatty acids produced in fermentation are absorbed (Milton et al. 1980; Milton and McBee 1982). In contrast, spider monkeys, which are strongly frugivorous at all times, turned over gut contents very rapidly. Mean time of first appearance (TFA) of markers in spider monkeys was 4.4 ± 1.5 h compared with 20.4 ± 3.5 h in howlers. This fast food passage permits spider monkeys to rid the gut rapidly of indigestible seeds present in fruit, their principal dietary item. Further, by turning over a large volume of fruit each day, they are apparently able to compensate for the low protein content of such foods.

The stomach, small intestine, and cecum of these two species are roughly similar in relative surface areas (Hladik 1967; Milton 1981), but colons of howlers have approximately double the capacity of those of spider monkeys. In effect, it would appear that each species is living on a diet that is not tenable for the other owing to differences in gut proportions and food passage rates. On a primarily fruit diet of the type eaten by spider monkeys, howlers with their voluminous colons and slow food passage rates presumably would not be able to meet their demands for protein. On the other hand, on a primarily folivorous diet of the type eaten by howlers, spider monkeys, with their relatively short narrow colons and fast food passage rates presumably would not be able to meet demands for energy. Thus gut morphology and food passage rates, in conjunction with other aspects of behavior and morphology characteristic of each species, appear to limit food choice so that each species is best nourished when eating a particular subset of the available plant resources. Further details of these experiments are found in Milton 1981.

These results suggested that similar features of gut morphology, as well as other factors related to food-processing efficiency, might also limit food choices by other primate species. To investigate this possibility I carried out the series of feeding experiments described next.

Feeding Experiments

In these experiments, individuals of 14 primate species were fed colored markers to obtain an estimate of food passage rates. Relevant data on all study subjects are presented in table 9.1. New World species were tested in the Republic of Panama at the Gorgas Memorial Laboratory, and at the Smithsonian Tropical Research Institute on Barro Colorado Island, and in Brazil at the Jardim Zoologico in São Paulo and the INCAP Jardim Zoologico in Manaus. Apes were tested at the San Francisco Zoo. Three human subjects also volunteered to eat markers to provide a data point for *Homo sapiens.*

Methods

Small colored plastic markers (about 4 mm wide, 1 mm thick) were concealed in foods offered to subjects. Subjects differed in the number of markers swallowed, but all animals listed in table 9.1 consumed enough markers that it appeared worthwhile to monitor fecal material for time of first appearance of markers. Either directly before or immediately after markers were swallowed, animals were fed their principal meal of the day, which in all cases consisted of a variety of fresh fruits, some leafy material such as lettuce, bread, generally soaked in milk and vitamins, and in a few cases pieces of commercial monkey chow. My objective was to get markers into the stomach along with a large, fresh bolus of food, since mixing in the stomach should distribute markers through the food. The time of first appearance of markers and the number of markers present in each set of feces could then be used to estimate the passage time of that meal.

There are a number of difficulties in using markers to estimate food passage rates (Alvarez and Freedlander 1924; Wiggins and Cummings 1976). Even under ideal conditions, using soluble and particulate markers simultaneously, it can be difficult to interpret the biological significance of data obtained. Since in these experiments all animals ate the same type of markers and approximately the same type of foods, data should be comparable, at least in rela-

Table 9.1. Time of First Appearance of Markers in Fourteen Primate Species

Species	No. Individuals, Sex and Age	Est. Mean Body Wt. (kg)[a]	No. Trials per Subject	Time of Feeding of Markers	No. of Hours for Time of First Appearance of Markers	Range of TFAs (when n = >1)
Cebus capucinus	3 AM[b]	3.0	2	0800; 1400	$\bar{X} = 3.5$	3.0–5.5
Cebus apella	1 AM	3.5	2	0800; 1400	3.5	
Pithecia monachus	1 AF	1.9	1	0900	20.0	
Chiropotes albinasa	1 AM, 1 SAM	3.0	1	0930	5.0	See text
Cacajao calvus	1 AM, 1 AF	3.0	1	0945	5.0	See text
Alouatta palliata	2 AM, 1 AF, 1 JF	6.9	[c]	0800; 1630	$\bar{X} = 20.4$	16–23
Ateles geoffroyi	1 AM, 6 AF, 2 JF	7.6	[c]	0800; 1400	$\bar{X} = 4.4$	2.75–7.75
Ateles paniscus	2 AM, 2 AF	7.8	1	1030	$\bar{X} = 5.25$	5.0–5.5
Lagothrix lagotricha	2 AM, 1 SAF	6.5	1	1000	$\bar{X} = 6.75$	6.5–7.0
Brachyteles arachnoides	1 AF	12.0	1	0900	8.0	
Pan troglodytes	3 AF	45.0	1	1600	36–38	See text
Pongo pygmaeus	1 AM, 1 AF	53.0	1	1600	36–38	36–38
Gorilla gorilla	1 AF, 1 JM	103.0	1	1600	36–38	See text
Homo sapiens	1 AF, 2 AM	$\bar{X}$ 67 ± 8.7	1	0730	26 ± 1.0	25–27

[a] Sources for weights as follows: All Pithecinae, *Cebus apella*, and *Ateles paniscus* from Fleagle and Mittermeier (1981) or Mittermeier, personal communication; *Alouatta palliata* from Milton, based on actual wts. of study subjects; *Ateles geoffroyi* from Querling (1950); *Lagothrix* and *Cebus capucinus* from Napier and Napier (1969); *Pan* and *Pongo* from Clutton-Brock and Harvey (1977). *Brachyteles*, female gorilla, and *Homo sapiens*, actual weights of study subjects. Juvenile gorilla, wt. = ca. 46 kg.

[b] A = adult; M = male; F = female; S = subadult; J = juvenile.

[c] Variable, see Milton (1981) for complete details.

tive terms. In my opinion these markers probably move through the gut in the same manner as indigestible food particles such as seed coats.

Once markers and food were swallowed, animals were checked at regular intervals to see if defecation had occurred. When feces appeared, the time was recorded and fecal material was examined for presence and quantity of markers. All primate facilities except Barro Colorado Island closed at 1700–1800 h and did not reopen until 0800 the following day. Therefore, during these hours animals could not be monitored. Time of first appearance of marker material in feces produced during this period could only be estimated based on appearance of the fecal matter. Because of this lack of continuity, data vary in quality. In some cases I was present when the first marker material was passed; in other cases, estimates are accurate to within 15 min; in other cases, TFA is simply a best estimate. When presenting results below, I give my opinion of the accuracy of the data. Further, though TFA in most cases is believed to be a good estimate, data from these trials cannot be used to indicate precisely how much food was passed through the tract per unit time; the data indicate only when markers first began to pass out of the tract and, in some cases, the percentage of marker material present. Moreover, animals were not fed natural diets. These results might change somewhat under free-ranging conditions owing to differences in the digestibility of wild foods or to different activity and feeding regimes of free-ranging animals.

Results

New World Monkeys

Cebus capucinus Animals ate good quantities of markers and food and were closely observed. Time of first appearance of markers averaged 3.5 h. Many markers were passed within 8 h after ingestion, but some were still appearing in the feces in small amounts 24 h later.

Cebus apella This animal ate good quantities of markers and food and was closely observed. Time of first appearance was 3.5 h. Many markers were passed within 8 h after ingestion, but some were still appearing in the feces in small amounts 24 h later.

Pithecia monarchus This animal ate moderate amounts of markers and food and was very closely observed during the first 8 h after ingestion during which time it is certain that no markers were passed. By 0800 the following day (approximately 23 h after the initial feeding), this animal had passed many markers. Feces containing markers appeared very fresh, and I estimate their passage at around daybreak (0530–0730). Time of first appearance in this animal is therefore placed at ca. 20 h. More fecal material containing markers was passed in my presence at 0800. The fecal material was produced in small pellets similar in appearance to those of a rabbit or goat.

Chiropotes albinasa Results of trials on both *Chiropotes* should be regarded as tentative. The adult male swallowed only three pieces of marker material and ate little food. Food passage rates were probably quite depressed. No markers were produced during the first 8 h of observation. One marker was recovered in feces at 0800 the following day and appeared to have been passed in ca. 20 h.

The subadult *Chiropotes* ate good quantities of both markers and food. Time of first appearance in this animal was ca. 5 h after ingestion and a considerable amount of marker material was passed within 8 h after ingestion. This animal had slightly liquid feces, however, and passage may therefore have been speeded up. Since conditions were not entirely normal in either case, results are inconclusive. Estimates from the subadult are believed to be more accurate because the animal did eat considerable marker material and food, and many markers were recoverd.

Cacajao calvus Each animal ate a moderate quantity of markers and food. Time of first appearance in both was ca. 5 h after ingestion, and a considerable amount of marker material was passed within 8 h after ingestion.

Alouatta palliata Animals ate good quantities of markers and food and were very well observed. Time of first appearance averaged 20.4

h—about 23 h if fed to animals before noon and about 16 h if fed after noon. Most markers were excreted within 30+ h after ingestion, but a few were still appearing in feces 72 h after the initial feeding. (See Milton [1981] for details of these feeding trials.)

Ateles geoffroyi Animals ate good quantities of markers and food and were very well observed. Time of first appearance averaged 4.4 h with a range of from 2.75 to 7.75 h. Most markers appeared to be excreted within 8+ h after ingestion but a few were still appearing in the feces more than 24 h after the initial feeding. (See Milton [1981] for details of these feeding trials.)

Ateles paniscus Animals ate good quantities of markers and food and were very well observed. Time of first appearance averaged 5.25 h. Most markers appeared to be excreted within 8+ h after ingestion.

Lagothrix lagotricha Animals ate good quantities of markers and food and were well observed. Time of first appearance averaged 6.75 h. More markers were seen in feces the following morning, ca. 22 h after the initial feeding.

Brachyteles arachnoides This adult female ate good quantities of food and markers and was very well observed. Time of first appearance was exactly 8 hours after ingestion. Considerably more marker material was seen in fecal material the following morning, presumably passed some 20–23 h after the initial feeding. A second experiment was carried out on this animal, but results could not be used, for she developed severe diarrhea with greatly accelerated passage rates that were decidedly abnormal.

Apes

All apes ate generous quantities of markers (ca. 150 each) and food and were well observed during daylight hours.

Pan troglodytes Good data are available for three adult females. One had a TFA of ca. 15 h, producing three markers in one set of feces. The first notable appearance of markers, however, occurred in this female around 36–38 h after ingestion. The other two females had

TFA's of around 36–38 h and produced many markers (25%) at this time.

Pongo pygmaeus The two adult orangs appeared to have approximately the same TFA, estimated at 36–38 h after ingestion. As for two of the three chimpanzees, TFA of markers occurred with the appearance of considerable (15%–25% of total) marker material.

Gorilla gorilla Markers were initially fed to five gorillas, but owing to logistical problems, accurate data could be collected only for two, a juvenile male and an adult female. The juvenile male had a TFA of ca. 17 h, producing two markers at this time in one set of feces. The first notable passage of marker material (ca. 25% of total) in this animal occurred approximately 36–38 h after ingestion. The adult female had a TFA of approximately 36–38 h and passed around 18% of the markers at this time. More than 84 h after the initial feeding, markers were still appearing in small amounts in her feces, as well as in those of two other adult animals whose passage rates could not be carefully monitored. This adult female and one other adult were observed to practice coprophagy (one incident for each animal) and one juvenile ingested feces of one adult female (one incident).

Homo sapiens The three human subjects, two adult men and one adult woman, swallowed 150 markers each along with considerable food and were very well observed. Time of first appearance averaged 26 h ± 1 h. As was characteristic of the Pongidae, considerable marker material was excreted at this time (33% of total in two male subjects, 10% in the female subject). In a series of detailed experiments on human transit times using glass beads as markers, Alvarez and Freedlander (1924) found that the average subject passed 15% of the beads within 24 h after ingestion, 40% within 48 h and 75%–80% within 96 h. Then days or even weeks could elapse before the remaining 20% were all finally recovered.

Discussion

Field studies show that different primate species often feed on different subsets of the available plant resources, but reasons for these species-specific patterns of food choice are not always apparent. It was suggested that features of gut morphology (and its effects on food passage rates), and other factors related to food-processing efficiency, might play an important role in determining what subset of plant foods a given primate species finds most suitable, i.e., that food choice might be dictated as much by internal constraints intrinsic to the digestive "strategy" of the feeder as by extrinsic factors such as nutrient content or relative availability.

Data on food passage rates show that smaller bodied species tend to pass food through the gut more rapidly than larger bodied species (fig. 9.1). Mehrtens (1971), in working with food transit times in ruminants, found that body size was the single variable showing the highest (negative) correlation with food passage rates. There seems

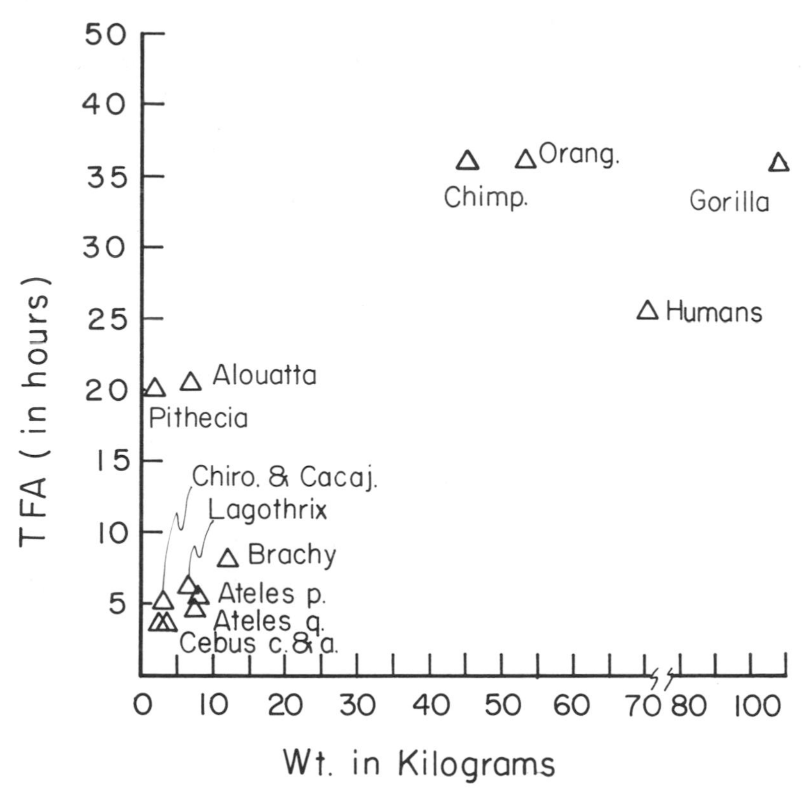

Figure 9.1. Time of first appearance of markers plotted against body weight for 14 primate species.

to be good physiological reason why smaller species should generally show faster food transit times. As pointed out by Parra (1978), as body size increases, metabolic costs per unit body weight decrease exponentially, while gut volume remains proportionate to body mass (see also Hungate et al. 1959). This implies that smaller homeotherms have disproportionately high energetic costs per unit body weight but no "extra" room in the gut to process a proportionately greater amount of food. One solution to this rather paradoxical problem is for smaller animals to turn over gut contents more rapidly than their larger counterparts (Hungate et al. 1959; Parra 1978). This fact helps to explain why smaller primate species generally seek out low-fiber, nutritionally concentrated resources, amenable to rapid digestion (Hladik 1977; Gaulin 1979).

Data show that of the smaller species examined, all except *Pithecia* had TFAs of 3.5 to 5 h and excreted considerable marker material within 8 h after ingestion. Field data show that all of these species except *Pithecia* avoid leaves as food (Fooden 1964; Oppenheimer 1968; Roosmalen, Mittermeier, and Milton 1980; Fleagle and Mittermeier 1981; Happel, 1981). The two *Cebus* species eat fruits and seeds from the first trophic level and insects and small vertebrates from the second (Oppenheimer 1968; Hladik and Hladik 1969; Milton, personal observation). The manual dexterity of this genus may have developed, at least in part, in association with its dietary habit of unrolling dead leaves to look for insect prey (personal observation).

Chiropotes is an unusual primate for its body size. It is generally reported to take all of its diet from the first trophic level, although Ayres and Nessimian (1982) indicate some use of food from the second trophic level. First-level foods reported for *Chiropotes* are fruit pulp, and immature and mature seeds, particularly those from members of the Lecythidaceae (Brazil nut family) and Bigoniaceae (Roosmalen et al. 1980; Ayres, personal communication). Such seeds can be a rich source of oils and protein and are low in indigestible bulk.

No data are available on the natural diet of *Cacajao,* but its food passage rates and dental morphology are similar to those of *Chiropotes,* suggesting that this genus may exploit a similar diet. Data are available on the relative surface area of different sections of the gut for most of these smaller species (including *Pithecia*); these show that the small intestine has by far the greatest relative surface area and that other sections of the gut are not notable in volume (Fooden 1964; Hladik 1967; Chivers and Hladik 1980). A well-developed small intestine is consistent with a diet of high-quality, volumetrically concentrated food resources, calling for a digestive strategy facilitating the rapid absorption of nutrients without the need for prolonged retention of food in other sections of the digestive tract.

The one *Pithecia* used in the trials had a strikingly long TFA in comparison with most other species tested. If future work confirms that long retention time is characteristic of the genus, *Pithecia* would appear to have a digestive strategy considerably different than those of other members of the Pithecinae or the *Cebus* species. The pelleted appearance of the feces also suggests a different strategy with respect to nutrient extraction. Data from Fooden (1964) suggest that *Pithecia* may show some reduction in the size of the small intestine and some increase in the size in the colon when compared with *Chiropotes chiropotes.* Free-ranging animals are reported to eat fruits of trees and vines (Fleagle and Mittermeier 1981). Happel (1981) noted fruit-eating in her observations of free-ranging animals but also commented on the fact that animals spent 16% of total feeding time eating leaves. In this context, it is interesting to note that *Pithecia* occurs sympatrically with both *Chiropotes* and *Cacajao,* but that members of the latter two genera do not appear to show geographical overlap (Mittermeier, personal communication; Ayres, personal communication). A different digestive strategy implies a different pattern of feeding that could lower competitive overlap between *Pithecia* and each of the other members of the Pithecinae, facilitating coexistence.

As primates increase in body size, metabolic costs per unit body weight become proportionally lower (Kleiber 1961), but absolutely more food is required (Bell 1971; Jarman 1974). Because larger body size confers greater energetic lability, however, larger species are more likely to show gut modifications and digestive strategies predicated on long retention time of food. The efficient digestion of plant cell wall material, particularly more lignified material, is a time-consuming process (Van Soest 1977, 1982). Van Soest (1982) has estimated that a body size of 10 kg or greater may be required for a digestive strategy predicated entirely on foregut fermentation. Most arboreal primates fall below this critical body mass, but because of lower energetic demands per unit body weight, larger primates should be able to exploit a wider range of fibrous materials more efficiently than smaller primates since they can "afford" somewhat slower food transit times.

Alouatta palliata is a moderate-sized arboreal primate (adult body mass 7–9 kg). As noted, members of this genus often live for long periods of time on diets consisting almost entirely of leaves, including some mature leaves. The relatively capacious hindgut and slow food passage rates provide conditions suitable for the efficient fermentation of plant cell wall material. Howler monkeys generally choose young leaves or mature leaves, which are unusually high in protein and relatively low in cell wall material (Milton 1979). Fermentation of such material should be more rapid than would be the case with most mature leaves (Van Soest 1977). Apparently because of limitations imposed by its body size and gut morphology, *Alouatta* must feed selectively, choosing leaves of high quality that provide maximal energetic and nutritional returns in exchange for the amount of time animals are able to invest in digesting them.

The three other moderate-sized primates examined in this study are all members of the subfamily Atelinae. *Ateles* falls in the middle of the three in body mass but has by far the most rapid turnover of ingesta. As noted, this genus is very strongly frugivorous. By passing a large volume of fruit through the gut each day and supplementing this basic fruit diet with selected leaf buds and young leaves, members of this genus are apparently able to satisfy all nutritional requirements. *Lagothrix* is somewhat smaller in body mass than *Ateles* and also has a somewhat slower food transit time. Field data show that animals eat fruits, leaves, and insects, probably more leaves and insects than is the case for *Ateles* (Izawa 1975; Kavanagh and Dresdale 1975). My data on captive animals suggest that *Lagothrix* may also be able to process harder fruits than *Ateles* (see appendix to this paper). *Lagothrix* occurs sympatrically with *Ateles,* but these data indicate that animals are feeding on somewhat different subsets of the available plant resources and that *Lagothrix* may also rely more heavily on foods from the second trophic level.

Brachyteles is the largest ateline and the largest neotropical primate. The adult female used in experiments weighed 12 kg (S. Rodriguez, personal communication), and free-ranging adults are reported to weigh some 15 kg (Aguirre 1971; J.C. Magalhães, personal communication). This large body size should confer sufficient energetic lability on animals so that they can exploit slower digesting foods efficiently. Data show that *Brachyteles* is highly folivorous (Milton, unpublished). In some months from 70% to more than 90% of total feeding time is spent eating leaves, including quantities of mature leaves. Animals also eat fruits and flowers when these are available but their dietary staple is foliage and animals are fully as folivorous as *Alouatta.* Like *Alouatta, Brachyteles* has a simple stomach and a capacious cecum and colon. Fecal material has the same curious pungent odor as that of *Alouatta,* and, similarly, shows a high content of endogenous fecal nitrogen (Milton, unpublished). These facts strongly suggest that *Brachyteles* routinely carries out fermentation of plant structural carbohydrates in the hindgut. The large body size of *Brachyteles* should enable it to be somewhat less selective than *Alouatta* when feeding. Faster food transit times with respect to TFA indicate that the digestive strategy of *Brachyteles* may be to take in copious quantities of leafy foods, holding such food in the gut only for sufficient time to extract the more accessible or degradable components, and then passing the more lignified material from the gut. In contrast, *Alouatta* is presumed to be a more selective feeder, taking in less lignified, higher quality leafy foods that can be more thoroughly degraded and then holding these foods in the gut for a longer fermentation treatment.

All of the Pongidae had approximately similar TFAs in spite of some considerable differences in adult body weight. Both chimps and orangs passed some 25% of the markers in the first 36–38 h after ingestion and 50%–67% within the first 48 h. My subjective impression was that chimps defecated smaller amounts more frequently than the orangs and had looser feces. Future trials may show that chimps pass more of a given meal within a 48-hour period than is the case for orangs. Field data indicate that chimps specialize more heavily on ripe fruits than is the case with orangs, who also eat quantities of unripe fruit as well as bark and leaves (Wrangham 1977; Rodman 1977, this volume). A strong dietary bias toward ripe fruit might well produce a pattern of frequent defecation of small amounts of fecal matter in chimps as is the case is *Ateles,* another ripe fruit specialist (Milton personal observation).

Adult gorillas appeared to retain a greater proportion of a given meal for a longer time than other pongids. They also show a somewhat greater relative volume in the colon than chimps and orangs (calculated from data in Chivers and Hladik 1980), and they are coprophagous (see also Harcourt and Stewart 1978). In the wild, gorillas eat both leaves and fruit, but leafy matter composes the greatest proportion of the diet; mountain gorillas may eat leafy material almost exclusively (Casimir 1975; Fossey and Harcourt 1977). The slow turnover time of meals and coprophagy strongly suggest that fermentation may provide gorillas with some of their required energy, and, by ingesting feces, they may also improve their protein and/or vitamin economies (McBee 1971).

The human subjects began to turn over food considerably more rapidly than either chimps or orangs, yet body weights were higher. The faster initial turnover rate in the human subjects may have been due to the more refined nature of their items of diet. Zoo-living pongids, however, are also eating somewhat refined diets. The slower TFA times of chimps and orangs could have been due to artificial feeding schedules characteristic of captive conditions. In general, however, it should be stressed than animal protein appears to be a more prominent dietary item for humans than for apes. Archaeological evidence indicates that hunting animal prey is in ancient trait in the hominid line (Isaac 1971). The small intestine of humans shows a greater relative volume than in apes, whereas apes show a greater relative volume in the area of the colon. Both the relatively large capacity of the human small intestine and the relatively rapid TFA times in my human subjects strongly support the inference that the human gut is particularly well adapted to process high quality dietary items that are volumetrically concentrated and rapidly digested. In a savanna-mosaic environment, higher quality dietary items are far more patchily distributed than in tropical forests (Milton 1981). In order to exploit such foods successfully, the ancestral line leading to modern hominids may have depended heavily both on meat protein in the diet and on overall increased food search efficiency to afford the costs associated with a diet based on hyperdispersed, high quality foods (see Milton 1981). Mobility is presumed to be a critical feature in the exploitation of such a diet in a savanna-mosaic setting. Evidence from field behavior of gorillas, orangutans, and chimpanzees (see, for example, Rodman, this volume) strongly indicates that in the hominoid line, a high intake of bulky, fibrous or slowly digesting food stuffs results in decreasing mobility and a limited day range (gorilla and orangutan vs. chimpanzee). The heritage features of the hominoid gut appear to be such that decreasing dietary quality sets energetic limitations on an active lifestyle. Thus I would postulate that the ancestral line leading to modern humans was more chimp-like in its dietary habits than orangutan- or gorilla-like and that this same trend toward exploitation of high quality dietary items was retained and intensified during hominid evolution in a savanna-mosaic setting and is still characteristic of the hominid line today.

Overview

The data presented above are often scant and represent single trials, only for TFA's and often with only one subject. Before any firm conclusions can be drawn from these results, similar experiments should be repeated on a larger scale and in more depth. Nonetheless, these results do suggest a few tentative conclusions.

The feeding repertoire of any animal is a combination of its behavior, morphology, and physiology. But, depending on the animal examined and its dietary regime, one of these features may assume special importance in foraging success. Since most anthropoids are strongly dependent on plant foods, features of their internal morphology, particularly their gut morphology, might show specializations to help them overcome many of the chemical problems potentially inherent in plant-based diets. Data show that in some cases (i.e. howler monkeys, spider monkeys, and possibly *Pithecia*), the structure of the gut and its effect on food transit time plays a very important role in helping animals compensate for or overcome nutritional problems that would otherwise be posed by their choice of dietary items. A knowledge of food passage rates is therefore critical in helping us understand the food choice patterns in these species.

In other species, though gut morphology is obviously important, other features too assume considerable importance in the successful exploitation of a particular set of dietary items. For example, the specialized dentitions of *Chiropotes* and *Cacajao* help animals open the hard fruits and seeds constituting much of their diet; the manual dexterity of *cebus* aids in searching for insects among leaves and fronds, and facilitates the exploitation of hard foods. The enormous body size of the gorilla may be viewed as a special dietary adaptation because it provides the body mass apparently required for an animal in this lineage to exploit a diet consisting primarily of fibrous bulky foods that must be eaten in huge quantities and retained in the gut for considerable time.

Though there seems to be a general trend for smaller-bodied primates to turn over food more rapidly than larger-bodied primates, there are notable exceptions. *Pithecia* is a small primate that shows a very depressed food transit time, whereas *Ateles* is a much larger animal with an extremely accelerated food transit time. It should not be surprising that there are exceptions to the general broad relationships among body size, food choice, and food transit time. Obviously more research is necessary to elucidate the finer details of food processing in these and other species, but some knowledge of food passage rates is necessary first. Finally, I view competition for limited resources as the ultimate factor underlying these proximate mechanisms of gut morphology and digestive processes which, in effect, allow a particular primate species to use a subset of the available plant resources to a degree not presumed possible by other primate primary consumers in the habitat.

Acknowledgements

The cooperation and assistance of many people were required to collect the data presented in this paper. I thank the Gorgas Memorial Laboratory in Panama and particularly Dr. James Harper for permission to work with their captive primates. Dr. Ladislau Deutsch, Chief of Mammals and Reptiles, Fundacion Parque Zoologico de São

Paulo, very graciously permitted me to work with the many rare primates under his supervision, and his assistant, Dr. Sergio Rodrigues, generously gave up many hours of his time to aid in this work. While at the Parque Zoologico in São Paulo, I was also fortunate to have the assistance of Ms. Denise Bretas, a student in primatology from UNICAMP, Campinas. In Manaus, Cap. Edino Camoleze of CIGS Jardim Zoologico ensured that I had the facilities required for my work. During my work in Brazil I was affiliated with the Instituto Nacional de Pesquisas de Amazonas (INPA), Manaus, under the sponsorship of José Marcio Ayres. While at the San Francisco Zoo I was aided by Mr. John Alcaraz, whose practical experience and knowledge of ape behavior was invaluable. I thank all of these individuals for their scientific curiosity and their generosity in giving me access to these primates, many of which are highly endangered in their natural habitats. My work in Panama and Brazil was partially funded by the Smithsonian Institution under the sponsorship of Dr. Ira Rubinoff, Director, Smithsonian Tropical Research Institute. Tom Milton and Kathy Troyer read an earlier draft of this manuscript and made many useful comments on the text.

Appendix

Other factors related to food processing efficiency can include the texture of foods eaten, mouth size and gape of the feeder, and features of the dentition. Below I present results of two experiments in which I investigated aspects of such influences on primate feeding patterns and food choice.

Feeding Rates in Howler Monkeys

Field data often show intraspecific as well as interspecific differences in food choice and feeding patterns. Such differences are typically attributed either to differences in body size or differences in the respective physiological states of given animals (Hladik 1977; Gaulin 1979). Though in many cases such explanations are doubtless correct, these differences also presumably reflect the fact that foods from different dietary categories are ingested at different rates and that intraspecific differences in body size exert an influence on the amount of food an animal can eat per unit time.

To determine the influence of food types and body size on feeding rates, three individually caged howlers (*Alouatta palliata*) were offered ad libitum access to a tared quantity of a particular food for a 5-minute period (table 9.2). After 5 minutes, the remaining food was removed and weighed to determine the amount ingested. This amount, when subtracted from the initial amount, gave the number of grams of food ingested per animal in the 5-minute trial period. This figure, divided by 5 gave the number of grams of food ingested per animal per minute.

Results of these trials, presented in table 9.2, show that adult how-

Table 9.2. Feeding Rates in *Alouatta palliata* [a]

Subject (Wt.)	Food	$\bar{X}$ Grams Eaten Min^{-1} (fresh wt)	No. Trials
Adult male (8.4 kg)	*Cecropia insignis*, flush	19.8 ± 3.0	Five 5-min. trials
	Ficus insipida, flush	21.0 ± 5.8	Four 5-min. trials
	Ficus insipida, ripe fruit	37.0	One 5-min. trial
Adult female (5.4 kg)	*Cecropia insignis*, flush	12.9 ± 2.7	Four 5-min. trials
	Ficus insipida, flush	13.3 ± 1.0	Four 5-min. trials
	Ficus insipida, ripe fruit	27.0	One 5-min. trial
Juvenile female (3.0 kg)	*Cecropia insignis*, flush	4.8 ± 0.9	Four 5-min. trials
	Ficus insipida, flush	4.8 ± 1.0	Four 5-min. trials
	Ficus insipida, ripe fruit	14.0	One 5-min. trial

[a] One other adult female, weight 7.3 kg., in one trial with ripe *F. insipida* fruit ingested 26.9 g min^{-1} (one 5-min. trial).

Cecropia insignis flush: 74% water, total cell wall material (CWM) = 54.7% dry wt. of sample; *Ficus insipida* flush: 79% water, CWM = 37.5% dry wt.; *Ficus insipida* fruit: 78% water; CWM = 34.1% dry wt.

lers ingest ripe fig fruit approximately twice as rapidly as young leaves by weight. The juvenile howler ingested ripe fig fruit approximately three times as rapidly as young leaves. Thus, regardless of body size, ripe fruit of this type is eaten much more rapidly than young leaves. However, the number of grams of food ingested per minute appears to be a function of the body size of the feeder. The adult male weighed 56% more than the female and was able to eat young leaves 56% more rapidly and fruit 37% more rapidly. This male weighed 180% more than the juvenile and ate leaves 392% more rapidly and fruit 164% more rapidly. The same pattern was found when the adult female was compared to the juvenile. The female weighed 80% more than the juvenile and ate young leaves 176% more rapidly and fruit 93% more rapidly. These data suggest that for young howlers even a minor increase in body size can confer a considerable processing benefit in food intake by weight.

Hard Foods

In working on feeding ecology of howlers and spider monkeys in Panama, I noticed that both species avoided eating fruits protected by a hard exocarp and seeds protected by a hard endocarp (e.g., *Apieba membranacea* fruit, seeds of Palmae). The seeds of many such fruits are routinely eaten by other animals, which suggests that toxicity is not a factor here. This raised the question of whether howlers and spider monkeys avoided such foods because they did not choose to eat them or whether there was actually some more fundamental reason why such foods were avoided. To answer this question I offered commercial walnuts in the shell to the following caged primates: one *Alouatta palliata,* two *Ateles paniscus,* two *Lagothrix lagotricha,* one *Cebus albifrons,* one *Cebus apella,* and one *Cacajao melanocephalus.* The howler, spider monkeys, and uakari sniffed and/or bit and/or licked at the nut shell for a second or two, then dropped the nut and lost interest in it. Each woolly monkey immediately placed the nut in the canine area of the jaw, cracked it with no apparent effort and ate the nut. All three cebus monkeys reacted in precisely the same way to walnuts. They alternately bit at the shell and pounded the nut forcefully on various surfaces in the enclosure, apparently trying to chip or crack the shell. Once a crack appeared, the teeth were used to open the shell further so that the nut could be eaten.

These results suggest that both dental morphology and manual dexterity play a role in determining what members of a particular species will regard as potential food. Apparently animals lacking teeth suitable for crushing hard objects and/or lacking in manual dexterity to compensate for their relatively less powerful dentition or smaller gape do not perceive items such as walnuts as food. Other species immediately recognize these same items as highly desirable and edible and can circumvent the problems involved at getting at the edible portion. I believe that none of these animals had ever seen a commercial walnut before these feeding experiments (although all may have seen nuts in the wild before capture), and so all were presumably equally naïve with respect to that potential food. These tests should be repeated with more animals and a greater variety of hard foods. The one *uakari* used in the experiments was a very small female, and I regard results with this animal as inconclusive.

References

Aguirre, A. C. 1971. *O mono Brachyteles arachnoides.* Rio de Janeiro, Brasil: Academia Brasileira de Ciencias.

Alvarez, W. C., and B. L. Freedlander. 1924. The rate of progress of food residues through the bowel. *J. Amer. Med. Assn.* 83:576–80.

Ayres, J. M., and J. L. Nessimian. 1982. Evidence for insectivory in *Chiropotes satanas. Primates* 23:458–59.

Bauchop, T., and R. W. Martucci. 1968. Ruminant-like digestion of the langur monkey. *Science* 161:698–700.

Bell, R. H. V. 1971. A grazing ecosystem in the Serengetti. *Sci. Amer.* 225:86–93.

Casimir, N. J. 1975. Feeding ecology and nutrition of an eastern gorilla group in the Mt. Kahuzi Region (Republique de Zaire). *Folia primatol.* 24:81–136.

Chivers, D. J., and C. M. Hladik. 1980. Morphology of the gastrointestinal tract in primates: Comparisons with other mammals in relation to diet. *J. Morphol.* 166:337–86.

Clutton-Brock, T. H. 1974. Primate social organization and ecology. *Nature* 250:539–42.

Clutton-Brock, T. H., ed. 1977. *Primate Ecology: Studies of Feeding and Ranging Behavior in Lemurs, Monkeys and Apes.* London: Academic Press.

Clutton-Brock, T. H., and P. H. Harvey. 1977. Sexual dimorphism, socionomic sex ratio and body weight in primates. *Nature* 269:797–800.

Feeny, P. 1971. Seasonal changes in oak leaf tannins and nutrients as a cause of spring feeding by winter moth caterpillars. *Ecology* 51:565–80.

Fleagle, J. G., and R. A. Mittermeier. 1981. Locomotor behavior, body size and comparative ecology of Surinam monkeys. *Amer. J. Phys. Anthrop.* 52:301–14.

Fooden, J. 1964. Stomach contents and gastrointestinal proportions in wildshot Guianan monkeys. *Amer. J. Phys. Anthrop.* 22:227–32.

Fossey, D., and A. H. Harcourt. 1977. Feeding ecology of free-ranging mountain gorillas (*Gorilla gorilla beringei*). In Clutton-Brock, ed. 1977:415–49.

Freeland, W. J., and D. H. Janzen. 1974. Strategies of herbivory in animals: The role of plant secondary compounds. *Amer. Natur.* 108:269–89.

Gaulin, S. J. C. 1979. A Jarman/Bell model of primate feeding niches. *Hu. Ecol.* 7:1–20.

Glander, K. E. 1975. Habitat and resource utilization: An ecological view of social organization in mantled howling monkeys. Ph.D. dissertation. Chicago: University of Chicago.

—— 1978. Howling monkey feeding behavior and plant secondary compounds: A study of strategies. In Montgomery, ed. 1978:561–74.

Happel, R. 1981. Natural history and conservation of *Pithecia hirsuta* in Peru. Unpublished manuscript.

Harcourt, A. H. and K. J. Stewart. 1978. Coprophagy by wild mountain gorillas. *E. Afr. Wildl. J.* 16:223–25.

Hespenheide, H. A. 1971. Food preferences and the extent of overlap in some insectivorous birds with special reference to the Tyrannidae. *Ibis* 113:59–72.

—— 1975. Selective predation by two swifts and a swallow in Central America. *Ibis* 117:82–99.

Hladik, A. 1977. Phenology of leaf production in rain forest of Gabon: Distribution and composition of food for folivores. In Montgomery, ed. 1978:51–72.

Hladik, C. M. 1967. Surface relative du tractus digestif de quelques primates. Morphologie des villosités intestinales et correlations avec le régime alimentaire. *Mammalia* 31:120–47.

—— 1977. A comparative study of two sympatric species of leaf monkey (*Lagothrix lagothricha*) in northern Colombia. *Primates* 1977:324–54.

—— 1978. Adaptive strategies of primates in relation to leaf-eating. In Montgomery, ed. 1978:373–95.

Hladik, A., and C. M. Hladik. 1969. Rapports trophiques entre vegetation et primates dans la forêt de Barro Colorado (Panama). *Terre et la Vie* 1:25–117.

Hungate, R. E., G. D. Phillips, A. McGregor, D. P. Hungate, and H. K. Buechner. 1959. Microbial fermentation in certain animals. *Science* 130:1192–94.

Isaac, G. 1971. The diet of early man: Aspects of archaeological evidence from lower and middle Pleistocene sites in Africa. *World Archaeology* 2:270–99.

Izawa, K. 1975. Foods and feeding behavior of monkeys in the upper Amazon basin. *Primates* 16:295–316.

Janzen, D. H. 1970. Herbivores and the number of tree species in tropical forests. *Amer. Nat.* 104:501–28.

Jarman, P. J. 1974. The social organisation of antelope in relation to their ecology. *Behaviour* 48:216–67.

Kavanagh, M., and L. Dresdale. 1975. Observations on the woolly monkey (*Lagothrix lagothricha*) in northern Colombia. *Primates* 16:285–94.

Kleiber, M. 1961. *The Fire of Life.* New York: Wiley.

Maynard, L. A., and J. K. Loosli. 1969. *Animal Nutrition.* New York: McGraw-Hill.

McBee, R. H. 1971. Significance of intestinal microflora in herbivory. *Ann. Rev. Ecol. Syst.* 2:165–75.

McKey, D., P. G. Waterman, C. N. Mbi, J. S. Gartlan, and T. T. Struhsaker. 1978. Phenolic content of vegetation in two African rain forests: Ecological implications. *Science* 202:61–64.

Mehrtens, D. R. 1971. Application of theoretical mathematical models to cell wall digestion and forage intake in ruminants. Ph.D. dissertation. Ithaca, N.Y.: Cornell University.

Milton, K. 1979. Factors influencing leaf choice by howler monkeys: A test of some hypotheses of leaf selection by generalist herbivores. *Amer. Natur.* 114:362–78.

—— 1980. *The Foraging Strategy of Howler Monkeys: A Study in Primate Economics.* New York: Columbia University Press.

—— 1981. Food choice and digestive strategies of two sympatric primate species. *Amer. Natur.* 117:476–95.

Milton, K. and R. H. McBee. 1983. Rates of fermentative digestion in the howler monkey, *Alonatta palliata* (Primates, Ceboidea). *Comp. Biochem. Physiol.* 74A:29–31.

Milton, K., P. J. Van Soest, and J. Robertson. 1980. The digestive efficiencies of wild howler monkeys. *Physiol. Zool.* 53:402–9.

Moir, R. J. 1967. Ruminant digestion and evolution. In C. F. Cole, ed. *Handbook of Physiology* 5:2673–94. Baltimore: Waverly Press.

Montgomery, G. C., ed. 1978. *The Ecology of Arboreal Folivores.* Washington, D.C.: Smithsonian Institution Press.

Napier, J. R., and P. H. Napier. 1969. *A Handbook of Living Primates.* London: Academic Press.

Oates, J. F., T. Swain, and J. Zantovska. 1977. Secondary compounds and food selection by Colobus monkeys. *Biochem. System. Ecol.* 5:317–21.

Oates, J. F., P. G. Waterman, and G. M. Choo. 1980. Food selection by the South Indian leaf-monkey, *Presbytis johnii* in relation to leaf chemistry. *Oecologia* 45:45–56.

Oppenheimer, J. R. 1968. Behavior and ecology of the white-faced monkey, *Cebus capucinus,* on Barro Colorado Island, Canal Zone. Ph.D. dissertation. Urbana, Ill.: University of Illinois.

Parra, R. 1978. Comparison of foregut and hindgut fermentation in herbivores. In Montgomery, ed. 1978:205–30.

Querling, D. P. 1950. *Functional Anatomy of the Vertebrates.* New York: McGraw-Hill.

Ricklefs, R. E. 1972. Dominance and the niche in bird communities. *Amer. Natur.* 106:538–45.

Rodman, P. S. 1977. Feeding behaviour of orang-utans of the Kutai Nature Reserve, East Kalimantan. In Clutton-Brock, ed. 1977:384–413.

Roosmalen, M. J. van, R. A. Mittermeier, and K. Milton. 1980. The bearded saki genus *Chiropotes.* In A. F. Coimbra-Filho and R. A. Mittermeier, eds. *Ecology and Behavior of Neotropical Primates,* pp. 419–41. Rio de Janeiro: Academia Brasileira de Ciencias.

Ryan, C. A., and T. R. Green. 1974. Proteinase inhibitors in natural plant protection. In V. C. Runckles and E. E. Conn, eds. *Metabolism and Regulation of Secondary Plant Products,* pp. 123–40. New York: Academic Press.

Schoener, T. W. 1965. The evolution of bill size differences among sympatric congeneric species of birds. *Evolution* 19:189–213.

Sussman, R. W. 1979. Ecological distinction in sympatric species of *Lemur.* In R. W. Sussman, ed. *Primate Ecology,* pp. 53–84. New York: Academic Press.

Van Soest, P. J. 1977. Plant fiber and its role in herbivore nutrition. *The Cornell Veterinarian* 67:307–26.

—— 1982. *Nutritional Ecology of the Ruminant.* Corvallis, Oregon: O and B Books.

Waterman, P. G., C. N. Mbi, D. G. McKey, and J. S. Gartlan. 1980. African rainforest vegetation and rumen microbes: Phenolic compounds and nutrients as correlates of digestibility. *Oecologia* 47:22–33.

Werner, E. E. 1977. Species packing and niche complementarity in three sunfishes. *Amer. Natur.* 111:553–78.

Westoby, M. 1974. An analysis of diet selection by large generalist herbivores. *Amer. Natur.* 108:290–304.

Wiggins, H. S., and J. H. Cummings. 1976. Evidence for the mixing of residue in the human gut. *J. Brit. Soc. Gastroent.* 17:1007–11.

Wrangham, R. W. 1977. Feeding behaviour of chimpanzees in Gombe National Park, Tanzania. In Clutton-Brock, ed. 1977:504–38.

10 Is Optimization the Optimal Approach to Primate Foraging?

DAVID G. POST

MATHEMATICAL MODELS of behavioral processes serve many functions, the most important of which may be to "clarify how much of a phenomenon has been comprehended" (Cohen 1972:417); a model is simply a "reconstruction of nature for the purposes of study" (Levins 1968:6) and thus embodies explicitly what we do, and what we do not, understand about the way nature operates. The model-building process can expose our ignorance in certain areas, prompting further investigation; it can underscore the critical position of certain untested assumptions within a deductive framework, assumptions that can then be examined in greater detail; it can illuminate old questions from a new perspective and point to new questions that, in the model's absence, might never have been asked (see Oster and Wilson 1978, chapter 8, for an extended discussion of these and related questions).

DAVID POST has studied feeding ecology of baboons of the Amboseli Game Park of Kenya with a particular interest in the relationship of sexual dimorphism in body size to sex differences in feeding patterns. In the following paper he examines the relationship of research to theory in the study of foraging, with a critical analysis of "optimal foraging theory" and its applicability to understanding food choices in free-ranging baboons. His analysis uncovers "hidden" assumptions of the theory and cautions against a tendency for untested hypotheses to be used as *post hoc* explanations of behavior. He concludes by advocating the Altmann-Wagner approach to optimal foraging, which takes the logical view that individuals will vary in their success at foraging optimally, and therein lies the opportunity to observe natural selection in operation.

It is in this spirit that this essay proceeds. It is neither a review of optimal foraging theory (for which the reader is referred to Krebs 1978 or Pyke et al. 1977) nor a summary of present knowledge concerning primate foraging (see Gaulin and Konner 1977 and the collections edited by Clutton-Brock 1977 and Sussman 1979). I assume, perforce, that the reader is familiar, in general outline at least, with recent currents of research in these two areas. Furthermore, it strikes me as unproductive at this time, for reasons that should become clear in what follows, to attempt a general comparison of theoretical predictions of optimal foraging theory with empirical observations on primate foraging. I intend, rather, to raise and address certain primarily conceptual issues regarding the logical structure of optimality arguments as they pertain to the foraging behavior of nonhuman primates. In the first flush of excitement surrounding the development of this body of theory, we are in danger of losing sight of many of its inherent limitations. I argue below that the potential of these theoretical developments will be fully realized only when we have gathered far more independent evidence bearing on the validity of the stringent assumptions underlying the optimization perspective and we have a better understanding of the "robustness" of these models in the face of violations of some of these simplifying assumptions. Although the tone of this essay is somewhat critical, I intend it to be constructively so. While I am concerned by what I regard as the improper use of the optimization perspective on foraging behavior in certain instances, its potential utility as an explanatory framework is profound. The problems discussed below are, in my view, more in the nature of interesting challenges than of insurmountable obstacles.

On the Nature of Optimal Foraging and Optimal Foraging Models

The Basic Assumptions

During foraging, animals are continually faced with decisions concerning, for example, which foods to eat and which to pass up, which habitat types to forage in and which to avoid, or when to leave particular habitat patches or feeding sites. Optimal foraging theory, as the term is used here, encompasses a variety of predictive decision models that specify how animals *should* behave when faced with such alternative courses of action (see Krebs 1978). As such, it is closely allied with other disciplines, particularly microeconomics, that treat individual decision strategies, a point to which I shall return at several junctures below (see also McCleery 1978; Rapport 1971; Rapport and Turner 1977).

All decision models begin with the assumption that the decision-maker is *rational* in the strict sense (see Edwards 1954; McCleery 1978; Simon 1955). Specifically, we envision an animal faced with a set of alternative *actions,* each of which produces a particular *outcome,* and we assume the following:

1. *All outcomes can be "weak-ordered" by the animal;* that is, given a choice between outcomes O_1 and O_2, either O_1 is preferred to O_2 ($O_1 > O_2$), O_2 to O_1 ($O_2 > O_1$), or they are equally desirable ($O_1 \sim O_2$).
2. *These preferences are transitive;* that is, if $O_1 > O_2$ and $O_2 > O_3$ then $O_1 > O_3$; if $O_1 \sim O_2$ and $O_2 \sim O_3$ then $O_1 \sim O_3$, etc.
3. *The decision-maker chooses among these alternative actions such that something, the so-called "currency" of the model* (*Schoener 1971*), *is maximized* (*or minimized*). Animals (or humans) are not *necessarily* rational in this sense, and indeed, the demonstration of rational preferences, even in simple laboratory experiments, has often proved surprisingly difficult (see, for example, Edwards 1954 and 1960; Simon 1957). Furthermore, it seems reasonable to suggest that this assumption will, in general, pose the most serious difficulties for the model-builder. What, in fact is being maximized?

Behavioral biologists here find refuge in the theory of natural selection. In McCleery's (1978) words:

> The thing that is maximized by a rational animal is called "utility," which can be thought of as equivalent to the net benefit

> derived from making a particular choice, and must be measurable in a currency common to the different options. . . . An appropriate common currency must be one related to the ultimate evolutionary advantage of indulging in certain behaviors . . . i.e. to the effects of performing these activities on the reproductive potential of the individual and his close kin.

Note that an explicit parallel is being drawn to microeconomic decision theory, where it is assumed that "individuals act in their various roles so as to maximize their own satisfaction or 'utility' " (Nicholson 1972:43). Utility, to the microeconomist, is indeed the "net benefit derived from making a particular choice," and the concept was first introduced by the mathematician Daniel Bernoulli, in the eighteenth century, for the purpose of explaining certain apparently paradoxical aspects of human behavior at the gambling tables (see below). The notion that humans are attempting to maximize utility during decision-making is fundamental to much of microeconomic and game theory (see Raiffa 1968; Rapoport 1969). The net utility of any particular outcome is (roughly) equivalent to the "happiness" or "satisfaction" derived from attaining that outcome, and, as such, it is a complex function of many variables (e.g., income, status, health and well-being), many of which can be measured with difficulty (if at all). The characteristics of a specific individual's utility function cannot be deduced from a set of axioms or laws governing human behavior but must be inferred from the decisions the individual makes. Furthermore, there is no absolute scale of utility that would allow meaningful interpersonal comparisons of utility to be made (see Edwards 1954; Raiffa 1968). Thus there is no way to know a priori whether a year's subscription to *Folia primatologica* has a higher utility value for me than a year's subscription to the Metropolitan Opera, nor is there any way to compare my utility assignments for these two outcomes to anyone else's.

In contrast, the biologist can deduce, from the theory of natural selection, that Darwinian fitness (or "inclusive fitness") is an animal's "utility," the quantity being maximized during decision-making. Our use of fitness in this context rests implicitly on several additional assumptions (Pyke et al. 1977):

4. *Foraging behavior shows heritable variation within populations.*
5. *Natural selection favors those individuals in a population contributing most genetic material to subsequent generations.* Hence,
6. *"Natural selection will result in a change with time of the average foraging behavior in the population towards that foraging behavior in the range of possible behaviors giving maximum fitness"* (Pyke et al. 1977:138).

Optimization models in behavior, then, require that the benefits and costs of alternative behavioral options be specified as increments or decrements to an individual's inclusive fitness, an extremely difficult task under the best of circumstances. In light of these difficulties, most models rely on some more readily measurable currency, common to the various options, that is, in Schoener's words, "commensurate with fitness" (Schoener 1971:369). In most (but not all) optimal foraging models the net rate of energy intake is chosen as this common currency, a choice that can be justified on a variety of grounds (see Pyke et al. 1977:138ff.; Schoener 1971). It can be shown, however (see below), that, in conditions likely to pertain in many foraging situations, models using *any* currency other than fitness may yield decision strategies that are, in fact, not truly "optimal."

Testing the Assumptions of Optimality Models

Optimal foraging theory is explicitly predictive, generating specific and falsifiable hypotheses concerning what animals should or should not do in particular situations. This is, of course a large part of its extraordinary appeal: it may yet transform this branch of behavioral biology into a truly deductive science. However, these predictions and hypotheses are necessarily *conditional,* a point that seems to be frequently overlooked. That is, optimal foraging models can generally be reformulated as an "if-then" statement (or a series of such statements) of the following form:

If a set of assumptions (such as the ones discussed in the previous section) is satisfied, and if a certain set of environmental conditions is encountered, then we should observe certain responses by the animals under observation.

Let us consider a similar, but much simpler, conditional statement: "If the sun shines brightly on Saturday, Stubby will go to the beach."

As stated, bright sunshine is a sufficient condition for Stubby's outing, and therefore, if we observe bright sunshine on Saturday and fail to observe Stubby at the beach, we have grounds for rejecting our "model" of Stubby's behavior.

But suppose, with this "model" in hand, we are told only that Stubby is at the beach on Saturday; can we then conclude that the sun is shining brightly? Clearly, such a conclusion confuses *necessity* for *sufficiency,* for we have not ruled out the possibility that Stubby will go to the beach if the sun is *not* shining brightly.

I would assert that this confusion of necessity for sufficiency is pervasive in much of the recent literature on optimal foraging (and "sociobiology" in general). It is often asserted, at least implicitly, that we can "test" a model's assumptions by verifying the congruence between its predictions and the behavior of the animals under study. But this is strictly true only if those assumptions form a set of necessary conditions for the occurrence of the specified outcomes, that is, if all alternative hypotheses can be rejected. This argument has been made by McCleery (1978:384–85) with specific reference to the optimization assumption (3 above):

> It is sometimes asserted, wrongly, that optimality is an hypothesis for testing, but a moment's reflection will show that this cannot be so. All that is tested in any given instance is whether a particular optimality model is correct; failure of a model does not mean that optimization as such is not occurring, since there may be another model which does fit the facts but which has not been thought of yet. Optimization is a weak postulate, *which is assumed to be true most of the time.* [my emphasis]

Or, as Maynard Smith (1978:35) has put it, "the essential point is that in testing (an optimization) model we are *not* testing the general proposition that nature optimizes, but the specific hypotheses about constraints, optimization criteria, and heredity." This is not simply a semantic quibble; optimization is not a hypothesis that is tested against competing alternatives (of nonoptimality), but rather it has been deduced from the dominant paradigm underlying investigations in this area, the theory of natural selection.

Consider, by way of example, Pyke's elegant analysis of the economics of territoriality in the golden-winged sunbird (Pyke 1979). He starts with the assumption (erroneously labeled a "hypothesis") that "an animal defending a territory will adapt a territory size (shape, etc.) and a time budget which maximize its individual Darwinian Fitness" (p. 132). Several alternative hypotheses are then proposed: in order to maximize fitness, animals may be

(a) maximizing net daily energy gain;
(b) maximizing time spent resting;
(c) minimizing total daily caloric expenditure; or
(d) maximizing the ratio of gross daily energy gain to total daily energetic cost.

Each hypothesis generates a set of predictions about sunbird behavior, which are then compared to empirical observations. Of the four hypotheses presented, only one (minimizing total daily caloric expenditure) generates predictions that are in close accord with the sunbirds' behavior. If none of the predictions had matched the birds' behavior, the four hypotheses could have been rejected (i.e., falsified) without affecting our notion of fitness maximization on the part of the sunbirds. That is, fitness maximization is not a "hypothesis," for it cannot be falsified within the context of the study; it is, rather, an assumption that allows us to generate falsifiable hypotheses, and just as deviations of observed from predicted behavior do not cause us to "reject" that assumption, so too is congruence between observed and predicted insufficient for us to "accept" that assumption.

We will, therefore, require *independent* evidence to support the assumptions underlying the optimization approach. Much research is presently underway along just these lines, and it is a good example of how the process of model-building opens up new or often unexplored areas for investigation. For example, the assumption of fitness maximization is itself built on the assumption that at least some components of the foraging process depend on heredity (assumption 4): "Natural selection cannot produce adaptation unless there is heredity" (Maynard Smith 1978:35). To my knowledge, however, there is as yet no documentation of a hereditary component of any aspect of the foraging behavior of nonhuman primates. I stress this point not to suggest that heredity plays no part in behavior of nonhuman primates but rather to point out the important role this assumption plays in the overall logical structure of the optimization perspective.

What If Our Assumptions Are Known To Be Invalid?

The preceding section should not be taken to imply that it is somehow illegitimate to invoke assumptions that cannot be adequately tested with available data. Indeed, this is often a valuable by-product of the model-building process, for it directs our attention to areas of research that may have previously been overlooked. In other circumstances, however, we may be well aware that certain of our model's underlying assumptions are clearly inappropriate for the animals under investigation. How can we then evaluate the congruence, or lack of it, between the model's predictions and the animals' behavior? To answer this question we need to have some idea how "robust" those predictions are, given violations of particular assumptions, and I suggest that this question has received far too little attention in the optimal foraging literature to date.

To take a specific example, certain characteristics of an "optimal diet" have been derived from models proposed independently by no fewer than nine authors (see Pyke et al. 1977:141, and references therein; also Maynard Smith 1978:44 ff.). In addition to assumptions 1–6 discussed earlier, these models all assume a fine-grained foraging context, that is, one in which resources (food types in this case) are encountered in proportion to their relative abundance in the environment. Under these conditions, the diet that maximizes the net rate of energy intake has the following important properties:

1. Whether or not a food type should be eaten is independent of the abundance of that food type and depends only on the absolute abundances of the food types of higher "rank" (foods might be ranked according to their ratios of caloric value to handling time).

2. Increasing food abundance should lead to increasing food specialization.

3a. A food type should be either completely included in the optimal diet or completely excluded from it; animals should not exhibit "partial preferences."

3b. In conjunction with the assumption that food types are *encountered* in proportion to their relative abundance (the "fine-grained" assumption), foods that are included in the optimal diet are also *eaten* in proportion to their relative abundance.

These predictions concerning the characteristics of an animal's "optimal" diet are slowly being transformed into dogma in the behavioral literature. This is regrettable in light of the facts that several of the underlying assumptions, in particular the assumption of a fine-grained foraging environment, are inappropriate for many animals and that little attention has been paid to the consequences of violating those assumptions (see below). Thus, to take two recent examples, we read that:

> The association between omnivory and open habitats can be understood by reference to Schoener's (1971) feeding strategy mode. If food abundance is lower in open than in forested habitats, open country animals should exploit a wider range of food types: they should be more omnivorous [prediction 2 above]. . . . Moreover, the relative contribution to the diet of any acceptable food item is proportional to its abundance in the habitat (Schoener, 1971) [prediction 3b]. . . . Of course, the e/t (energy per time) (or, put more generally, the benefit/cost ratio) of a particular kind of food must be above some threshold value or that food will never be eaten regardless of its abundance [prediction 3a] (Gaulin 1979:15–16).

The feeding behavior of generalists has been the subject of much theoretical discussion (Schoener 1971; Emlen 1973; Freeland and

Janzen 1974). Their characteristic strategies may be summarized as follows:

> (1) They eat a potentially wide range of food types, but at any one time concentrate on the most familiar and readily available (Freeland and Jantzen 1974). Diversity in the diet is a response to declining availability of preferred foods [Prediction 2].
> (2) Food items are eaten according to their relative availability [prediction 3b] (Homewood 1978:385).

I am suggesting, at the very least, that considerable caution be exercised in the application of these predictions to the foraging behavior of organisms without first examining whether *all* underlying assumptions are appropriate *or* whether the models are sufficiently robust in the face of violations of those assumptions. Let us turn to a consideration of these questions in more detail.

Some Considerations of Coarse-Grain Foraging

My claim that most optimal foraging models assume a fine-grained foraging context deserves more careful examination. Consider a general optimal diet model attempting to predict which of k available food types a predator should consume (or, alternatively, which of k patch types a predator should visit). Such a model will include some parameter (say $P(i)$, $i = 1, 2, \ldots, k$) representing the probability, at any point in the foraging process, that the next prey (or patch) type encountered is type i. How can this probability distribution be estimated? In most models, including those that generate the predictions about the characteristics of the optimal diet discussed earlier (see above and Pyke et al. 1977:141), $P(i)$ is assumed to be proportional to the relative abundance of prey (patch) type i in the environment. This is equivalent to assuming a fine-grained foraging context.

While it is clear that this assumption greatly increases the conceptual and mathematical tractability of these models by providing a means of estimating these encounter probabilities with reference to readily measurable quantities (relative abundances), the implications of this assumption should be clearly understood.

Thus:

(a) The $P(i)$s are constant through time. In particular,

(b) they are independent of the behavior of the animals themselves

and thus they are not in the decision-makers' "control" (see especially Charnov 1976).

(c) As a result of (a) and (b), "foraging behavior at one point in time does not alter the optimal foraging behavior at a later point in time" since the $P(i)$s are constant and uninfluenced by the animal's foraging behavior; thus "the predictions from the optimization are independent of time scale and any time scale can be used" (Pyke et al. 1977:139).

Consider, however, the foraging environment depicted schematically in figure 10.1. This figure depicts the spatial distribution of three habitat types (of 24 such habitat types recognized) within the home range of a single social group of yellow baboons (*Papio cynocephalus*) in the Amboseli National Park, Kenya (see Post 1978, fig. 27, from which this figure has been adapted). We can assume, for purposes of this discussion, that the ranges of food items available in the three habitat types are nonoverlapping (see table 10.1 and Post 1978, table 62). Fig. 10.1 also shows the location of all available sleeping groves in which the group spends the night; these define the beginning and end points of all foraging day journeys, each of which averages approximately 5 km (Post 1978).

As the baboons traverse this area in search of food, it is clear that the probability of encountering any particular food type ($P(i)$) is fluctuating throughout the day. Furthermore, those fluctuations are, to a certain extent, a function of the group's previous foraging decisions (e.g., which direction away from the sleeping groves to move). For example, the probability of encountering a sapling fever tree (*Acacia xanthoploea*), a highly preferred food type found in habitat type B, will be low for the entire day once the decision to forage in the *Acacia tortilis* woodland in the southern portion of the home range

Figure 10.1. A map showing the distribution of "patches" of three habitat types within the home range of Alto's Group in the Amboseli National Park, Kenya. A = *Acacia tortilis* woodland, B = regenerating *Acacia xanthophloea* woodland, and C = bare "pans" of "black cotton" soil, largely devoid of vegetation. See table 10.1 for a list of foods available in each of these habitat types. The solid circles represent the locations of sleeping groves, and the solid horizontal bar represents a distance of 1 km.

Table 10.1. Summary of Foods Available in the Three Habitat Types Depicted in Fig. 10.1 (from Post 1978)

Habitat Type	Foods Available
A (*Acacia tortilis* woodland)	*Acacia tortilis* flowers and seed pods *Salvadora persica* leaves and berries *Cynodon plectostachyus* corms *Lycium europaeum* leaves
B (regenerating *Acacia xanthophloea* woodland)	*Acacia xanthophloea* gum *Sporobolus spicatus* seedheads, blades
C (black-cotton soil pans)	None

has been made. In this manner the animals' foraging decisions at one time constrain the foraging options that will be available at a later time. The degree to which decisions at one time constrain future options depends, of course, on a complex interaction among a number of variables: the size of the foraging area, the size of individual patches, the nonrandomness in the distribution of patches, and the locomotor capabilities of the animals. In the nonhuman primates, there appear to be systematic patterns in the proportion of the total home range that can be traversed during a single day journey. By using data on home range area and day-journey length of 39 primate species compiled by Clutton-Brock and Harvey (1977), the following least-squares regression equation was obtained:

$$\text{day-journey length} = (1.04)\,(\text{home-range area})^{.312}$$
$$r = .676, N = 39, p < .01$$

In other words, the ratio of day-journey length to home-range area is decreasing in species with larger home ranges. Note also that the ratio of day-journey length to (home-range area)$^{\frac{1}{2}}$, which expresses the two variables in equivalent linear units, also decreases as home ranges get larger.

These considerations may have numerous implications for our understanding of the "optimal" decision strategy of the baboons, two of which will be touched upon here.

The Role of Uncertainty

Optimization models in ecology predict that an animal, when faced with alternative options whose outcomes are known with certainty in advance, will choose that option whose return, calculated in units of Darwinian fitness, is highest. As discussed above, because of the difficulties inherent in specifying the fitness costs or benefits of behavioral acts, most optimal foraging models rely on some more readily measurable currency (e.g., net rate of energy intake). If fitness is a nondecreasing function of this currency, the option that maximizes the net rate of energy intake is necessarily the optimal (i.e., fitness maximizing) option as well.

What, then, of a choice among options whose outcomes are *not* known with certainty in advance, as in the case of, say, the evaluation of distant food patches by an animal with less-than-perfect information about the food abundance in each? Uncertainty of this kind is likely to be a component of most foraging processes as a consequence both of the imperfect ability of animals to gather, store, and recall all of the relevant information concerning the spatial distribution of resources and of temporal variability in that resource distribution (see Post, in press). Although uncertainty may characterize fine-grained foraging contexts (e.g., Pulliam 1974), it is likely to be most pronounced in highly patchy environments where there are substantial difficulties in "monitoring" the availability of resources over a wide area.

It is clear that the optimal strategy in such circumstances is to choose the option that *maximizes the expected fitness return.* That is, suppose we present our forager with an oversimplified choice between two options (1 and 2). Option 1 can result in either outcome A (with probability P_{A1}) or outcome B (with probability P_{B1}), as can option 2 (with probabilities P_{A2} and P_{B2}, respectively). If, in turn, outcomes A and B result in fitness increments F_A and F_B, respectively, then the expected fitness return of the two options can be computed as:

$$\text{Expected Fitness Return, Option 1: } P_{A1}F_A + P_{B1}F_B$$
$$\text{Expected Fitness Return, Option 2: } P_{A2}F_A + P_{B2}F_B$$

The option with the highest expected fitness return should be chosen by a fitness-maximizing forager.

As mentioned above, in foraging situations *without* this kind of uncertainty, if fitness is a nondecreasing function of some currency, then the option that maximizes that currency is the fitness-maximizing, and hence optimal, option as well. It might be assured that the optimal strategy in the uncertain foraging context is to maximize the expected return of that currency (see, e.g., Charnov 1976; Pulliam 1974). It can, however, be shown that this is not necessarily the case.

Assume that fitness is a nondecreasing but convex function of the rate of energy intake (fig. 10.2); this would represent, in Schoener's words, "the decreasing or at least limited ability to convert energy into additional protoplasm or offspring" (1971:373). We present our forager with two options:

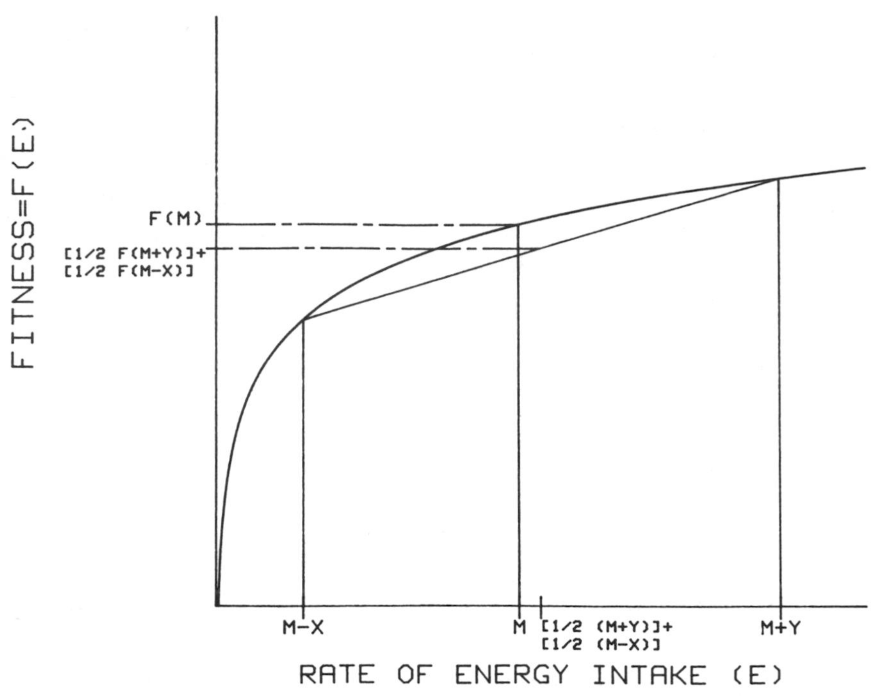

Figure 10.2. A hypothetical curve showing a convex relationship between fitness and the rate of energy intake. $F(M)$ represents the fitness gain associated with a guaranteed return of M energy units. The expected fitness return of a "gamble" giving the animal $M + Y$ energy units (with probability 1/2) or $M - X$ units (with probability 1/2) is $1/2\ F(M + Y) + 1/2\ F(M - X)$; this can be found graphically as the midpoint of the line connecting $F(M + Y)$ and $F(M - X)$. Clearly the latter option yields lower fitness than the former.

Option 1 guarantees (i.e., with probability = 1) a return of M energy units.

Option 2 will yield $M + Y$ units (with probability = ½) or $M - X$ units (also with probability = ½).

If $Y > X$ then the expected energetic return of option 1 (M) is less than the expected *energetic* return of option 2, ($\frac{1}{2}(M+Y) + \frac{1}{2}(M-X)$). However, the expected *fitness* return of option 1 ($F(M)$) is greater than that of option 2 ($\frac{1}{2}F(M+Y) + \frac{1}{2}F(M-Y)$), a simple consequence of the convexity of the fitness vs. energy intake curve. Option 1 is therefore the optimal (fitness-maximizing) choice, despite the fact that option 2 yields a higher expected rate of energy intake. In simple terms, option 2 is too "risky"; a berry in the hand (option 1) may indeed be worth more than two berries possibly present in some distant bush (option 2).

This simple decision model was derived from one initially formulated to account for certain seemingly paradoxical aspects of human gambling behavior (see, e.g., Raiffa 1968). Most people are "risk-aversive" in that they would prefer a guaranteed payment of $100 (option 1 with $M = 100$) to a gamble yielding $10,200 with probability ½ and −$9,900 with probability ½ (option 2 with $Y = 10{,}100$ and $X = 10{,}000$) despite the greater expected monetary return of the latter ($150 vs. $100). This choice is consistent with the notion that people are maximizing expected utility (= fitness), not expected monetary returns (= rate of energy intake), and that utility is a convex function of money (this is the economists' "diminishing marginal utility of money").

In short, even given the basic premise that animals are choosing the fitness-maximizing alternatives, one cannot predict a priori their choice when confronted with alternative actions unless (a) one can specify the costs and benefits of each outcome in terms of increments or decrements to the individual's fitness or (b) the surrogate currency of the decision model (e.g., rate of energy intake) is known to be *linearly* related to fitness.

I suspect that uncertainty characterizes the foraging environments of many primates, feeding, as they often are, on unpredictably occurring, widely scattering foods. I am hesitant, therefore, to base my predictions of how such animals should behave on models that do not incorporate considerations of risk and uncertainty in their predictive framework. Consider, for example, the question of "partial preferences" (prediction 3a above). Just as an individual's response to a "risky" decision at the gambling table will be contingent upon his or her present income (or previous winnings), so, too, should an animal's energy reserves (i.e., its "previous winnings") influence its response to risky foraging decisions. As a baboon heads away from its previous night's sleeping grove, it may "pass up" certain foods owing to its expectation of higher returns elsewhere. At the end of the day, with its belly perhaps only half full and little time remaining before darkness falls, it may be far more reluctant to pass up a *cer-*

tain, albeit small, return in favor of an option with higher expected return but high risk. Although this argument needs to be formulated in considerably greater detail, I would suggest that it may be relevant to the observation that, among primates, "there appears to be a common tendency for fruit [high risk, high return?] to be eaten relatively more than foliage [low risk, low return?] early in the activity period" (Clutton-Brock 1977:531).

The Role of Food Abundance in Patchy Environments

Consider the prediction that the inclusion of any particular food item (say, the fruit of *Acacia tortilis* trees) in the diet is independent of its abundance (prediction 1 above). If the baboons are to include this item in their diet, they must enter the *tortilis* woodland (habitat type A in fig. 10.1). If baboons are choosing foraging patches "rationally," their decision to forage in the *tortilis* woodland is a function of the "expected return" from foraging there, which is in turn a function of the amount of fruit available. In other words, the abundance of *A. tortilis* fruit is a determinant of the amount of time spent foraging in the *tortilis* woodland, and hence of the amount of *A. tortilis* fruit in the diet, in contradiction to prediction 1. The model is not, of course, "wrong"; it is simply inappropriate for a coarse-grained foraging environment *where the animals' decisions themselves determine the encounter probabilities with various food items.* Clearly, in these circumstances, I have no reason to expect the baboons to behave in accordance with this prediction.

Furthermore, coarse-grained foraging environments have an additional dimension of "risk" that is related to the "traveling salesman problem" (see Altmann 1974) in that the "expected return" from foraging in the *tortilis* woodland cannot be the sole determinant of the baboons' optimal foraging path given the highly patchy distribution of their food supplies. Consider a salesman who has a fixed time period within which to make sales presentations in a series of Northeastern cities and, say, Houston. In deciding where to "forage," he must consider both the "expected return" (expected dollar value of sales/day) in each city *and* the location of each city relative to other "foraging" areas. The probability of coming up "empty handed" may be uniformly low in all cities; however, he is more reluctant to risk such failure in Houston than in Boston because there are no nearby "foraging areas" where he can recoup his losses before his resources (time and money) run out. Thus, his expected return from the Houston excursion must be relatively high before the trip will be undertaken; more importantly, however, we can imagine two salesmen, *each behaving optimally,* coming to very different conclusions about the desirability of such a trip to Houston. Salesman A, already living beyond his means, with high mortgage payments and three offspring at expensive private universities, looks at the nonzero probability of failure (and subsequent bankruptcy) and cancels the trip. Salesman B, on the other hand, is swayed by the nonzero probability of extravagant profits and books the first flight out. Without knowing their respective attitudes toward this risk, there is no way we, as observers, could have predicted their decisions.

For the baboons, too, each foraging decision constrains future options and decisions, and the "cost" of a wrong decision has to be measured with reference to the diminution of present *and future* returns. The decision to forage in the *tortilis* woodland must include a consideration of the cost that foraging in this area imposes on the baboons' subsequent foraging, as a consequence of the particular spatial distribution of resources within their home range. It may be that the baboons can ill afford to come up "empty handed" in the *tortilis* woodland because they are then poorly situated with respect to the availability of other foods. Predictions based solely on the "expected value" of that foraging decision are unlikely, therefore, to fully account for their observed foraging behavior and food choice. This may help explain, as I have suggested elsewhere (Post, in press), why the baboons foraged within the *tortilis* woodland only when the abundance of *A. tortilis* fruits (or flowers) was at its highest, and hence, why the consumption of these items was restricted to extremely brief periods. By contrast, the fruits of the other tree species in the area, *A. xanthophloea* (the fever tree), which is much less "patchily" distributed, were eaten throughout the period when they were available.

An Alternative Approach

My remarks above have been directed toward a particular class of optimization models that are designed to predict the course of animal decision-making. There exists, however, another approach to the "optimality" of animal behavior that deserves mention, however brief (see Altmann and Wagner 1978). Suppose that, on the basis of a thorough knowledge of the nutritional requirements of an animal and the nutritional composition of available foods, we can specify that combination of foods that is "optimal" (i.e., fitness maximizing) without regard to the actual decisions required to achieve that precise mix of foods. Thus, with n available foods, the optimal diet can be represented by a point in n-dimensional space. The feeding behavior of a group of animals is then observed, and each animal's diet is similarly placed within this n-dimensional space. The expectation is not that each animal will have achieved this optimal diet but rather that the "distance" between the point representing each animal's diet and the optimum should, *ceterus paribus* and assuming that our identification of the optimum is correct, correlate with that individual's ability to survive and reproduce.

Note that this approach implicitly anticipates (and, in a sense, requires) intrapopulation variability in the degree to which individuals depart from "optimal" behavior. It attempts to investigate the action of natural selection directly and thereby avoids the "adaptationist" (*sensu* Gould and Lewontin 1979) assumptions inherent in the static, decision-oriented perspective (i.e., the assumption that natural selection has already produced optimally designed organisms). Deviations from the optimum are thus regarded as grist for natural selection's mill, rather than as cause for the model's rejection. Analysis of complex primate diets necessary to determine reasonable optimal points in the dietary space and to place individual diets relative to these optima will be difficult, though not impossible. Such an approach, which places emphasis on what individuals *do* rather than on what, for a variety of more or less justifiable reasons, they are *expected* to do, holds promise for effective analysis of food choice by primates.

Summary and Final Comments

Optimal foraging theory proceeds from the assumption of "rational preferences," an assumption itself built on a series of assumptions concerning heredity, intrapopulation variability, and the "guiding hand" of natural selection. We may accept that assumption a priori (although cogent arguments have been presented to suggest, at the very least, that some caution be exercised at this point [see Gould and Lewontin 1979]). I have tried to argue that, since this assumption is not *verified* by observing the congruence between theoretical predictions and empirical reality, we will require independent confirmation of the premises underlying the optimization assumption. The optimization assumption is a strong one in the sense that it asserts not merely that animals choose those options that yield *maximum* net fitness benefits. There are many reasons why animals may forage suboptimally, and, again, much research is presently proceeding along just these lines, a good example of the ways in which the process of model-building opens up new and largely unexplored areas for investigation. For example, while most optimal foraging models implicitly assume that the animals "know" the values of all parameters appearing in the optimization equations (e.g., the density and/or caloric value of all available foods, the amount of time available for foraging, etc.), it is recognized that this assumption will frequently conflict with the necessarily limited capabilities for information storage and retrieval possessed by most organisms (e.g., Gill and Wolf 1977; Pyke et al. 1977; Smith and Sweatman 1974). Furthermore, the actual identification of the optimal course of action may require prodigious computational abilities far in excess of what most animals are likely to possess (see Oaten 1977 for a particularly trenchant commentary on this problem).

Furthermore, our models of this decision-making process necessarily involve further simplifying assumptions. They should, therefore, be invoked as predictors only in those cases where either these simplifying assumptions are known to be appropriate or where the model's predictions are not significantly affected by their violation. I have tried to illustrate this point with reference to both the assumption of fine-grained foraging environments and the use of surrogate currencies other than fitness in foraging contexts characterized by less than perfect information ("uncertainty").

At its simplest, optimal foraging theory is a logical structure embodying a series of assumptions that generates, by deduction, falsifiable hypotheses about decision-making during foraging. It thus has an enormously important role to play in the process of hypothesis formation and hypothesis testing in this area of behavioral ecology. That role will be best served within primatology (as elsewhere) when both the strengths and limitations of the theory are well understood.

Acknowledgments

Financial support during the preparation of this paper was provided by a grant from the National Science Foundation (BNS 79-12810). My fieldwork in Amboseli was made possible by grants from the National Science Foundation, the National Institute of Mental Health, the Sigma Xi Society, and the Boise Fund, and by the generous cooperation of the Kenyan Ministry of Tourism and Wildlife. Although I take full responsibility for all the ideas expressed above, I would like to thank S. Altmann and G. Hausfater for stimulating me to think about many of these problems and J. Cant and P. Rodman for comments on an earlier draft.

References

Altmann, S. A. 1974. Baboons, space, time, and energy. *Amer. Zool.* 14:221–48.

Altmann, S. A., and S. Wagner. 1978. A general model of an optimal diet. In D. Chivers and J. Herbert, eds. *Recent Advances in Primatology,* 1:407–11. London: Academic Press.

Charnov, E. L. 1976. Optimal foraging: The marginal value theorem. *Theoret. Pop. Biol.* 9:129–36.

Clutton Brock, T. H. ed. 1977. *Primate Ecology: Studies of Feeding and Ranging Behaviour in Lemurs, Monkeys, and Apes.* London: Academic Press.

Clutton-Brock, T. H. and P. H. Harvey. 1977. Species Differences in feeding and ranging behaviour in primates. In Clutton-Brock 1977:557–84.

Cohen, J. 1972. Aping monkeys with mathematics. In R. Tuttle, ed. *The Functional and Evolutionary Biology of Primates,* pp. 415–36. Chicago: Aldine.

Edwards, W. 1954. The theory of decision-making. *Psych. Bull.* 51:380–419.

—— 1960. Measurement of utility and subjective probability. In H. Gulliksen and S. Messick, eds. *Psychological Scaling: Theory and Applications,* pp. 114–48. New York: Wiley

Emlen, J. M. 1973. Ecology: An Evolutionary Approach. Reading, Mass.: Addison-Wesley.

Freeland, W., and D. Janzen. 1974. Strategies of herbivory by mammals: The Role of plant secondary compounds. *Amer. Natur.* 108:269–89.

Gaulin, S. J. C. 1979. A Jarman-Bell model of primate feeding niches. *Hum. Ecol.* 7:1–20.

Gaulin, S. J. C., and M. Konner. 1977. On the natural diet of primates, including humans. In R. Wurtman and J. Wurtman, eds. *Nutrition and the Brain,* 1:244–61. New York: Raven Press.

Gill, F., and L. Wolf. 1977. Nonrandom foraging by sunbirds in a patchy environment. *Ecology* 58:1284–96.

Gould, S. J., and R. C. Lewontin. 1979. The spandrels of San Marcos and the Panglossian paradigm: A critique of the adaptationist programme. *Proc. Roy. Soc. Lond.* 205:581–98.

Homewood, K. 1978. Feeding strategy of the Tana mangabey (*Cercocebus galeritus galeritus*). *J. Zool., Lond.* 186:375–91.

Krebs, J. R. 1978. Optimal foraging: Decision rules for predators. In J. R. Krebs and N. B. Davies, eds. *Behavioural Ecology,* pp. 23–63. Sunderland, Mass.: Sinauer.

Levins, R. 1968. *Evolution in Changing Environments.* Princeton, N.J.: Princeton University Press.

Maynard Smith, J. 1978. Optimization theory in evolution. *Ann. Rev. Ecol. Syst.* 9:31–56.

McCleery, R. 1978. Optimal behaviour sequences and decision-making. In J. R. Krebs and N. B. Davies, eds. *Behavioural Ecology,* pp. 377–410. Sunderland, Mass.: Sinauer.

Nicholson, W. 1972. *Microeconomic Theory: Basic Principles and Extensions.* Hinsdale, Ill.: Dryden Press.

Oaten, A. 1977. Optimal foraging in patches: A case for stochasticity. *Theoret. Pop. Biol.* 12:263–85.

Oster, G., and E. O. Wilson. 1978. *Caste and Ecology in the Social Insects.* Princeton, N.J.: Princeton University Press.

Post, D. G. 1978. Feeding and ranging behavior of the yellow baboon. Ph.D. dissertation. New Haven, Conn.: Yale University.

—— In press. Feeding behavior of yellow baboons in the Amboseli National Park, Kenya. *Int. J. Primatol.*

Pulliam, H. R. 1974. On the theory of optimal diets. *Amer. Natur.* 108:59–74.

Pyke, G. H. 1979. The economics of territory size and time budget in the golden-winged sunbird. *Amer. Natur.* 114:131–45.

Pyke, G. H., H. R. Pulliam, and E. L. Charnov. 1977. Optimal foraging: A selective review of theory and tests. *Quart. Rev. Biol.* 52:137–54.

Raiffa, H. 1968. *Decision Analysis.* Reading, Mass.: Addison-Wesley.

Rapoport, A. 1969. *Two-Person Game Theory: the Essential Ideas.* Ann Arbor, Mich.: University of Michigan Press.

Rapport, D. 1971. An optimization model of food selection. *Amer. Natur.* 105:575–87.

Rapport, D., and J. E. Turner. 1977. Economic models in ecology. *Science* 195:367–73.

Schoener, T. W. 1971. Theory of feeding strategies. *Ann. Rev. Ecol. Syst.* 2:369–404.

Simon, H. 1955. A behavioural model of rational choice. *Quart. J. Econ.* 69:99–118.

—— 1957. *Models of Man.* New York: Wiley.

Smith, J. N. M., and H. P. A. Sweatman. 1974. Food-searching behavior of titmice in patchy environments. *Ecology* 55:1216–32.

Sussman, R. W. 1979. *Primate Ecology: Problem-Oriented Field Studies.* New York: Wiley.

11 A Conceptual Approach to Foraging Adaptations in Primates

JOHN G. H. CANT AND
L. ALIS TEMERIN

PRECEDING CHAPTERS serve two functions. First, they provide analyses of a broad range of specific issues in the study of primate foraging and illustrate the current state of the art. Second, several of them, most explicitly those by Kay and by Post, discuss conceptual approaches for investigating certain classes of adaptations for foraging.

Our aim in this chapter is to present a general conceptual framework for investigating primate foraging. We have found that preparing this has been useful to us as a means of organizing what

JOHN CANT is currently studying aspects of ecology and behavior of several species of monkeys and apes in the Gunung Leuser National Park, northern Sumatra, as a means of understanding catarrhine evolution. In the past he has studied behavioral ecology of spider and howler monkeys at Tikal, Guatemala. ALIS TEMERIN is completing her doctoral dissertation on theoretical, historical, and comparative perspectives on the cercopithecoid-hominoid differentiation. She has worked in the field in East Kalimantan and at Tikal.

In the following paper, Cant and Temerin develop a framework for research on foraging adaptations and discuss the relationship of preceding chapters to that framework. Though the paper concerns primates, it is of broader utility on the zoological spectrum. The paper provides a reasoned basis for ordering priorities in observation and analysis of foraging of animals by breaking down the general problem of finding and harvesting food into component problems and showing how various aspects of the whole organism contribute to solutions to these subproblems.

seemed at first to be highly diverse and in many cases weakly interrelated material. We hope that the result will prove useful to others in a similar way, by helping identify issues and connections between them and thereby facilitating the formulation and testing of hypotheses.

We define *foraging* in a very general and inclusive way as the location, acquisition, and assimilation of food. Thus all behavioral activities and physiological processes leading to nutritional gain fall under this rubric. There has been debate recently about whether or not the principle of natural selection is a tautology and whether it is possible to test hypotheses about adaptation (e.g., Maynard Smith 1978; Gould and Lewontin 1979; Brady 1979). For present purposes we define an adaptation simply as a trait that contributes to inclusive fitness because it helps fit the organism to its environment. We adopt Maynard Smith's (1978) view: what is to be tested is not the theory of natural selection or the thesis that nature optimizes but specific hypotheses about the ways that selective forces operate on phenotypes.

A common practice of biologists studying adaptations has been to divide animals into different categories of traits defined by academic disciplines, e.g., morphological, physiological, behavioral, and then analyze traits of one kind separately from attempts to understand other sorts of traits. This carving up of animals into categories of traits—a "trait-category orientation"—may fail to take account of the integrative nature of adaptation and evolutionary change. For example, mastication and digestion operate together to deal with qualities of food items (Kay, this volume), but how many empirical studies analyze teeth and digestion in a unified way?

In place of a trait-category approach, we prefer to look at foraging adaptations from the perspective of *problems* faced by a primate in finding and acquiring food. Traits of several different kinds are likely to be involved at any time in the solution to a problem, and initial focus on foraging problems and their solutions can help us deal with the interdependence of traits.

This focus on problems and their solutions is not new in biology and is basic in optimal foraging studies (e.g., Pyke et al. 1977). Our discussion and the optimal foraging approach are complementary:

we attempt a broad outline of problems and the interactions of traits and environmental factors, and optimal foraging modeling is an analytical tactic that is highly useful in elucidating solutions to certain kinds of foraging problems (see Post, this volume).

Examples of a problem-oriented approach in primate studies are uncommon. A notable one is Cartmill's (1974) explanation of the evolution of stereoscopic vision and prehensile appendages in primates. By taking account of the obstacles to mobility and food acquisition posed by structural features of the arboreal habitat, he showed how aspects of vision must have coevolved with traits of locomotor morphology to facilitate predation on mobile prey in a small-branch milieu.

In the following pages we begin with a summary of our conceptual approach. The approach is then illustrated by delineating the kinds of problems primates face on a daily basis when they forage. We conclude with a brief discussion of the use of this framework in understanding foraging adaptations.

The Conceptual Framework

As we have noted, the crux of our approach is a focus on problems and solutions in foraging. Several different classes of variables constitute what we term "the foraging universe," and we first state briefly the relations between these variables.

Synopsis

Our conceptual framework is illustrated in figure 11.1 The goal of a foraging animal is to meet its metabolic demands for nutrients and water. Thus we begin with an animal who must satisfy a particular set of nutritional needs. Environmental factors (both biotic and physical) determine the circumstances under which needs must be met, and these two elements combine to produce problems that the foraging animal has to solve.

Traits of morphology and physiology determine the sets of behaviors open to an animal, or its behavioral capacities. These capacities enable an animal to solve the problems encountered during forag-

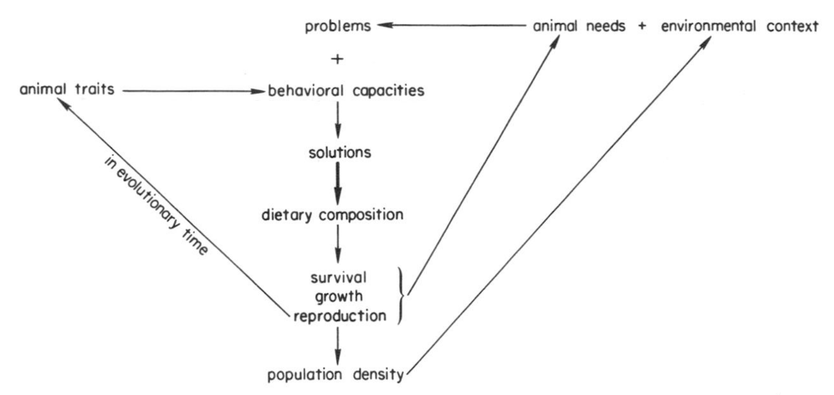

Figure 11.1. A model of the foraging process. Arrows indicate deterministic relations between classes of variables.

ing, and also determine the precise manner in which solutions are achieved. A particular dietary composition (as measured over a meaningful interval of time) is the eventual outcome of this process.

Dietary intake is intimately related to an animal's survival, growth, and reproduction through its role in determining the allocation of nutritional gain to different categories of metabolic needs. The consequent impact on animals' life history traits is eventually manifested at the population level by variations in population density.

These last components of the foraging universe influence other groups of variables contributing to the foraging process and thus introduce feedback. First, at any time an animal's allocation of nutrient gain to different requirements is influenced by the degree to which previous dietary intake has satisfied metabolic demands. Second, changes in population density will contribute to alterations in the environment of a foraging animal. Third, fitness, most closely approximated in our scheme by "reproduction," affects the evolution of animal traits.

Further Definition

An abstract model of the foraging process may convey the relations between elements of the foraging universe, but it does not portray adequately the often varied nature of these elements. Here we discuss them in more detail as they appear in figure 11.1.

Animal Needs. Dietary intake must satisfy a primate's protein, calorie, vitamin, mineral, and water requirements. Specific needs vary temporally according to animal activity and the availability of nutrient stores deposited when gains exceeded metabolic demands. Size, age, and sex differences contribute to variation among individuals in their nutritional requirements.

Environmental Context. Many different aspects of the environment may influence a foraging primate. We describe nine categories of environmental factors and suggest ways these contribute to the problems that must be solved.

1. Climate. Ambient temperature and rainfall are probably the most significant climatic factors affecting primates. Scheduling of activity may be directly influenced by the association between temperature and animal heatload, or between rainfall and ease of activity (e.g., many primates cease foraging during heavy rainstorms). Scheduling may also be indirectly influenced through the impact of ambient temperature on the activity patterns of ectothermic prey.

2. Physical structure of the habitat. Ground topography and the structure and density of vegetation are principal variables. At ground level, surface configuration and vegetation density influence travel routes and speeds. Structural features of arboreal habitats, including branch sizes, resilience, inclination, and continuity, play a critical role in determining patterns of travel and food harvest activity (see chapters by Crompton, Garber, Grand, this volume). A habitat's physical structure can also be important in determining the risk of predation by affecting predator visibility and opportunities for escape from attack.

3. Predators. Risk of predation is important in the lives of most primates, though different species and age/sex classes vary in susceptibility to predation. Predator habits and densities may influence both the location and scheduling of activity.

4. Co-consumers. Generally, several different animals have some foods in common in a given habitat. These co-consumers can belong to the same species as the foraging primate or to other species and may or may not feed in a food patch concurrently. Concerted feeding activity, which often characterizes primates in a social group, has

an impact on the amount of food a patch yields to a consumer. For example, the density of food items will be decreasing at the moment the animal is attempting to acquire them, and dominance relationships are likely to influence rates of food acquisition also. When food consumption occurs in the absence of the focal primate, the spatiotemporal distribution of food in the habitat is altered. Co-consumers can also influence access to food patches (e.g., coordinated defense against intruders), and the location of patches or the items contained within them (e.g., when animals are foraging for insect prey).

5. Spatiotemporal distribution of food patches. We define a food patch as a localized aggregation of food items (fruit, leaves, invertebrates, etc.) that is separated from other such aggregations by regions of markedly lower food density. Patches are distributed in both space and time, and although it is theoretically possible that they are evenly dispersed, it is more likely that primates must deal with clumped and/or random distributions. The resulting variation in interpatch distances and resource predictability influences search and travel activities (Ghiglieri, this volume; Rodman, this volume).

6. Patch size. Food patches contain variable numbers of food items. Size represents a quantitative measure of the items present in a patch at a certain time. Patch sizes vary with location, and also with species identity in cases where trees, shrubs, or vines constitute patches. The same patch can vary in size through time as well. These factors help determine the gain a primate can achieve by feeding in a particular patch.

7. Arrangement of foods within a patch. The distribution of plant and animal foods within a patch is variable. Microhabitat variation within patches and the growth forms of trees and other plants probably contribute to uneven spatial distribution of particular food items in most situations.

8. Physical characteristics of food items. Foraging primates are attracted to foods by sensory cues such as color, odor, texture, and, in the case of animal prey, behavior. The location of potential food items relative to substrates influences their discovery, as well as the type of effort needed to procure them. Some plant foods may be located below ground (rhizomes) or along surfaces (gum). Others

hang (fruit, leaves) or project (grasses) from substrates while still others are substrates (bark, clay). Similarly, animal prey may be found below, on top of, or independent of surfaces. The size, shape, and physical structure of food items are critical variables as a primate readies them for ingestion. Mechanical qualities of an item are determined by the proportions, as well as absolute quantities, of its structural elements (chitin, collagen, silica, lignin, etc.). These affect the item's resistance to the shearing, tensile, and compressive forces that animals can exert during food preparation (e.g., opening a fruit husk or separating pulp from seed) and mastication.

9. Constituents of food items. Primates encounter foods that vary in their content of water and potentially digestible constituents, including cell wall material, nonstructural carbohydrates, protein, lipids, vitamins, and minerals. Plant and animal toxins, another class of food constituents, are variously present and at the very least, may negatively affect digestive processes.

Problems. The problems a foraging primate must solve can be identified initially without reference to particulars. There are eight major ones, including "when to forage," "how to move within a food patch," and "how to ready items for ingestion." Others are listed in table 11.1. We present an extensive discussion of the problems primates confront in the next section of the paper.

Animal Traits. Each primate displays a particular assemblage of morphological and physiological characteristics that influence its foraging capacities. These characteristics are of course functionally integrated, and we separate them into aspects of structure and function for explanatory purposes only. Similarly, it is most convenient to organize traits along functional (or systemic) lines. Major categories include body size, musculoskeletal anatomy of the locomotor system, dentition, digestive processes, metabolic rates, and neurophysiological processes. A more comprehensive list is contained in table 11.1. Age- and sex-dependent variation in various aspects of morphology or physiology (e.g., tissue and body proportions, metabolic rates) contributes to patterns of intraspecific variation in traits.

Table 11.1. Some Components of the Foraging Universe

I. Environmental Factors Climate Physical structure of habitat Predators Co-consumers Spatiotemporal distribution of food patches Patch size Arrangement of foods within patch Physical characteristics of food items Constituents of food items	II. Foraging Problems When to forage How to locate a food patch What food patch to enter How to move within a food patch How to acquire food items How to ready items for ingestion When to leave a patch How to travel between patches
III. Animal Traits Morphological Body size and proportions Tissue proportions (skin, fat, muscle, skeleton, gut, etc.) Musculoskeletal structure of limbs, trunk, jaw, neck, and face Dentition Digestive tract structure and volume Brain size and structure Sense organ structure Physiological Contraction rates and fatigue resistance of muscle fibers Metabolic pathways Rates of metabolism Digestive processes Neurophysiological function	IV. Behavioral Capacities Cognitive and perceptual Learning and problem-solving abilities Abilities to respond to sensory stimuli Travel Locomotor abilities Thermoregulatory abilities Feeding Postural abilities Manipulatory abilities Masticatory abilities Digestive abilities

Behavioral Capacities. A primate's traits of morphology and physiology place intrinsic limits on what it can do in particular circumstances. At the same time these traits influence strongly the actual form of animal-environment interactions (see Temerin 1983). Thus the potential behaviors of a primate have a distinctive character, and they are always limited in scope.

Three major categories of behavioral capacities are important to a foraging animal: cognitive and perceptual, travel, and feeding. Each can be broken down into component abilities (see table 11.1), which may or may not be strictly behavioral. For example, locomotor abilities, a principal component of travel capacities, are expressed behaviorally as progression modes (e.g., quadrupedal running, armswinging, bipedal leaps) and speeds (Crompton, this volume; Garber, this volume). On the other hand, digestive abilities, one set of factors contributing to feeding capacities, encompass physiological variables such as assimilation efficiency and food passage rate (Milton, this volume). A brief characterization of each set of behavioral capacities follows.

1. Cognitive and perceptual capacities. These play an integral role in all aspects of the foraging process. Features such as memory, problem-solving abilities, and relative importance of the various sensory modes (visual, olfactory, auditory, etc.) jointly contribute to how a primate locates a patch, searches for food items within it, and tests the suitability of a potential food, for example.

2. Travel capacities. Progression from one food patch to another or from place to place within a patch depends on an animal's travel capacities. Locomotor abilities help determine progression modes, pathways chosen, speed, and endurance. Thermoregulatory abilities influence certain travel variables, including endurance and speed, through their impact on an animal's susceptibility to heat or cold stress during activity.

3. Feeding capacities. Both the acquisition of food items and the eventual assimilation of nutrients are in the domain of feeding capacities. Postural abilities help determine how a primate stabilizes itself while procuring food, and manipulatory abilities (including use of hands, lips, and teeth) contribute to food acquisition and preparation. Masticatory abilities influence prepation and of course the- manner in which food is reduced into pieces for swallowing (Kay, this volume). Finally, digestive abilities bear critically on rates of nutrient gain. In combination, all these components of feeding capacities help determine whether and how quickly foods can be eaten.

Solutions. The behaviors of a foraging primate—selection of foods, pathways traversed in travel from one patch to another, postures used during food harvest, timing of foraging, and so on—constitute solutions to the problems encountered. At any one time these solutions usually represent a subset of all possible solutions. This is due to two factors: primates generally inhabit complex and variable environments, resulting in problems of an analogous nature; and their behavioral capacities are typically extensive. Solutions are often complex entities themselves, incorporating multiple behavioral variables. For example, when a primate travels arboreally from one food tree to another, its behavior (i.e., the solution to the problem of how to travel between patches) will incorporate several variables, including locomotor mode(s), route taken (partly a function of locomotor mode), and speed of progression (in part a function of locomotor mode and pathway). The problem-solving process is also complicated. Solutions to one problem are frequently interdependent (see above example), and they also influence and are influenced by solutions to other problems. We elaborate upon this aspect of solutions in the next section.

Dietary Composition. The foraging process leads to dietary intake. Primate diets are ecletic; even over the period of a single day animals typically utilize several different food sources. Diets comprise a variety of food types (e.g., invertebrates, fruit, gum, young leaves), and they also exhibit particular species composition. Both aspects of dietary composition vary through time and among individuals of different age and sex classes.

Survival, Growth, and Reproduction. Dietary intake determines whether a primate is able to maintain metabolic homeostasis and, in the case of immature or reproductively active individuals, whether the additional needs generated by growth or reproduction can be satisfied as well. Furthermore, an animal's nutrient condition may play an initial role in determining when the physiological processes leading to growth or reproduction begin or, once they have started, whether they continue. For example, ovulation will be influenced by previous nutritional intake, and outcome of a pregnancy (abortion or full term) may depend on concurrent dietary conditions.

Because some nutrients can be stored when intake exceeds demands (e.g., excess calories stored as fat), food intake must satisfy certain needs only "on the average." That is, dietary requirements need not be continually satisfied in all cases, and gains may fall below needs for brief intervals of time. (How long these intervals may be is species dependent.) This permits short-term lability both in food intake and in the allocation of nutrients to different life history categories. For example, environmental conditions may prevent a primate from getting enough food, a male may devote foraging time to agonostic behavior instead, or a lactating female may have greater needs than she can satisfy. In these situations animals must have reserves to call upon, and abilities to accumulate reserves can contribute to long-term cycles in growth and reproduction.

Population Density. Animal allocation of nutrient gain to the competing demands of homeostasis, growth, and reproduction eventually influences rates of survival and reproduction, proximate determinants of population size. Temporal variation in these parameters may lead to variations in population density.

Feedback Relations. The foraging process results in a particular pattern of dietary intake, and as we have noted, this eventually influences life history and population parameters. The latter variables have in turn an impact on certain other elements of the foraging universe and thus create a feedback network. This network operates in both ecological and evolutionary time.

In ecological time two separate feedback relations contribute to temporal variation in the problems a foraging primate must solve. The first, which takes place over comparatively shorter time intervals, is generated by the impact of prior food intake on subsequent dietary needs. If a primate has not been able to satisfy all of its nutrient requirements and its condition is poor as a consequence (survival is in jeopardy), then subsequent needs will reflect this imbalance. For example, protein or calcium intake may have been below necessary levels, leading to an intensified demand for foods containing relatively high proportions of these nutrients. An analogous situation results if previous intake has triggered reproductive activity or increased growth rates, augmenting nutritional requirements. Alternatively, if dietary intake exceeds needs, and the excess is stored, then at some future time nutrient gain can be less than the concurrent expenditures. The second feedback relation is produced by the effect population density has on a foraging primate's environment. Numbers of co-consumers vary directly with local population density, and food availability will vary concomitantly.

A single feedback relation operates in evolutionary time. Reproductive activity, influenced by dietary intake, is an important determinant of fitness. Hence it bears on the evolution of the morphological and psychological traits related to foraging. When evolutionary change in these traits occurs, there is a corresponding alteration in behavioral capacities. This has, of course, an eventual impact on solutions and the consequences thereof.

Illustration of the Approach: Problems Faced by a Foraging Primate

Each day a foraging primate faces a series of problems in food acquisition. We have identified eight problems (see table 11.1) and in this section discuss their specific natures. We view the foraging universe of primates as it exists in ecological time and select one day as the unit of analysis. Thus both a primate's traits and environmental circumstances may be considered relatively invariant. This approach is, however, taken for purposes of illustration only. Short- and long-term variation in life history processes and environmental factors leaves no question that periods longer than a day are critical for the foraging process, and these must eventually be considered.

As a first approximation, we think of a primate setting out in the morning with the basic need of finding and consuming enough of the proper food to maintain metabolic homeostasis. Problems arise the moment the animal awakens with "when to forage?" and then quickly multiply. We treat each one in logical sequence but stress that solutions to one problem are rarely separable from solutions to another. This is conveyed in part through figures that summarize for each problem the patterns of organism-environment interaction that may be involved in its solution (see figs. 11.2–11.9). In each figure the problem under discussion is contained in a central circle. This is surrounded by concentric bands, representing components of the foraging universe: abilities, behavioral capacities, environmental variables, and where applicable, solutions to other problems. Functionally allied elements are grouped into clusters, and when there is more than one cluster, each constitutes a particular aspect of the problem. Design considerations have led us to leave important elements out of the figures: morphological and physiological traits are not included, and abilities are specified only for feeding and locomotor capacities because it is here that very distinct abilities contribute to a single class of behavioral capacities.

Our aim is to set forth for each problem all the kinds of organism-environment interactions that may be occurring in primates. We limit this treatment to the immediate context of foraging, however, and are aware that except for mention of co-consumers, we have largely excluded from discussion relationships between foraging and sociality. Those relationships are the subject of intense interest in primatology, and some major questions are: How many others "should" an individual forage with? What are the benefits of associating with different kinds of conspecifics (kin, non-kin, age-sex classes)? If grouping is beneficial for whatever reason, how might an individual minimize adverse effects of intragroup competition for food? How and why do foraging patterns of different members of a group differ? With respect to the last question, Post (this volume) provides a very useful discussion of modeling optimal diet and then determining how the diet of each group member differs from the theoretical optimum. Other chapters in this book by Ghiglieri, Rodman, and Waser examine important relationships between foraging behavior and social organization, particularly ways in which the demands of foraging may limit the degree to which individuals can be social. Full understanding of foraging requires that we elucidate not only foraging itself but also its effects on other behaviors.

Problem 1: When To Forage?

Primates typically alternate foraging with other activities, and the first problem we consider is the scheduling of foraging (figure 11.2). In most cases activity must be organized over roughly a 12-hour period, night or day. This is because the option of nocturnal vs. diurnal activity (or of continuous activity) is generally not open to primates in ecological time. (Exceptions are found among some Malagasy prosimians, e.g., *Lemur mongoz* [Harrington 1978], which may be active over the entire diel cycle.) Aspects of the physical environment,

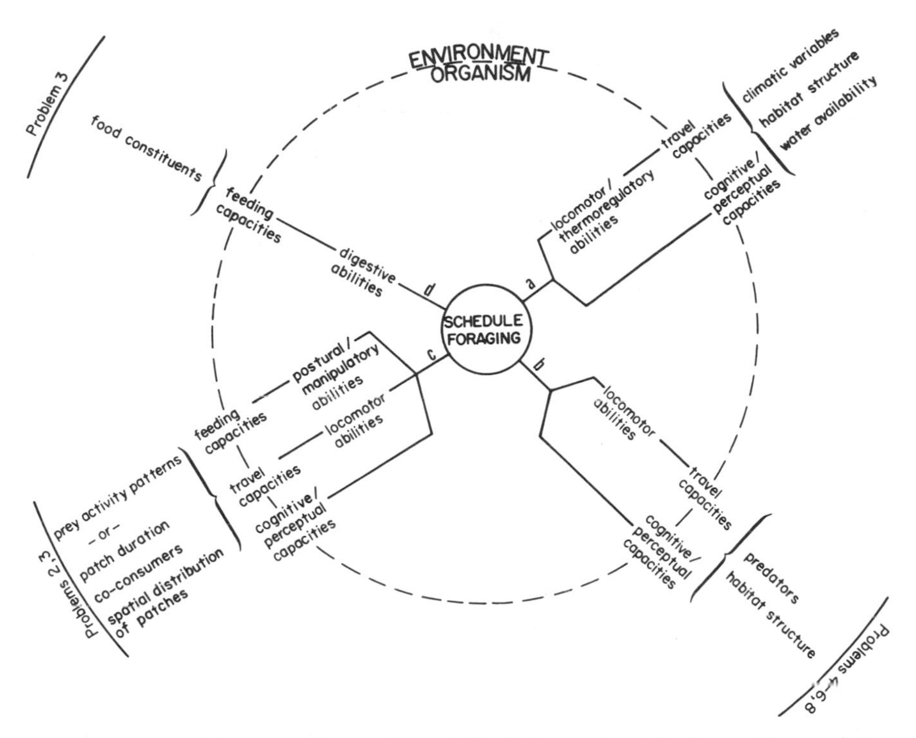

Figure 11.2. Problem 1: When to forage? Variables pertinent to the problem and its solution are organized in concentric bands about a central circle. Proceeding outwards, the bands comprise animal abilities, capacities, environmental variables, and other problems whose solutions are relevant in the present context. Functionally allied elements are grouped into separate clusters (a–d), and the dashed line represents the interface between organism and environment. See text for further discussion.

predators, co-consumers, and the nature of food resources may all influence when a primate forages.

Climatic variables, including temperature and rainfall, and habitat structure contribute to the problem of scheduling foraging, particularly insofar as they influence the maintenance of thermal homeostasis (cluster a, figure 11.2). For example, heat stress may underlie patterns of activity in many diurnal primates. Daily temperature regimes, in combination with site of activity (shaded or exposed microhabitats) and accessibility of water sources, help determine the potential for thermal stress. Whether foraging activity is actually constrained by this factor will depend in large part on a primate's travel capacities: in particular, how far an animal is able to travel—or how long it can remain active—before gaining an insupportable heat load. These behaviorable variables will be influenced by how much heat a primate absorbs from the environment, the quantity it generates by muscular activity, and abilities to dissipate excess heat; these features are in turn influenced by animal traits such as pelage density and coloration, body size and proportions, circulatory patterns, nasal structure, and so on.

A primate's sensory evaluation of its environment while foraging is also influenced by climatic variables. Rainfall especially may hamper visual, olfactory, auditory, and tactile functions and thus lead to a halt in foraging activity. During heavy rainfall prey items may be more difficult to locate, or progression may entail added risk of falls owing to slippery substrates.

Susceptibility to predation may influence when a primate forages (cluster b, fig. 11.2). Predator traits and habitat structure help determine the chances of predation. Various primate traits also contribute to this risk. Details of appearance, including pelage coloration and the degree of stealth in foraging, affect a predator's discovery of its potential prey, while the perceptual and cognitive capacities of primates bear on the detection of their potential predators. The latter traits, in combination with locomotor abilities, help determine modes of escape (and the success thereof). Agile arboreal species may move rapidly in flight while slow-moving species might attempt concealment, for example. If predation is significant for a primate population, risk could be reduced by not foraging at certain times or by pursuing activity in high-risk locations only when the perceptual abilities of predators are most limited. Gautier-Hion et al. (1981) have suggested, for example, that *Cercopithecus cephus* limits foraging in certain habitat zones to times of reduced light as a method of reducing risk from avian predators, and Crompton (this volume) analyzes the ways galagos of differing body size and patterns of habitat use deal with the threat of predation.

Foraging activity must be closely tied to when, for how long, and where food patches exist. The distributional patterns of food species, as well as the presence of co-consumers, are relevant here (cluster c, fig. 11.2). Food patches may appear and disappear or fluctuate in size over short time intervals (ca. 12 hours). Where animal prey is consumed, the activity patterns of the prey are a major source of temporal variation in the existence of a patch and the potential gain derived from foraging in it. Similarly, patches containing plant foods may vary on a short time scale, and both plant characteristics and the presence of co-consumers can account for this. For example, the neotropical species *Caseria corymbora* bears ripe fruit for only a few hours of the morning, during which many consumers deplete the supply (Howe 1977). The tree then ceases to be a patch or at least becomes a radically smaller one for the rest of the day. Nocturnal ripening of more fruit leads to a renewal of the patch the next morning. The spatial distribution of patches also influences a primate's foraging schedule, inasmuch as the time needed for travel between food patches affects how much time is allocated to foraging. Thus animals exploiting highly dispersed patches might begin foraging earlier (or halt later) than otherwise.

The impact of such spatiotemporal variation in food sources on a primate's foraging schedule is mediated by traits contributing to all three major categories of behavioral capacities. Important ones are those influencing food harvest rates (postural and manipulatory abilities, perceptual and cognitive abilities, and the proximate determinants thereof) and travel speed (locomotor abilities and their determinants).

Food constituents, in combination with primate feeding capacities, constitute a final source of constraints on the timing of foraging (cluster d, fig. 11.2). Whenever rates of food intake exceed digestion rates, animals face the problem of having full stomachs. Under such circumstances feeding activity must cease until digestive processes create more room for food. These conditions probably occur most often when primates ingest substantial amounts of plant fiber, bulky material that moves slowly through the digestive tract. Gut volume and food passage rates are important determinants of digestive abilities and thus must bear on the need to interrupt foraging. A more subtle influence on foraging schedules is also possible. Over a day many primates eat plant foods varying in fiber content, and when particular items are eaten may be partly influenced by whether there is a subsequent need to halt food intake. Consumption of foods with high fiber content near the end of the activity cycle will not affect subsequent foraging activity as much as it might at other times. (Post [this volume] offers an alternative hypothesis incorporating considerations of uncertainty and risk-taking to explain the "fruit-early" and "leaves-late" phenomenon.)

The last two aspects of problem 1 (clusters c and d) depend most clearly on the identity of the food items a primate consumes. This is determined largely in part by the solution to problem 3 (what food patch to enter), and it is thus appropriate to consider the latter solution as a major source of constraints on the scheduling of foraging. (Note that this consideration holds for the other problems as well [see figs. 11.3, 11.5–11.9]). At the same time, however, a primate's capacity to forage on a certain schedule—and thus its ability to exploit highly ephemeral food patches or consume large quantities of fibrous foods, for example—can bear on which food patches are exploited. This factor contributes to the interdependence of the two problems and their solutions. Solutions to other problems may also influence when a primate forages: the manner in which food patches are located (problem 2) may help determine the most profitable time to forage, and the speed with which animals are able to harvest food (problems 4–6) and travel between patches (problem 8) in particular circumstances can affect how long they risk predation and hence the timing of such activities.

Problem 2: How To Locate a Food Patch?

As we noted earlier, the food patches of primates are probably uneven in distribution. Animal search patterns must deal with this by reducing the random element of search as much as possible. There are several complementary methods: use memory and learning while choosing search paths, and use sensory cues to locate patches before gaining their immediate vicinity and also to monitor the temporal dimensions of patch existence.

Four sets of environmental factors and related behavioral capacities operate in locating food (fig. 11.3). First, foods differ in spati-

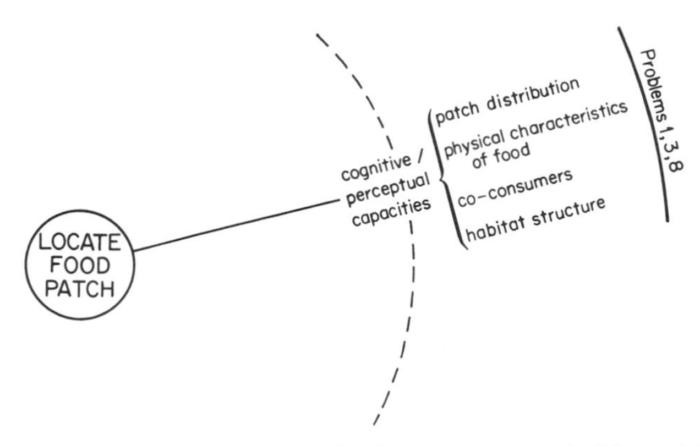

Figure 11.3. Problem 2: How to locate a food patch? See legend of figure 11.2 and text for further discussion.

otemporal distribution, and the difficulty of locating patches should vary concomitantly. The cognitive capacities of animals—based on abilities to remember past search routes, the location of patches that have been exploited or monitored previously, etc.—help determine predictability of patch location and thus the effectiveness of search patterns. Studies of the relation between encephalization (a morphological index of cognitive capacities) and dietary content in several groups of mammals, including primates (Clutton-Brock and Harvey 1980), bats (Eisenberg and Wilson 1978), and rodents (Mace et al. 1981) have demonstrated, for example, that species eating more evenly distributed foods, such as foliage, have smaller relative brain sizes. In this volume, Ghiglieri suggests that the advanced mental abilities of chimpanzees may have evolved in response to patchiness of food.

Second, food items may provide sensory cues to the location of patches. The color and odor of plant foods may signal the presence of a patch to searching primates, as may characteristics of potential animal prey, such as color, noise of activity, and possibly odor. A primate's perception of cues depends in part on the development of the different sensory modalities: visual acuity, olfactory sensitivity, auditory range, and so on.

A third environmental factor, the presence of co-consumers, can possibly furnish cues that enable animals to find food patches. Co-

consumers that have located a patch may signal this to others by their simple presence (e.g., Andersson et al. 1981); by a change in their activity, as when certain members of a dispersed social group stop moving and begin to feed; by active vocalization (e.g., male chimpanzees; Ghiglieri, this volume); and by some mode of information transfer, as when individuals aggregate between bouts of foraging activity (e.g., Menzel 1979; Krebs 1974; Waltz 1982).

Habitat structure, a fourth aspect of the environment, may either impede or facilitate a primate's reception of sensory cues. Small-scale topographic relief and foliage density, for example, are likely to affect the visual fields of animals and the attenuation of sound waves.

Solutions to several other problems also impinge on the ways primates locate food sources. Foraging schedules (problem 1) help determine the environmental circumstances under which activity occurs, e.g., light and sound conditions, or the presence of co-consumers belonging to other species. Dietary choice (problem 3) establishes the nature and distribution of food items, and patterns of travel between patches (problem 8) contribute to opportunities for perceiving cues and signals identifying food patches.

From the investigator's viewpoint, the observable results of search and travel are travel routes and patch visitation patterns. One of the few studies to address these topics is Waser's comparison, in this volume, of the ranging behavior of two species of mangabeys. *Cercocebus galeritus* backtracks more than *C. albigena,* apparently returning to recently used areas more frequently, and Waser suggests that this behavioral contrast reflects habitat differences in rates of invertebrate patch renewal.

Problem 3: What Food Patch To Enter?

Primates must choose among many potential food sources when they decide which to exploit, and each decision is represented by entry into a particular food patch. The problem of food choice depends to a large extent on how an animal solves the other foraging problems (see fig. 11.4). Only occasionally are elements of this problem unique to it, and we consider these in greatest detail, leaving

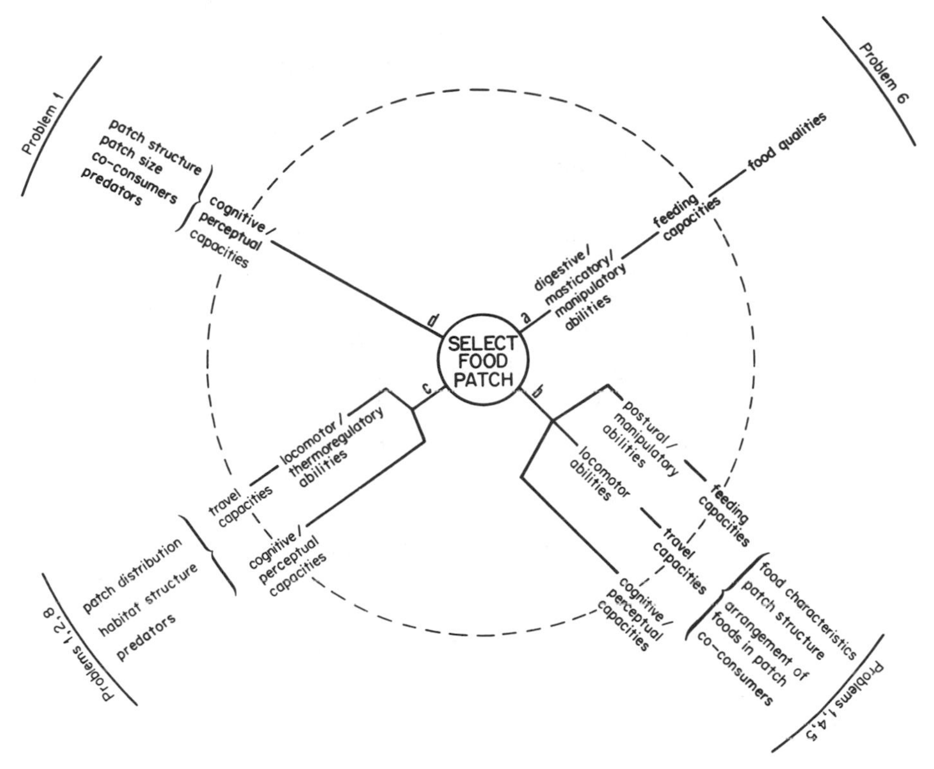

Figure 11.4. Problem 3: What food patch to enter? See legend of figure 11.2 and text for further discussion.

extensive discussions of the other foraging problems to the appropriate sections to avoid undue repetition. Because so many variables are involved we emphasize that our discussion does not adequately convey how the process of reaching a solution must actually work. (The same consideration also applies to the other foraging problems.) Primates presumably do not confront the problem of food choice by progressing linearly through its component parts, as represented here. Rather, they somehow combine, in more or less simultaneous fashion, an ongoing appraisal of environmental circumstances with past experience and learning in order to choose among behavioral options.

Several sets of variables bear on patch choice, and each is a source of constraints on which foods an animal eventually consumes. We begin with factors that determine the nutritional benefits of ingesting a particular item (cluster a, fig. 11.4). The constituents of food items are the nutrients that are potentially available to a consumer. Actual availability—what a primate assimilates and how quickly it does so—is influenced by animal abilities to extract nutrient gain through the process of digestion and by abilities to ready items for ingestion (problem 6). Gut structure and volume, and digestive physiology, are determinants of assimilation efficiencies and food passage rates (Milton, this volume). The latter variables are also influenced by solutions to problem 6, which determine the nature of the material actually ingested. For example, an ability to chew finely the plant cell wall material or chitin enhances digestive efficiencies (Kay and Sheine 1979; Kay this volume), and rejection of fibrous fruit pulp after chewing can lead to higher rates of food passage.

An item should be excluded from a primate's diet if the animal is unable to derive any nutrient benefit from it, either alone or in combination with other foods. Thus the elements comprising cluster a establish rigid boundaries to the domain of possible food sources. Choice among the remaining possibilities will be strongly influenced by factors bearing on food procurement (clusters b–d, fig. 11.4).

A number of variables relate to how animals gain possession of the food items they eventually ingest (cluster b, fig. 11.4). Certain food characteristics, including the ways plant items are attached to substrates and the nature of prey microhabitats, help determine the techniques of food procurement a consumer must employ. This constitutes problem 5, and a primate's postural and manipulatory abilities contribute to its solution (see below). Details of patch structure, such as the disposition and sizes of the branches of a food tree, the presence of co-consumers, and the distribution of foods within a patch, contribute to the problem of how a primate moves about while feeding (problem 4). Travel capacities are particularly important in this context.

The solutions to the two problems of food procurement and within-patch movement have both a direct and an indirect impact on the central problem of food choice. First, they determine whether access to foods is possible. Second, they help determine how efficiently a patch can be exploited and hence its value relative to other patches. The solution to another foraging problem, when to forage (problem 1), also has some significance for cluster b (and for clusters c and d) since the amount of time allocated to foraging helps establish which rates of food harvest will ultimately result in nutritional gains.

Because primates must usually exploit several different food sources over a day (or night) to satisfy their nutritional requirements, travel from one patch to another—problem 8—bears on food choice (cluster c, fig. 11.4). The solution to this problem is influenced by the spatiotemporal distribution of patches, habitat structure, and the presence of predators, on the one hand, and by the travel and perceptual and cognitive capacities of the forager, on the other. It is also affected by how animals locate food patches (problem 2), inasmuch as search modes may help determine travel speeds or routes. As in the previous context (cluster b), both access and behavioral efficiency are pivotal variables contributing to the relative value of different food patches. A primate obviously cannot exploit a patch it does not reach in the time it has available (which is influenced by problem 1), and travel incurs costs (risk, time, and energy, for example) that must be measured against the potential gains a patch offers.

A primate's appraisal of this variable—the potential gains offered by a patch—is the final aspect of the process leading to selection of a food patch (cluster d, fig. 11.4). Pertinent environmental variables include patch size (or richness) and structure and the presence of co-consumers and possible predators. The quantity of food items each patch contains determines the maximum number of items a consumer can harvest (i.e., potential yield). If co-consumers feed concurrently in the patch, its effective yield will be less than this, so that from a forager's point of view a patch becomes smaller with increasing numbers of co-consumers. But the simultaneous presence of co-consumers may also enhance the attractiveness of a patch if detection of predators is improved as a result. Patch structure is also important with respect to predation, for it helps determine the likelihood of discovery by a predator and the opportunities to escape from an attack.

An animal's cognitive and perceptual capacities contribute to methods of appraising food patches, e.g., the sensory modes used to evaluate environmental variables, how information is processed, and the extent to which it is then integrated with past experience. These capacities also contribute in large part to the accuracy of the appraisal, and this of course affects the ultimate success of the foraging process.

Choice among accessible food patches by foraging primates should be strongly related to the efficiency with which each can be exploited. The latter will be determined by the interaction between variables determining the gains (clusters a and d) and variables determining the costs of achieving those gains (clusters b and c). Patches will as a consequence differ in their value to an animal, but they need not represent simple "yes or no" options (e.g., Krebs 1978). The diverse diets of most primates, in combination with their complex environments, strongly suggest that patches represent spectra of benefits to consumers and thus fall along a varying scale of preference. This increases the importance of abilities to monitor food availability.

Of the eight foraging problems we discuss, it is to this one, food choice, that the optimal foraging approach has been applied most often. We defer to Post's chapter on optimal foraging in this volume and simply wish to draw attention to his emphasis on the fact that most primates probably experience coarse-grained foraging environments. Post explains how coarse grain complicates efforts to understand food choice.

Finally, body size is an important factor underlying many of the relationships we have been considering, e.g., composition of food items and their benefits to consumers, methods of moving within patches and of acquiring food items, and travel between patches. Four other chapters in this volume deal with body size in greater depth than is appropriate here: Temerin et al. present a general analysis of body size and foraging, Garber investigates how small body size of tamarins influences dietary requirements and positional behavior on widely differing substrates, Crompton examines the effects of body size on foraging problems and solutions in two species of galagos, and Rodman elucidates relationships between sexual dimorphism in body size and foraging in two species of great apes.

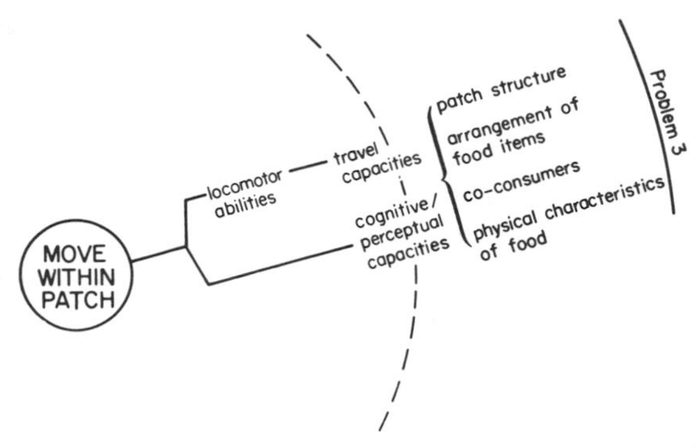

Figure 11.5. Problem 4: How to move within a food patch? See legend of figure 11.2 and text for further discussion.

Problem 4: How To Move Within a Food Patch?

Once a patch has been entered the foraging primate must gain access to food items. Patterns of movement within the patch—locomotor modes, travel speeds, and routes used—depend both on abilities to locate food items and on locomotor abilities (see fig. 11.5).

If a consumer must locate food while it moves through a patch, i.e., if olfactory cues are important or foods are hidden from view, for example, then both travel speeds and routes will be influenced. Relevant factors include the physical characteristics of food items, their distribution within the patch, patch structure, co-consumers, and an organism's perceptual and cognitive capacities. Each contributes to the problem of food location in much the same way that it (or its counterpart) does in the wider context of patch location (see discussion of problem 2), with one exception. Co-consumers may hinder rather than facilitate the location of food items by making progression along certain routes unattractive or impossible. A low-ranking member of a social group may avoid close encounters with higher ranking individuals (e.g., Ripley 1970).

The environmental variables just enumerated, excepting food characteristics, influence how an animal moves about a patch whether or not it must search simultaneously for food items. Details of patch structure, including surface characteristics and vegetation densities at ground level, and the distribution, orientation, texture, and size of substrates located above ground, establish opportunities for the use of particular locomotor modes and/or speeds. Traits influencing locomotor abilities (e.g., anatomy of the locomotor apparatus and body size) help determine the actual manner of progression. Forelimb—dominated abilities in hylobatids, for example, allow them to "climb" among the small substrates of their food trees a major portion of the time (Fleagle 1980), whereas dusky leaf monkeys (*Presbytis obscura*), which do not share these abilities, most often progress quadrupedally along supports in feeding trees (Fleagle 1978).

Relationships among food distribution within the patch, patch structure, and the presence of co-consumers, help determine an animal's pathways as it moves about. The spatial distribution of food items, including locations relative to substrates (e.g., where gum sources or fruits are located with respect to the supports in a tree), establishes the spots to which animals must move. Patch structure bears on the nature and location of substrates and interacts with animal locomotor abilities (see above) to set bounds on the set of possible pathways to foods. As illustration, male howler monkeys (*Alouatta palliata*) are heavier than females and use thin supports in the outer regions of tree crowns less frequently (Mendel 1976); and a small primate with claws is more likely to climb a large, smooth-barked trunk when exploiting a food patch than an equal-sized one without claws (Cartmill 1979; also see Crompton, this volume; Garber, this volume).

An animal's perceptual and cognitive capacities constitute the final determinant of travel routes within a patch. They enable it to recognize, evaluate, and choose among the potential routes established by environmental factors and locomotor abilities.

Problem 5: How To Acquire Food Items?

Food items must of course be procured—plant parts harvested or prey captured—before they can be eaten. There are two aspects to the problem of gaining possession of food (see fig. 11.6). First, an animal must stabilize its body while acquiring items (cluster b, fig.

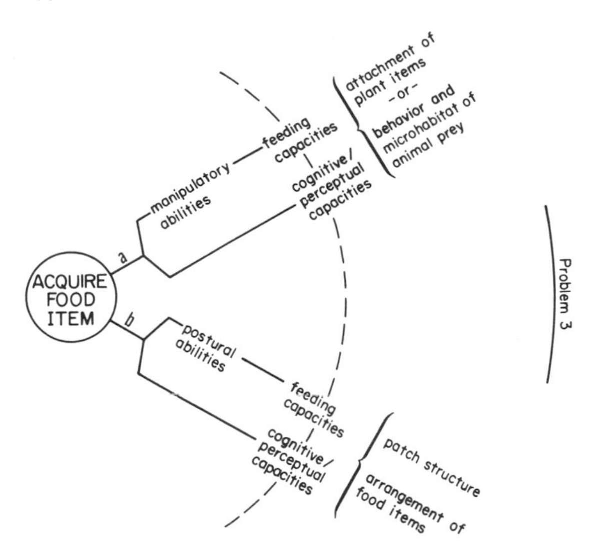

Figure 11.6. Problem 5: How to acquire food items? See legend of figure 11.2 and text for further discussion.

11.6). Characteristics of the available substrates and their position relative to the location of food items contribute to the circumstances under which a primate must secure its position. The manner of accomplishing this rests on an animal's abilities to distribute its mass on, beneath, or among supports. Many traits underlie these postural abilities, including body size and proportions (each of which comprises several variables), limb joint flexibility and the accompanying skeletal and musculo-ligamentous structure, development of flexor musculature in hands and feet, fiber composition of limb muscles, and the pattern of blood supply to these muscles.

Second, an animal must actually gain possession of each food item (cluster a, fig. 11.6). Mode of attachment of plant parts and the nature of prey microhabitats and/or escape behavior operate here. A primate's ability to detach plant parts or catch prey is influenced by its manipulatory abilities, particularly the prehensile component of these; i.e., abilities to use teeth, lips, hands, and possibly feet to grasp and remove (or immobilize) food items. A wide variety of animal traits are involved in the use of these prehensile organs. Relative thumb length and opposability, in combination with hand musculature and innervation, contribute to the ability to manipulate small objects (e.g., Napier 1961); a rhinarium limits upper lip mobility and thus restricts the use of lips in prehension (Maier 1980); underbites enhance the incisal stripping of plant items (Walker, in press); and so on. Perceptual and cognitive capacities are also critical, e.g., eye-hand coordination in the capture of mobile prey, and problem-solving abilities in the use of tools to obtain otherwise inaccessible items.

Problem 6: How To Ready Items for Ingestion?

Most of the food items that primates acquire must be readied for ingestion (see fig. 11.7). This consists of two activities: preparation and mastication (Kay, this volume). Many items include edible and inedible (or undesirable) parts, and in some cases these act as barriers to ingestion as well. There may be soil adhering to the bases of grass stems, tough husks on fruit, large seeds, or dense hairs on caterpillars, for example. Both manipulatory and masticatory abilities, comprising the use of hands, lips, and teeth, help endow a consumer with the capacity to extract the portions it wishes to eat and discard the rest. Problem-solving abilities also contribute to this behavioral capacity.

The size, shape, and physical construction of the remaining food material pose further obstacles to ingestion and to subsequent diges-

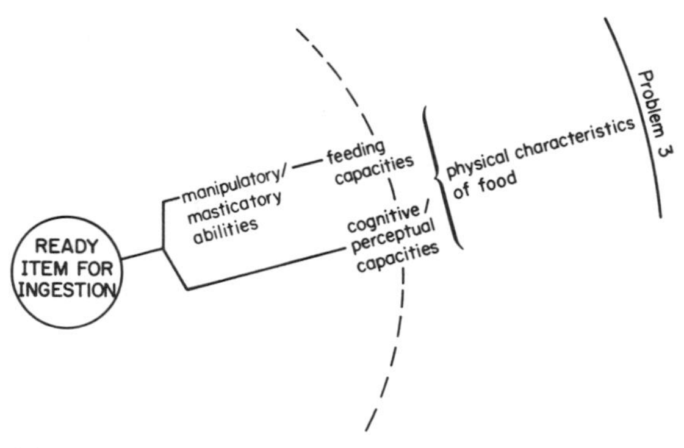

Figure 11.7. Problem 6: How to ready items for ingestion? See legend of figure 11.2 and text for further discussion.

tion. Mastication helps deal with these obstacles, and it depends partly on tooth, jaw, and cranial morphology. Molars with long shearing blades contribute to capacities to grind finely such items as leaves and arthropod exoskeletons (Walker and Murray 1975; Sheine and Kay 1977), for example, and a fused mandibular symphysis—found in all anthropoid primates but absent in many prosimians—is associated with the generation of greater masticatory forces (Hylander 1979). Production and composition of saliva are also relevant because the mixture of saliva with food particles may, in addition to aiding deglutition, initiate digestion.

Problem 7: When To Leave a Patch?

A primate's decision to leave a patch and forage elsewhere, or how long it feeds in a patch, may be a response to patch depletion, interference from other consumers that prevents access to foods, or satiation (either actual or relative to desired intake levels; see fig. 11.8).

Foraging activity produces a rate of patch depletion, and this, coupled with knowledge about the availability of food elsewhere, influences an animal's decision to leave before satiation or the disappearance of all food (e.g., Charnov 1976; Pyke 1981). Depletion rates are a function of focal primate harvest rates (determined by solutions to problems 4, 5, and 6), the number of other animals also exploiting

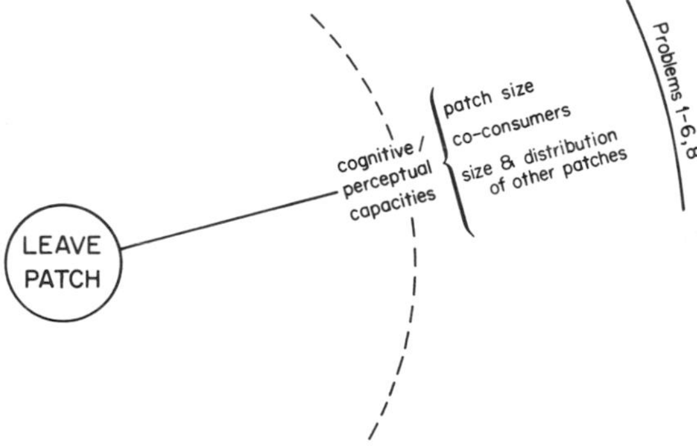

Figure 11.8. Problem 7: When to leave a patch? See legend of figure 11.2 and text for further discussion.

the patch, and their respective rates of food acquisition. Co-consumers will also be the source of an apparent reduction in food availability if they restrict access to food items, such as by enforcing social dominance. The spatiotemporal distribution and size of other food patches help determine the attractiveness of a patch as it becomes depleted. An animal's ability to locate other potentially more profit able patches (influenced by the solution to problem 2), the costs of traveling to a new patch (determined by the solution to problem 8), and the time it has available (determined by the solution to problem 1) are all critical factors in this context.

If a primate remains in a patch only until all food items have been harvested, or until it has eaten as much as it wants, length of stay is determined simply by harvest rates and quantity ingested. Other factors, including scheduling of foraging activity and pursuit of activities unrelated to foraging, may lead to greater duration of patch occupation.

Problem 8: How To Travel Between Patches?

Primates must get from one patch to another (see fig. 11.9). This is a multifaceted problem, involving choice of progression route, locomotor mode(s), and travel speed. Interpatch travel is influenced by much the same variables that bear on animal movement within a

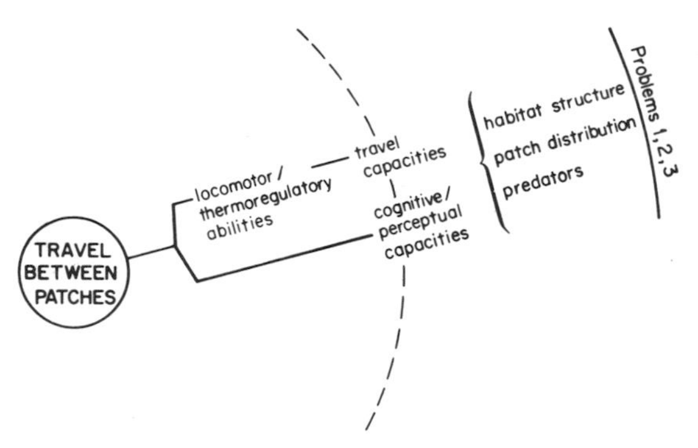

Figure 11.9. Problem 8: How to travel between patches? See legend of figure 11.2 and text for further discussion.

patch (problem 4), the primary distinction being the larger scale of the present problem and, as a result, its greater complexity.

Locus of travel is an option in primates typically exhibiting both arboreal and terrestrial progression. (Otherwise we consider the arboreal/terrestrial variable as fixed in ecological time.) Three aspects of the environment deserve attention: habitat structure, spatial distribution of patches, and predators. Patch distribution determines the shortest possible distance between two patches, and habitat structure frequently obstructs travel and thereby increases actual path length beyond the shortest possible distance. Contrasts in the availability of arboreal and terrestrial substrates suggest that we should expect the difference between actual path length and shortest distance to be greater for arboreal than terrestrial travel. For animals that feed in trees, the savings of terrestrial travel increase with increasing interpatch distance, because the investment in descending and ascending feeding trees becomes a smaller fraction of the total route traveled. Differences in habitat structure at ground and tree levels affect exposure to predators, and a primate's locomotor abilities influence success of escape from predator attack in each situation. Locomotor abilities also bear on choice of travel route and hence on the degree of disparity between arboreal and terrestrial path length.

The pathway an animal traces between two patches, whether it be arboreal or terrestrial, is strongly influenced by capacities to circumvent possible barriers to progression presented by habitat structure. This is particularly so for arboreal travel, and Grand (this volume) emphasizes the importance of discontinuities and unstable substrates in forest canopy. Success in dealing with barriers to progression rests on cognitive/perceptual and travel capacities. Travel paths are also influenced by how a primate locates food patches (the solution to problem 2).

Most primates have the option of using different locomotor modes, especially in arboreal locales where several may be displayed in sequential fashion. Habitat structure interacts with locomotor abilities to help determine how an animal moves in particular situations. For example, the field studies by Garber and by Crompton (this volume) describe relationships between locomotor modes and substrate properties. Locomotor abilities are founded largely on details of locomotor morphology and physiology but may also be influenced by behavioral variables, such as carrying infants or objects.

Locomotor speeds are influenced by mode of progression—brachiation or leaping allows faster arboreal travel than quadrupedalism (Cant 1977), for example, and by the nature of the substrates traversed, e.g., the size and inclination of the branches forming an arboreal travel path. Substrate characteristics determine the circumstances under which propulsive forces must be generated and balance maintained and are thus a source of limits on the maximum travel speeds for a locomotor mode. The spatiotemporal distribution of patches may also be related to speed. If patches are highly ephemeral, rapid rates of travel may be necessary to exploit them successfully. But as distance between patches increases, the possibility of heat stress, as well as muscle fatigue, increases. Reduction in travel speed will limit these risks. Travel speeds may be further influenced by risk of predation. For example, an animal might increase speed to reduce probability of attack before reaching refuge. Alternatively, an animal might travel slowly to reduce the risk of detection, depending on habitat structure.

A primate's foraging schedule (influenced by the solution to problem 1) is related to speeds of progression. It establishes the time available for foraging and thus may influence how quickly travel is performed. Also, it sets the time of foraging activity and thus affects exposure to thermal regime and risk of predation.

Concluding Discussion

Foraging is an ongoing process, involving repeated solutions to problems that are constantly changing in nature. Over an interval of time a primate eats a number of different foods, producing a certain pattern of environmental exploitation, represented by the identity, quantity, and spatiotemporal distribution of the items consumed. When we study foraging our eventual goal is to explain why animals exploit the environment in the ways they do. This involves inquiry into both proximate and ultimate causes. In the first context, we aim to determine how the foraging process actually works or why particular solutions to problems occur. In the second, evolutionary, context, we try to determine why a particular foraging strategy has evolved or what selection pressures have led to the evolution of the traits producing solutions.

The conceptual framework we have described may help achieve these goals in two complementary ways: by suggesting how to design initial studies of the foraging process and thus provide a foundation for subsequent analyses of causation, and by suggesting what to focus on when formulating and testing hypotheses about proximate and ultimate causes. We elaborate briefly on these applications of our conceptual approach.

Documentation of the Foraging Process

In accordance with the framework's primary emphasis on problems and solutions in foraging, the first (often difficult) task should be documentation of the behaviors that constitute solutions to the problems we have outlined and the associated environmental factors. Numerous aspects of behavior and environment must be dealt with, and there are two general considerations in the choice of variables.

First, the need to juggle multiple variables is exacerbated when we study primates because they generally inhabit complex environments. Habitat structure is often three dimensional and uneven, there is substantial variation in the relative abundance and distribution of potential food sources, conspecifics and other co-consumers often exploit the same patches, and so on. The problems primates face thus tend to be complicated and also to exhibit considerable short-term variation as well. By the same token, the behavioral repertoires of these animals are generally large. It is accordingly important to document animal-environment interactions with respect to the *immediate* contexts of problem-solving, lest important components of problems and solutions be missed or causal relationships be obscured. Aggregate variables (*sensu* Orians 1980), such as the proportion of the day spent foraging or day-range length, provide little insight into the particulars of resource exploitation given the wide range of variation in patch sizes and distribution of the food items constituting even one class (e.g., fruit). Furthermore, the possibility that such variables are produced by more than one sort of causal pathway must be entertained.

Second, as we have tried to illustrate (see figs. 11.2–11.9), solutions to one foraging problem usually influence solutions to other problems as well. Some problems such as food acquisition and interpatch travel are relatively narrow, and progress can be made by examining them more or less in isolation. Broader and more interconnected problems, e.g., foraging schedule and patch chioce, are more demanding. For example, the solution to interpatch travel includes use of particular substrates and locomotor modes, aspects of environment and animal, but knowledge of these variables alone would not permit evaluation of a primate's capacity to exploit patches of differing spatiotemporal distribution, part of the patch choice problem. For that we would also need data on travel speeds, length of pathways traversed, and distances between patches, variables "emerging" in the solution to interpatch travel.

Analysis of Causation

Proximate Causes. Our conceptual approach emphasizes deterministic relations among the elements of the foraging universe (see fig. 11.1). It is thus readily apparent that to investigate the proximate causes of foraging—why a primate has a particular diet or why sympatric species forage differently—we need to focus on the elements that precede solutions in our representation of the foraging process: problems and their determinants, and behavioral capacities and their determinants.

We suggest that analysis of proximate causes can be a three-part task. First, define the problems through measures of environmental variables and of animal needs (i.e., nutritional requirements). Second, elucidate behavioral capacities by determining animal abilities in specific environmental circumstances and also the morphological and physiolgoical foundations of those abilities. Third, combine the results of the first two parts in order to identify the critical interactions that produce a particular solution in specific environmental conditions. This entails examining costs and gains associated with different behavioral options (potential solutions), as well as the sorts of compromises reached between conflicting animal needs and/or solutions.

Assessment of proximate causes is an undeniable challenge in studying one population of primates, not to speak of studies of several species. We think, however, that our emphasis on dealing with the complete foraging process (i.e., all eight problems) does not create insurmountable obstacles. The conceptual framework permits two complementary approaches: investigation of solutions on an hour-to-hour or day-to-day scale, as in much optimal foraging research; and delineation of foraging strategies, summary characterizations of foraging patterns (i.e., overall dietary composition, ranging patterns, positional behavior, etc.). The latter approach compensates for the lack of time depth associated with a focus on short-term aspects of foraging and hence can contribute significantly to our appreciation of the influence of temporal variation in environmental factors on primate foraging. Some detail and analytical precision are lost in the second approach, for it partly sacrifices thorough understanding of the actual processes leading to solutions, but it permits more rapid progress and provides an opportunity to explore a larger empirical sample than is possible with the first approach. We stress that the two approaches are complementary: summary characterizations cannot be a substitute for understanding particular solutions.

Second, certain features of our conceptual framework and the animals we study facilitate analysis. Organisms are viewed in their entirety, by emphasizing integration of traits and resultant behavioral capacities. Despite—or perhaps because of—conflicts and compromises in the problem-solving process, there has to be some consistency among different traits and capacities and among the interdependent solutions to different problems. Thus it should be possible to focus initially on variables that have particularly well-defined roles in constraining behavioral capacities, in order to construct a preliminary explanation of proximate causes. Digestive, masticatory, and locomotor abilities, and the associated environmental variables, are good candidates in this respect. Consistency of traits and interdependence of solutions will then allow robust predictions about unexplored elements, leading to a more nearly complete understanding of the foraging process.

Ultimate Causes. Within the domain of our framework it is possible to tackle two kinds of issues in the evolution of foraging: the nature of the selection pressures that operated during the evolution of foraging adaptations and the sources of intrinsic constraints on possible pathways of evolutionary change.

With respect to selection pressures, we wish to determine the environmental circumstances under which the changes resulting in specified foraging traits were selectively advantageous (or at least, not selected against). Because the paleontological record is inadequate for this purpose, we must rely heavily on inferences drawn from analyses of foraging in extant forms. Investigation of the foraging problems and solutions displayed by a population allows us to identify environmental factors that appear to be the source of significant selection pressures. If the animals studied are adapted to their environments, there is a good chance that these selection forces were associated with the evolution of the specified traits. Additional insight gained from comparisons with other populations and from information on phylogeny and paleoenvironments enables reconstruction of the past events that resulted in the origin (or intensification) of the selection pressures.

The special combination of traits that characterizes a primate species is an important source of limits on opportunities for evolutionary change (e.g., Gould and Lewontin 1979). Evaluation of how traits are functionally integrated—the manner in which they jointly contribute to needs, capacities, and eventually solutions—will help us understand the second issue in ultimate causation. This task involves the analysis of proximate causes in order to establish how animal traits set constraints on behavioral options and hence determine which problems can be solved with enough success to ensure survival and reproductive success.

Final Comments

Research tactics have been discussed only in the broadest sense, and we expect their choice to be determined by the questions a researcher chooses to investigate. Nonetheless, one consideration merits expression: comparative research is essential. Nature has provided a tremendous number of adaptive experiments, and it is usually differences between populations or species that enable us to understand characteristics of any one species (Hamilton 1973; Maynard Smith 1978). Comparisons help us notice distinctions and thus stimulate inquiry, and they also furnish a source of analytical control when investigating causes. Several chapters in this volume (Crompton, Grand, Kay, Milton, Rodman, Temerin et al., Waser) demonstrate the value of interspecific comparisons in the study of foraging adaptations.

The importance of comparative research is further emphasized when we consider the purpose of studying foraging. In the most immediate context, it is to gain an understanding of a critical aspect of animal life. Because foraging provides the foundation for so many other aspects of an animal's life history, its study also has value in the larger context of understanding primate diversity. Thus in the broader view, the investigation of foraging adaptations will be an important source of insight into the factors limiting the distribution and abundance of populations and into the evolutionary foundations of diversity.

Acknowledgments

We are indebted to Peter S. Rodman for valuable discussions during the past several years and for suggestions and editorial assistance with respect to this chapter. Leo Berenstain and John C. Mitani made helpful comments on the penultimate version.

References

Andersson, M., F. Gotmark, and C. G. Wiklund. 1981. Food information in the Black-Headed Gull, *Larus ridibundus*. *Behav. Ecol. Sociobiol.* 9:199–202.

Brady, R. H. 1979. Natural selection and the criteria by which a theory is judged. *Syst. Zool.* 28:600–621.

Cant, J. G. H. 1977. Ecology, locomotion, and social organization of spider monkeys (*Ateles geoffroyi*). Ph.D. dissertation. Davis, Calif.: University of California.

Cartmill, M. 1974. Rethinking primate origins. *Science* 184:436–43.

—— 1979. The volar skin of primates: Its frictional characteristics and their functional significance. *Amer. J. Phys. Anthrop.* 50:497–509.

Charnov, E. L. 1976. Optimal foraging, the marginal value theorem. *Theoret. Pop. Biol.* 9:129–36.

Clutton-Brock, T. H., and P. H. Harvey. 1980. Primates, brains, and ecology. *J. Zool., London.* 190:309–23.

Eisenberg, J. F., and D. E. Wilson. 1978. Relative brain size and feeding strategies in the Chiroptera. *Evolution* 32:740–51.

Fleagle, J. G. 1978. Locomotion, posture, and habitat utilization in two sympatric, Malaysian leaf-monkeys (*Presbytis obscura* and *Presbytis melalophos*). In G. G. Montgomery, ed. *The Ecology of Arboreal Folivores,* pp. 243–51. Washington, D.C.: Smithsonian Institution Press.

—— 1980. Locomotion and posture. In D. J. Chivers, ed. *Malayan Forest Primates,* pp. 191–207. New York: Plenum.

Gautier-Hion, A., J. P. Gautier, and R. Quris. 1981. Forest structure and fruit availability as complementary factors influencing habitat use by a troop of monkeys (*Cercopithecus cephus*). *Rev. Ecol. (Terre et Vie)* 35:511–36.

Gould, S. J., and R. C. Lewontin. 1979. The Spandrels of San Marco and the Panglossian paradigm: A critique of the adaptationist programme. *Proc. Roy. Soc. London.* B205:581–98.

Hamilton, W. J. III. 1973. *Life's Color Code.* New York: McGraw-Hill.

Harrington, J. E. 1978. Diurnal behavior of *Lemur mongoz* at Ampijoroa, Madagascar. *Folia primatol.* 29:291–302.
Howe, H. F. 1977. Bird activity and seed dispersal of a tropical wet forest tree. *Ecology* 58:539–50.
Hylander, W. L. 1979. Mandibular function in *Galago crassicaudatus* and *Macaca fascicularis:* An in vivo approach to stress analysis of the mandible. *J. Morph.* 159:253–96.
Kay, R. F., and W. S. Sheine. 1979. On the relationship between chitin particle size and digestibility in the primate *Galago senegalensis. Amer. J. Phys. Anthrop.* 50:301–8.
Krebs, J. R. 1974. Colonial nesting and social feeding as strategies for exploiting food resources in the Great Blue Heron (*Ardea herodias*). *Behaviour* 51:99–134.
—— 1978. Optimal foraging: Decision rules for predators. In J. R. Krebs and N. B. Davies, eds. *Behavioural Ecology,* pp. 23–63. Sunderland, Mass.: Sinauer Associates, Inc.
Mace, G. M., P. H. Harvey, and T. H. Clutton-Brock. 1981. Brain size and ecology in small animals. *J. Zool., London.* 193:333–54.
Maier, W. 1980. Nasal structures in Old and New World monkeys. In R. L. Ciochon and A. B. Chiarelli, eds. *Evolutionary Biology of the New World Monkeys and Continental Drift,* pp. 219–41. New York: Plenum.
Maynard Smith, J. 1978. Optimization theory in evolution. *Ann. Rev. Ecol. Syst.* 9:31–56.
Mendel, F. 1976. Postural and locomotor behavior of *Alouatta palliata* on various substrates. *Folia primatol.* 26:36–53.
Menzel, E. W. 1979. Communication of object-locations in a group of young chimpanzees. In D. A. Hamburg and E. R. McCown, eds. *The Great Apes,* pp. 359–71. Menlo Park, Calif.: Benjamin/Cummings.
Napier, J. R. 1961. Prehensility and opposability in the hands of primates. *Symp. Zool. Soc. London.* 5:115–31.
Orians, G. H. 1980. Micro and macro in ecological theory. *BioScience* 30:79.
Pyke, G. H. 1981. Honeyeater foraging: A test of optimal foraging theory. *Anim. Behav.* 29:878–88.
Pyke, G. H., H. R. Pulliam, and E. L. Charnov. 1977. Optimal foraging: A selective review of theory and tests. *Quart. Rev. Biol.* 52:137–54.
Ripley, S. 1970. Leaves and leaf-monkeys: the social organization of

foraging in gray langurs *Presbytis entellus thersites.* In J. R. Napier and P. H. Napier, eds. *Old World monkeys: Evolution, Systematics, and Behavior,* pp. 481–509. New York: Academic Press.

Sheine, W. S. and R. F. Kay. 1977. An analysis of chewed food particle size and its relationship to molar structure in the primates *Cheirogaleus medius* and *Galago senegalensis* and the insectivoran *Tupaia glis. Amer. J. Phys. Anthrop.* 47:15–20.

Temerin, L. A. 1983. Evolutionary sources of primate diversity. Ph.D. dissertation. Davis, Calif.: University of California.

Walker, P. L., in press. An evaluation of incisor function in folivorous and frugivorous primates based on observations of *Alouatta villosa* and *Ateles geoffroyi. Amer. J. Phys. Anthrop.*

Walker, P. L., and P. Murray. 1975. An assessment of masticatory efficiency in a series of anthropoid primates with special reference to the Colobinae and Cercopithecinae. In R. H. Tuttle, ed. *Primate Functional Morphology and Evolution,* pp. 135–50. The Hague: Mouton.

Waltz, E. C. 1982. Resource characteristics and the evolution of information centers. *Amer. Natur.* 119:73–90.

INDEX

Activity patterns, 316 (*See also* Foraging problems and solutions; various taxa); influence of diet, 87-88; orbital size as predictor, 40-41

Adaptations, 2, 5 (*See also* Foraging adaptations; Causation); analysis of causation, 336-38; classes of, 2; definition, 305; problem of phylogenetic inertia, 213-14; reconstruction in extinct species, 21-49; study of problems and solutions, 305; trait-category orientation, 305

Alouatta (howler monkey), 6, 56, 267; body weight, 56; function of prehensile tail, 65, 67; locomotion across terminal branches, 57; locomotor descent, 61, 65, 67

Alouatta palliata: diet, 266; digestive processes, 266; feeding rates, 272-73; food choice, 256; food passage rates, 255-56, 258, 260-61; gut morphology, 256, 269; response to hard foods, 273; sex difference in support use, 328

Anatomy (*See also* Dentition; Eyes; Gut Morphology; various taxa): difficulty of predicting activity locale, 42; experimental and observational studies, 6; information on adaptations, 22; information on foraging behavior, 24-25

Anatomy, comparative: and functional significance of structural variation, 5; history of study in primates, 4, 5, 7, 8; and interpretation of fossil primates, 5; isolation from natural settings, 5; and primate classification, 3-4; and primate phylogeny, 4, 5

Animal needs, 308

Animal traits, 310, 311

Anthropoidea (*See also* various taxa): anterior dentition, 31; incisor size and diet in Old World forms, 43

Aotus trivirgatus, 40, 41

Arboreal supports, 54-55. *See also* Locomotion; various taxa

Arctocebus calabarensis, 33

Ateles (spider monkey), 6, 7, 268, 269, 270; diet, 266; energy costs of locomotion, 228; food choice, 256; food passage rates, 266; function of prehensile tail, 65-67

Ateles geoffroyi, 56; body weight, 56; food passage rates, 256, 258, 261; gut morphology, 256; locomotion across terminal branches, 57, 58, 60, 61; locomotor ascent, 67; locomotor descent, 65, 67; suspension, 64-68

Ateles paniscus: food passage rates, 258, 261; response to hard foods, 273

Australopithecines, 10

Avahi laniger, 40, 41

Baboons. See *Papio; Theropithecus*

Behavior (*See also* Behavioral capacities; Behavioral ecology and sociobiology; Feeding rates; Food choice; Foraging; Locomotion; Optimal foraging models; Social Systems; Socioecology); early naturalistic studies of primates, 6, 9-11; mathematical models, 280; observational techniques, 13; optimization models, 281-300; rationale for study of comparative socioecology, 8-9; social patterns, 10, 12, 13; socio-

Behavior (*Continued*)
ecology, 11-12; theory of feeding strategies, 12
Behavioral capacities, 319, 336; feeding, 312, 319; perceptual and cognitive, 312, 318, 321, 325-26, 328, 330; travel, 312, 318, 324, 325
Behavioral ecology and sociobiology, 12–14
Binturong *(Arctictis binturong)*, 55
Body size, 326. (*See also* Diet: Locomotion; various taxa); and ability to exploit widely distributed food species, 231-32; and ability to withstand food shortage, 113, 231; adaptive significance of, 240-41; and arboreal locomotion, 55; and diet, 27-30, 104, 221-40; and dietary trends, 237-39; and digestive strategies, 265-66; and energy budgets, 102-3; and energy metabolism, 28, 221; and evolutionary potential, 239-40; feedback relations among size-related parameters, 241; and folivory, 28-29; and food processing capacities, 221; and frequency of use of locomotor modes, 101-2; and gut volume, 103-4, 221; and harvest rates, 222-24; influence on energy costs of locomotion, 102-3, 226-27; influence on evolution of diets, 233-40; influence on feeding rates, 272-73; influence on food choice, 264; influence on mobility, 227-29, 240; influence on physical access to food items, 229-30; influence on travel speeds, 224-26; and insectivory, 27-28, 103; and nutrient constraints on choice of food species, 230-31; and rates of digestion, 104; relation to food passage rates, 262-68; relation to life history parameters, 241; size distribution of folivorous primates, 29-30; size distribution of frugivorous primates, size distribution of insectivorous primates, 29-30; and support use, 100
Brachiation. *See* Locomotion; various taxa
Brachyteles arachnoides (woolly spider monkey); body weight, 267; diet, 267; digestive processes, 267; food passage rates, 258, 261, 267; gut morphology, 267
Brain size, 321

Cacajao (uakari), 4, 265, 270; dental morphology, 264; food passage rates, 264
Cacajao calvus, food passage rates, 258, 260
Cacajao melanocephalus, response to hard foods, 273
Callitrichidae (marmosets and tamarins) (See also *Saguinus*): digital morphology, 114-15; group sizes, 127; incisors, 44
Capuchin monkey. See *Cebus*
Carnivores, 55
Catarrhini (Old World monkeys and apes), enamel thickness of molars, 36-38
Causation: proximate, 1, 334, 336-38; ultimate, 1-2, 335, 338
Cebidae: food passage rates and diet, 264-67; gut morphology, 264-65
Cebus (capuchin), 38, 265, 270; diet, 264; food passage rates, 264
Cebus albifrons, response to hard foods, 273-74
Cebus apella, 38; food passage rates, 258, 260; response to hard foods, 273-74
Cebus capucinus, food passage rates, 258, 259
Cercocebus (mangabey): comparison between *albigena* and *galeritus*, 197-212; ranging behavior, 322
Cercocebus albigena, 34, 35, 38, 45, 46, 322; day range, 201; diet, 200-201; diversity of food species, 201; group composition, 196-97; group size, 205, 212; habitat, 197-98; home range size, 204, 211; intergroup dispersion, 207; isolated males, 206; motor patterns during foraging, 200; phenology of important food species, 198; population density, 199; rates of movement,

202, 204, 209, 210; search patterns, 202-4, 209; subgroups, 205-6; substrates exploited during foraging, 200, 209; time spent foraging, 199-200; vertical distribution, 197
Cercocebus galeritus, 322; day range, 201; diet, 200-201, 210; diversity of food species, 201, 210; group composition, 197; group size, 205, 209-10, 212; habitat, 198, 199; habitat productivity, 211, 212; home range size, 204, 211; intergroup dispersion, 207, 211; intragroup dispersion, 205-7, 209, 210; isolated males, 206-7; motor patterns during foraging, 200; phenology of important food species, 198; population density, 198-99, 211; rates of movement, 202, 204, 209; search patterns, 202-4, 209, 211-12; subgroups, 205-6; substrates exploited during foraging, 200, 209; time spent foraging, 200, 210; vertical distribution, 197
Cercopithecidae (Old World monkeys); diets, 34-35, 46; molar shearing quotients, 34-35, 45, 46; selection of foods within diet categories, 47; structural carbohydrate in diet, 47
Cercopithecus, 35, 199
Cercopithecus ascanius, 34, 35, 167
Cercopithecus mitis, 34, 35
Cheirogaleus medius, 239
Chimpanzee. See *Pan*
Chiropotes (saki), 270; diet, 264
Chiropotes albinasa, 258, 260, 264
Chiropotes chiropotes, 265
Chiroptera (bats), 321
Claws, behavioral significance of, 328. *See also* Tegulae
Climate, 308, 317-18. *See also* various taxa
Co-consumers, 308-9, 318-19, 321-22, 325, 327, 332
Colobus badius, 34, 35, 45, 46, 164
Colobus guereza, 34, 35, 45, 46
Colobus satanas, 34, 35
Comparative approach, 134, 339
Craniofacial morphology and diet, 23-24
Crocuta crocuta (spotted hyena), 163
Darwinian paradigm, 4, 5, 9, 13
Daubentonia, 4, 43
Day range. *See* Ranging behavior; various taxa
Dentition: bilophodonty, 23; conflicting functional demands, 31; correlation with diet, 30-39; enamel microwear and diet, 39; functional convergence of molar systems, 32; functional differences in molars of Old World monkeys and apes, 45-48; function of anterior teeth, 31, 42-44; function of cheek teeth, 31; incisor size and diet in Old World anthropoids, 43; incisors of callitrichids, 44; incisors of lemurs and lorises, 43; information on feeding ecology, 22; molar enamel thickness, 23, 36; molar enamel thickness and diet, 35-38; molar enamel thickness and locale of activity, 37; molar enamel thickness and shearing crest development, 37; molar enamel thickness, functional significance, 37, 38; molar function, 31-32; molar shearing capacities, 331; molar shearing quotients, 33, 35; molar shearing quotients and diet, 32; molars of folivores, 32; molars of frugivores, 32; molars of insectivores, 32; selection pressures on, 44; traits associated with eating hard and tough objects, 37; traits associated with folivory, 37; underbites, 330
Diet, 25-27 (*See also* Food choice; Foraging; Optimal foraging models; various taxa); basis for selection of food items, 322-26; and body size in arboreal forms, 240; and brain size, 321; classification of, 25; and dentition, 30-39; energetic models of, 27-28, 233-39; evolution, 30 (*See also* under Body size); of human ancestors, 268-69; influence of gut volume, 103-4; influence on locomotion, 100-101; and life history tactics, 113-14; model for extinct species, 30; model of evolutionary change in, 233-39; relation to ranging behavior, 228, 269; role of

Diet (*Continued*)
secondary compounds in food choice, 253-54; seasonal changes in, 75, 78; temporal pattern of food intake, 296, 320

Digestion, 254-56, 319 (*see also* various taxa); circumvention of plant defenses, 254; contributory role of mastication, 32; food passage rates, 255-56; 257-68, 324; food passage rates and body size, 263-68; physiology of, 324; of plant foods, 28, 252, 265-66; rate and efficiency, 28-29, 324; saliva, role of, 331; strategies of, 255, 265-66

Digestive abilities, 312, 319, 324, 337

Digestive efficiency. *See under* Digestion

Energy budgets: influence of body size, 102-3; influence on diet, 103

Energy metabolism and body size, 221. *See also* Body size

Evolutionary change, intrinsic constraints, 338

Extinct species, 74; body weight predictions, 30; inferences on foraging behavior, 48-49; interpretation of morphology, 22-24; living models for assessing diets of, 44; predictions of activity pattern, 40-41; predictions of diet, 30; predictions of vertical distribution, 41-42

Eyes: correlation between orbital size and activity pattern, 40, 41; structure in nocturnal versus diurnal species, 40

Feeding rates, 272-73. *See also* Harvest rates; various taxa

Folivore, definition of, 25

Food acquisition, 328-30. *See also* various taxa

Food choice, 249-50, 269-70. *See also* Diet; Foraging; various taxa bounds on dietary quality set by size, 220-29; bounds on nutrient content of diet set by size, 231; constraints on, 323-24; factors affecting recognition of food items, 274; feeding trials, 273-74; influence of body size on species composition of diets, 229-32; nutrient balance, 29, 251-52; role of digestive system, 256

Food patches. *See* Patches

Food processing, relation to body size, 221

Foods (*See also* various taxa): digestibility of, 32, 254-55 (*See also* Digestion); location of, 320-22, 325; nutrient content, *see* physical and chemical composition; nutritional benefits, 324; physical and chemical composition, 25-26, 28, 29, 32, 35, 47, 81-82, 124, 251-54, 309-10, 330; plant defenses, 250-54; preparation and mastication, 330-31; procurement, 324; quality of, 220; variation in attributes of, 221

Foods, source of sensory cues, 321

Foraging, 2, 305, 334 (*See also* Diet; Food choice; Optimal foraging models; various taxa); allocation of time for, 316-20, 325; analysis of causation, 336-38; complementary approaches to study, 337-38; documentation of foraging process, 335-36; feedback systems, 314-15; interdependence of problems and solutions, 320, 322-23, 336; purpose of study, 334-35, 339; time and energy parameters of, 233-38

Foraging adaptations: conceptual approach, 306-15; why study primates, 2-3

Foraging and sociality, 316. *See also* Behavior; Social systems; various taxa

Foraging problems and solutions, 305, 310, 311, 312-13, 335-36; how to acquire food items, 320, 324, 328-30, 331; how to locate food patch, 320-22, 325, 327, 332, 333; how to move within food patch, 320, 324, 327-28, 331, 333; how to ready items for ingestion, 320, 324, 330-31; how to travel between patches, 320, 322, 324, 332-34; what food patch to enter, 320, 322-26; when to forage, 316-20, 322, 325, 332, 334; when to leave patch, 331-32

Foraging strategy, 337. *See also* various taxa
Foraging universe, 306, 311, 336
Forest structure. *See* Arboreal supports: Locomotion; various taxa
Fossils. *See* Extinct species
Frugivore, definition of, 25

Galago: arboreal zones, 82-84; comparison between *crassicaudatus* and *senegalensis,* 75-77, 81-82, 84-93, 105-8; distribution of food, 82-84; distribution of supports, 82-84; exposure to predation, 82-84; habitat structure, 82-84; influence of size on diet, 104; influence of size on locomotion, 104; lack of sympatry, 107; locomotor modes, 91-92; prey-capture techniques, 101; relation between locomotion and diet, 101; relation between locomotor mode and height (stratum), 102; relation between size and locomotor modes, 101-2; seasonal variation in food supply, 93-97; support use, 100
Galago alleni, 33, 34, 108
Galago crassicaudatus: activity profiles, 87-88, 97, 99; adaptive significance of locomotor patterns, 104-5; body weight, 76; competitors, 76-77; diet, 75, 76-77, 78, 99; distribution, 76-77; foraging strategy, 105-8; frequency use locomotor modes, 92-93, 98, 99; gum sources, 81; habitat, 78, 79; heights of locomotion, 84-87, 98, 99; location of food sources, 89; postures during feeding, 88-89; predators, 76-77; seasonal changes in behavior, 97-99; support diameters, 90-91, 98, 99; support orientations, 89-90, 98, 99
Galago demidovii, 33, 108
Galago elegantulus, 31, 33, 34, 43
Galago senegalensis, 28; activity profiles, 87-88, 94, 95, 96, 97; adaptive significance of locomotor patterns, 104; body weight, 76; competitors, 76; diet, 75, 76, 78, 94, 95, 96, 97; distribution, 76; foraging strategy, 105-8; frequency use locomotor modes, 92-93, 94, 95, 96, 97; gum sources, 81; habitat, 78, 79; heights of locomotion, 84-87, 94, 95, 96, 97; location of food sources, 89; postures during feeding, 88-89; predators, 76; seasonal changes in behavior, 94-97 support diameters, 90-91; support orientations, 89-90, 94, 95, 96, 97
Gibbon. See *Hylobates;* Hylobatidae
Gorilla gorilla, 6, 10, 270; diet, 268; food passage rates, 258, 262, 268
Gut morphology, 324. *See also* various taxa

Habitat, 308
Habitat structure, 322, 333. *See also* Arboreal supports; various taxa
Harvest rates, 332 (*See also* various taxa): components of, 222-23; relation to body size, 222-24, 272-73
Heat stress, 317-18, 334
Heterohyrax brucei (hyrax), 39
Home range size. *See* Ranging behavior; various taxa
Hominoidea (apes and man): diet and ranging behavior, 269; evolution of suspension, 238
Homo, 4
Homo sapiens: diet, 268; food passage rates, 258, 262, 268; gut morphology, 268
Howler monkey. See *Alouatta*
Hylobates (gibbon), 6, 47, 56, 240; digestive abilities, 47; food particle sizes, 47; functional significance of forelimb morphology, 65
Hylobates lar, 45, 46, 56: body weight, 56; diet, 46; harvest rates, 224; locomotion across terminal branches, 57-59; locomotion across terminal branches with descent, 59-61; locomotor ascent, 63, 67; locomotor descent, 61-63, 65, 67; locomotor speed, 62; molar shearing quotients, 45, 46; response to locomotor accident, 63; suspension, 64-68; travel rates, 240

Hylobatidae (lesser apes): evolution of brachiation, 48, 238; locomotion, 328; selection of foods within diet categories, 47; structural carbohydrate in diet, 47

Insectivore, definition of, 25
Insectivory: energetic model of, 27-28; maximum body size, 28, 103

Jarman-Bell principle, 237

Lagothrix lagotricha (woolly monkey); diet, 266; food passage rates, 258, 261, 266; response to hard foods, 273
Leaf monkey. See *Presbytis*
Lepilemur, 40, 41
Lepilemur mustelinus, 29
Life history parameters, 241
Life history strategies, 167
Locomotion (*See also* Arboreal supports; Body size; Positional behavior; Posture; Travel; various taxa); anatomical basis of, 65-67, 69; appropriate comparative sample, 56; arboreal travel in large species, 55; arboreal travel in medium-sized species, 55; arboreal travel in small species, 55; and body size, 55, 68; brachiation. *See* suspension; categories of, 7; conceptual approach, 64; control of space (economy), 67-70; definition of, 113-14; energy costs of, 102-3, 104, 226-27; experimental and observational studies of, 6; factors influencing, 327-28; influence on diet, 100-101; pathways traced, 328; selection pressures, 69-70; semibrachiation, 7; speed, 334; subleties of performance, 55; support characteristics, 54-55, 334; suspension, 64-68, 238; suspension, biomechanics of, 65
Locomotor abilities, 311-12, 318, 327, 328, 333, 334, 337
Locomotor modes, 333-34
Locomotor repertoires, arboreal species, 55
Locomotor studies, methodology, 80-81

Lorisidae, 40, 41 (*See also* various species); diets, 33-34; molar shearing quotients, 33-34
Loxodonta africana (African elephant), 163

Macaca (macaques), 6; digestive abilities, 47; food particle sizes, 47
Macaca fascicularis, 45, 46, 48, 218-19; access to food sources, 230; activity profile, 224, 225; body weight, 218; comparison with *Pongo,* 221-22, 224-26, 228-29, 230-32; day range, 225-26; diet, 221, 222, 231; distribution of food sources, 232; feeding unit size and biomass, 225, 228; habitat, 219; number feeding bouts per day, 225, 226; time at food sources, 225, 226; travel speeds, 224-25
Macaca sinica, 56; body weight, 56; locomotion across terminal branches, 57, 58, 59, 60; locomotion along branch, 66; locomotor ascent, 67; locomotor descent, 62-63, 65, 67; tail morphology and function, 67
Man. See *Homo*
Mandibular symphysis, 331
Manipulatory abilities, 312, 324, 329-30, 331, 337
Masticatory abilities, 312, 3330, 331
Mating systems, 148-49 (See also *Pan, Pongo*); model based on effective reproductive success, 151-54
Microcebus murinus, 239
Miopithecus talapoin, 127
Mobility. *See* Body size; various taxa
Muscle fatigue, 334

Optimal foraging models, 282, 298, 299-300, 305-6, 326, 337; assumptions of, 282-84, 287, 289; implications of coarse-grained foraging, 289-92; predictions, 288, 289, 290, 293-95; robusticity of, 287-89, 296; significance of food abundance in patchy environment, 296-97; testing assumptions of, 284-87; uncertainty and risk, 292-97
Orangutan. See *Pongo*

Orbit size, 40-41
Oppossums, 55
Organismal biology, 1, 2; integration of animal traits and behavioral capacities, 337-38

Pangolins, 55
Pan troglodytes (chimpanzee), 4, 6, 10; activity profiles, 141, 168-70; activity profiles, sex differences in, 168-70; body weight, 135; comparison with *Pongo,* 141-48, 190, 238-39; constraints on male body size, 150; day ranges, 145-46, 239; density of important food species, 180-81; diet, 142-43, 170-72, 238-39, 267; distances between food sources, 144, 145; distribution of important food sources, 181; effects of provisioning, 162; female dispersion, 138-39, 189; female exogamy, 136, 154-55; food calls, 151, 174-76; food passage rates, 258, 261-62, 267-68; food species, 171; grouping patterns, 139; habitat, 163-64; history of naturalistic study, 161-62; home range size, 146-47; intercommunity aggression, 136; life history strategy, 167; male communal activity, 150-51; male competition for offspring, 154; male dispersion, 140; male effective reproductive success, 151-54; male relatedness, 152; mating behavior, 137; mating system, 151-54; mobility, 147-48; phenology of food species, 181-83; relation between patch sizes and feeding aggregation sizes, 176-80; relation of food supply to local population density and travel party size, 182, 184; reproductive behavior, 167; sex ratio of groups, 152; sexual dimorphism, 135-36, 148, 154; social system, 136-40, 155, 162; spacing of foragers, 172-73; time at food source, 143-44; travel costs, 148; travel party composition, 186-89; travel party size, 185-86; travel speed, 239
Papio (baboons), 10, 43, 196
Papio cynocephalus: day range, 290; foods available, 291; habitat types, 290, 291; "optimal" decision strategies in foraging, 295-97; probability of food encounter, 290, 292
Patches, 309; arrangement of food items within, 309, 328; depletion rates, 331-32; distribution, 309, 318-19, 332, 333; exploitation efficiency of, 326; influence of structure on locomotion, 327; relative value, 325, 326; size, 309, 319, 332; structure, 324, 328
Peccaries, 38
Perodicticus potto, 33-34
Phaner furcifer, 43
Phylogenetic inertia, 213-14
Piltdown Man, 5, 9
Pithecia (sakis), 264, 269, 270; diet, 265
Pithecia monarchus, food passage rates, 258, 260, 265
Pithecinae, competition among, 265
Plant defenses, 250-51 (*See also* Foods: physical and chemical composition); indigestible materials, 252-53; nutrient content, 251-52; secondary compounds, 253-54
Pongidae (great apes) (See also *Gorilla, Pan Pongo*): food passage rates and diet, 267-68; gut morphology, 268
Pongo pygmaeus (orangutan), 38, 55, 218-19: access to food sources 230; activity profiles, 141-42, 224, 225; activity profiles, sex differences in, 170; body weights, 135, 218; comparison with *Pan troglodytes,* 141-48, 190, 238-39; comparison with *Macaca fascicularis,* 221-22, 224-26, 228-32; constraints on male body size, 149-50; constraints on travel, 228-29; day ranges, 145-46, 225-26, 239; diet, 142-43, 221-22, 231, 238, 267; distances between food sources, 144, 145; distribution of food sources, 232; feeding unit size and biomass, 225, 228; food passage rates, 258, 262, 267, 268; habitat, 219; home range size, 146-47; locomotor morphology, 239; male classes, 137; male dispersion, 137-38, 140; male

Pongo pygmaeus (*Continued*)
mating strategy, 149; mobility, 147-48, 240; number feeding bouts per day, 225, 226; sex differences in travel, 218; sexual dimorphism, 135-36, 149; social system, 137, 155; time at food sources, 143, 225, 226; travel speed, 224-25, 239
Population density, 314. *See also* various taxa
Positional behavior, 113-14 (*See also* Locomotion; Postural abilities; Posture; various taxa); and forces acting on musculoskeletal systems, 114
Postural abilities, 312, 324, 329
Posture (*See also* Locomotion; Positional behavior; various taxa): definition of, 114; during feeding, 328-29
Predation risk, 308, 318, 325, 333, 334. *See also* various taxa
Presbytis (leaf monkeys), 6, 8
Presbytis melalophos, 45, 46
Presbytis obscura, 45, 46; locomotion, 328
Problem-solving abilities, 312, 328, 330
Procavia johnstoni (hyrax), 39
Propithecus (sifakas), 31, 43
Pseudocheirus peregrinus (phalangeroid marsupial), 29

Ranging behavior, 228 (*See also* Travel; various taxa); relation between day range and home range, 292; relation to diet, 269
Reproduction. *See* Life history strategies; various taxa
Rhinarium, 330
Rodentia, 31, 55, 321

Saguinus (tamarins), 113
Saguinus oedipus: activity profile, 117-18; body weight, 117; diet, 117-18; food distribution and abundance, 118, 126, 127; foraging activity, 122; foraging strategy, 129; group size, 127; habitat, 115-16; heights of locomotion, 118; home range, 115; insect-hunting tactics, 128; locomotion during foraging, 123-24; postures during feeding, 119-22; relation between positional behavior and food type, 124, 125; significance of exudates in diet, 128; significance of insects in diet, 124, 126-27; support use, 118; support use and feeding postures, 118; support use during feeding, 119-22; support use during foraging, 123-24; tegulae, function of, 118, 128
Saimiri, 127
Sexual dimorphism, 149. *See also* various taxa
Siamang. See *Symphalangus*
Social behavior. *See also* Behavior; Sociobiology; Socioecology; Social systems; various taxa; methodological advances in study of, 13-14
Social systems (*See also* various taxa); female strategies, 140; male cohesiveness, 163; male strategies, 140, 148-49; and sexual dimorphism, 149
Sociobiology, development of, 13
Socioecology: failings of, 12; focus of inquiry, 12; history of study, 8-12
Spider monkey. See *Ateles*
Squirrels, 55
Strepsirhini (prosimians), toothcomb, 31, 43
Suspension. *See also* Locomotion; various taxa
Symphalangus syndactylus (siamang), 45, 46, 48, 240; diet, 46; harvest rates, 224; molar shearing quotients, 45, 46; travel rates, 240

Tails: nonprehensile, function, 67; prehensile, function, 65, 67
Tamarin. *See* Callitrichidae; *Saguinus*
Tarsius (tarsier), 40, 41, 108
Tegulae, functional significance, 100, 118, 128. *See also* Claws
Theory of feeding strategies, 12. *See also* Optimal foraging models
Thermoregulatory abilities, 312, 318
Theropithecus (gelada baboon), 43
Travel (*See also* Locomotion; Ranging behavior; various taxa): arboreal versus terrestrial, 227-28, 333; between

patches, 325, 332-34; energy costs of, 226-27, 228, 333; path lengths, 333; speeds, 224-26; within food patch, 324, 327-28

Underbites, 330

Woolly monkey. See *Lagothrix*
Woolly spider monkey. See *Brachyteles*